T0351088

Graduate Texts in Mathematics **18**

Springer
New York
Berlin
Heidelberg
Barcelona
Hong Kong
London
Milan
Paris
Singapore
Tokyo

Graduate Texts in Mathematics

(continued after index)

Paul R. Halmos

Measure Theory

P.R. Halmos
Department of Mathematics
Santa Clara University
Santa Clara, CA 95053

Mathematics Subject Classification (2000): 28-01, 28A05, 28A10, 28A12, 28A20, 28A25

Library of Congress Cataloging in Publication Data

Halmos, Paul Richard, 1914–
Measure theory.

(Graduate texts in mathematics; 18)
Reprint of the ed. published by Van Nostrand, New York in series: The University series in higher mathematics,
Bibliography: p.
1. Measure theory. I. Title. II. Series.
[QA312.H26 1974] 515'.42 74-10690

Printed on acid-free paper.

Printed and bound by R. R. Donnelley and Sons, Harrisonburg, VA.

Printed in the United States of America.

9 8

ISBN 0-387-90088-8
ISBN 3-540-90088-8 SPIN 10830847

Springer-Verlag New York Berlin Heidelberg
A member of BertelsmannSpringer Science+Business Media GmbH

PREFACE

My main purpose in this book is to present a unified treatment of that part of measure theory which in recent years has shown itself to be most useful for its applications in modern analysis. If I have accomplished my purpose, then the book should be found usable both as a text for students and as a source of reference for the more advanced mathematician.

I have tried to keep to a minimum the amount of new and unusual terminology and notation. In the few places where my nomenclature differs from that in the existing literature of measure theory, I was motivated by an attempt to harmonize with the usage of other parts of mathematics. There are, for instance, sound algebraic reasons for using the terms "lattice" and "ring" for certain classes of sets—reasons which are more cogent than the similarities that caused Hausdorff to use "ring" and "field."

The only necessary prerequisite for an intelligent reading of the first seven chapters of this book is what is known in the United States as undergraduate algebra and analysis. For the convenience of the reader, § 0 is devoted to a detailed listing of exactly what knowledge is assumed in the various chapters. The beginner should be warned that some of the words and symbols in the latter part of § 0 are defined only later, in the first seven chapters of the text, and that, accordingly, he should not be discouraged if, on first reading of § 0, he finds that he does not have the prerequisites for reading the prerequisites.

At the end of almost every section there is a set of exercises which appear sometimes as questions but more usually as assertions that the reader is invited to prove. These exercises should be viewed as corollaries to and sidelights on the results more formally expounded. They constitute an integral part of the book; among them appear not only most of the examples and counter examples necessary for understanding the theory, but also definitions of new concepts and, occasionally, entire theories that not long ago were still subjects of research.

It might appear inconsistent that, in the text, many elementary notions are treated in great detail, while, in the exercises, some quite refined and profound matters (topological spaces, transfinite numbers, Banach spaces, etc.) are assumed to be known. The material is arranged, however, so that when a beginning student comes to an exercise which uses terms not defined in this book he may simply omit it without loss of continuity. The more advanced reader, on the other hand, might be pleased at the interplay between measure theory and other parts of mathematics which it is the purpose of such exercises to exhibit.

The symbol ▮ is used throughout the entire book in place of such phrases as "Q.E.D." or "This completes the proof of the theorem" to signal the end of a proof.

At the end of the book there is a short list of references and a bibliography. I make no claims of completeness for these lists. Their purpose is sometimes to mention background reading, rarely (in cases where the history of the subject is not too well known) to give credit for original discoveries, and most often to indicate directions for further study.

A symbol such as $u.v$, where u is an integer and v is an integer or a letter of the alphabet, refers to the (unique) theorem, formula, or exercise in section u which bears the label v.

ACKNOWLEDGMENTS

Most of the work on this book was done in the academic year 1947–1948 while I was a fellow of the John Simon Guggenheim Memorial Foundation, in residence at the Institute for Advanced Study, on leave from the University of Chicago.

I am very much indebted to D. Blackwell, J. L. Doob, W. H. Gottschalk, L. Nachbin, B. J. Pettis, and, especially, to J. C. Oxtoby for their critical reading of the manuscript and their many valuable suggestions for improvements.

The result of 3.13 was communicated to me by E. Bishop. The condition in 31.10 was suggested by J. C. Oxtoby. The example 52.10 was discovered by J. Dieudonné.

P. R. H.

CONTENTS

CHAPTER X: LOCALLY COMPACT SPACES

CHAPTER XI: HAAR MEASURE

CHAPTER XII: MEASURE AND TOPOLOGY IN GROUPS

§0. PREREQUISITES

The only prerequisite for reading and understanding the first seven chapters of this book is a knowledge of elementary algebra and analysis. Specifically it is assumed that the reader is familiar with the concepts and results listed in (1)–(7) below.

(1) Mathematical induction, commutativity and associativity of algebraic operations, linear combinations, equivalence relations and decompositions into equivalence classes.

(2) Countable sets; the union of countably many countable sets is countable.

(3) Real numbers, elementary metric and topological properties of the real line (e.g. the rational numbers are dense, every open set is a countable union of disjoint open intervals), the Heine–Borel theorem.

(4) The general concept of a function and, in particular, of a sequence (i.e. a function whose domain of definition is the set of positive integers); sums, products, constant multiples, and absolute values of functions.

(5) Least upper and greatest lower bounds (called suprema and infima) of sets of real numbers and real valued functions; limits, superior limits, and inferior limits of sequences of real numbers and real valued functions.

(6) The symbols $+\infty$ and $-\infty$, and the following algebraic relations among them and real numbers x:

$$(\pm\infty) + (\pm\infty) = x + (\pm\infty) = (\pm\infty) + x = \pm\infty;$$

$$x(\pm\infty) = (\pm\infty)x = \begin{cases} \pm\infty & \text{if } x > 0, \\ 0 & \text{if } x = 0, \\ \mp\infty & \text{if } x < 0; \end{cases}$$

$$(\pm\infty)(\pm\infty) = +\infty,$$

$$(\pm\infty)(\mp\infty) = -\infty;$$

$$x/(\pm\infty) = 0;$$

$$-\infty < x < +\infty.$$

The phrase **extended real number** refers to a real number or one of the symbols $\pm\infty$.

(7) If x and y are real numbers,

$$x \cup y = \max\{x,y\} = \tfrac{1}{2}(x + y + |x - y|),$$

$$x \cap y = \min\{x,y\} = \tfrac{1}{2}(x + y - |x - y|).$$

Similarly, if f and g are real valued functions, then $f \cup g$ and $f \cap g$ are the functions defined by

$$(f \cup g)(x) = f(x) \cup g(x) \quad \text{and} \quad (f \cap g)(x) = f(x) \cap g(x),$$

respectively. The supremum and infimum of a sequence $\{x_n\}$ of real numbers are denoted by

$$\bigcup_{n=1}^{\infty} x_n \quad \text{and} \quad \bigcap_{n=1}^{\infty} x_n,$$

respectively. In this notation

$$\limsup_n x_n = \bigcap_{n=1}^{\infty} \bigcup_{m=n}^{\infty} x_m$$

and

$$\liminf_n x_n = \bigcup_{n=1}^{\infty} \bigcap_{m=n}^{\infty} x_m.$$

In Chapter VIII the concept of metric space is used, together with such related concepts as completeness and separability for metric spaces, and uniform continuity of functions on metric spaces. In Chapter VIII use is made also of such slightly more sophisticated concepts of real analysis as one-sided continuity.

In the last section of Chapter IX, Tychonoff's theorem on the compactness of product spaces is needed (for countably many factors each of which is an interval).

In general, each chapter makes free use of all preceding chapters; the only major exception to this is that Chapter IX is not needed for the last three chapters.

In Chapters X, XI, and XII systematic use is made of many of the concepts and results of point set topology and the elements of topological group theory. We append below a list of all the relevant definitions and theorems. The purpose of this list is not to serve as a text on topology, but (a) to tell the expert exactly which forms of the relevant concepts and results we need, (b) to tell the beginner with exactly which concepts and results he should familiarize himself before studying the last three chapters, (c) to put on record certain, not universally used, terminological conventions, and (d) to serve as an easily available reference for things which the reader may wish to recall.

Topological Spaces

A **topological space** is a set X and a class of subsets of X, called the **open sets** of X, such that the class contains 0 and X and is closed under the formation of finite intersections and arbitrary (i.e. not necessarily finite or countable) unions. A subset E of X is called a G_δ if there exists a sequence $\{U_n\}$ of open sets such that $E = \bigcap_{n=1}^{\infty} U_n$. The class of all G_δ's is closed under the formation of finite unions and countable intersections. The topological space X is **discrete** if every subset of X is open, or, equivalently, if every one-point subset of X is open. A set E is **closed** if $X - E$ is open. The class of closed sets contains 0 and X and is closed under the formation of finite unions and arbitrary intersections. The **interior,** E^0, of a subset E of X is the greatest open set contained in E; the **closure,** $\bar{E}$, of E is the least closed set containing E. Interiors are open sets and closures are closed sets; if E is open, then $E^0 = E$, and, if E is closed, then $\bar{E} = E$. The closure of a set E is the set of all points x such that, for every open set U containing x, $E \cap U \neq 0$. A set E is **dense** in X if $\bar{E} = X$. A subset Y of a topological space becomes a topological space (a **subspace** of X) in the **relative topology** if exactly those subsets of Y are called open which may be obtained by intersecting an open subset of X with Y. A **neighborhood** of a point x in X [or of a subset E of X] is an open set containing x [or an open set containing E]. A **base** is a class $\mathbf{B}$ of open sets such that, for every x in X and every neighborhood U of x, there exists a set B in $\mathbf{B}$ such that $x \in B \subset U$. The **topology of the real line** is determined by the requirement that the class of all open intervals be a base. A **subbase** is a class of sets, the class of all finite intersections of which is a base. A space X is **separable** if it has a countable base. A subspace of a separable space is separable.

An **open covering** of a subset E of a topological space X is a class $\mathbf{K}$ of open sets such that $E \subset \bigcup \mathbf{K}$. If X is separable and $\mathbf{K}$ is an open covering of a subset E of X, then there exists a countable subclass $\{K_1, K_2, \cdots\}$ of $\mathbf{K}$ which is an open covering of E. A set E in X is **compact** if, for every open covering $\mathbf{K}$ of E, there exists a finite subclass $\{K_1, \cdots, K_n\}$ of $\mathbf{K}$ which is an open covering of E. A class $\mathbf{K}$ of sets has the **finite intersection property** if every finite subclass of $\mathbf{K}$ has a non empty intersection. A space X is compact if and only if every class of closed sets with the finite intersection property has a non empty intersection. A set E in a space X is **σ-compact** if there exists a sequence $\{C_n\}$ of compact sets such that $E = \bigcup_{n=1}^{\infty} C_n$. A space X is **locally compact** if every point of X has a neighborhood whose closure is compact. A subset E of a locally compact space is **bounded** if there exists a compact set C such that $E \subset C$. The class of all bounded open sets in a locally compact space is a base. A closed subset of a bounded set is compact. A subset E of a locally compact space is **σ-bounded** if there exists a sequence $\{C_n\}$ of compact sets such that $E \subset \bigcup_{n=1}^{\infty} C_n$. To any locally compact but not compact topological space X there corresponds a compact space X^* containing X and exactly one additional point x^*; X^* is called the **one-point compactification** of X by x^*. The open sets of X^* are the open subsets of X and the complements (in X^*) of the closed compact subsets of X.

If $\{X_i: i \in I\}$ is a class of topological spaces, their **Cartesian product** is the set $X = \times \{X_i: i \in I\}$ of all functions x defined on I and such that, for each i in I, $x(i) \in X_i$. For a fixed i_0 in I, let E_{i_0} be an open subset of X_{i_0}, and, for $i \neq i_0$, write $E_i = X_i$; the class of open sets in X is determined by the requirement that the class of all sets of the form $\times \{E_i: i \in I\}$ be a subbase. If the function ξ_i on X is defined by $\xi_i(x) = x(i)$, then ξ_i is continuous. The Cartesian product of any class of compact spaces is compact.

A topological space is a **Hausdorff space** if every pair of distinct points have disjoint neighborhoods. Two disjoint compact sets in a Hausdorff space have disjoint neighborhoods. A compact subset of a Hausdorff space is closed. If a locally compact space is a Hausdorff space or a separable space, then so is its one-point compactification. A real valued continuous function on a compact set is bounded.

For any topological space X we denote by $\mathfrak{F}$ (or $\mathfrak{F}(X)$) the class of all real valued continuous functions f such that $0 \leqq f(x) \leqq 1$ for all x in X. A Hausdorff space is **completely regular** if, for every point y in X and every closed set F not containing y, there is a function f in $\mathfrak{F}$ such that $f(y) = 0$ and, for x in F, $f(x) = 1$. A locally compact Hausdorff space is completely regular.

A **metric space** is a set X and a real valued function d (called **distance**) on $X \times X$, such that $d(x,y) \geqq 0$, $d(x,y) = 0$ if and only if $x = y$, $d(x,y) = d(y,x)$, and $d(x,y) \leqq d(x,z) + d(z,y)$. If E and F are non empty subsets of a metric space X, the distance between them is defined to be the number $d(E,F) = \inf\{d(x,y): x \in E, y \in F\}$. If $F = \{x_0\}$ is a one-point set, we write $d(E,x_0)$ in place of $d(E,\{x_0\})$. A **sphere** (with **center** x_0 and **radius** r_0) is a subset E of a metric space X such that, for some point x_0 and some positive number r_0, $E = \{x: d(x_0,x) < r_0\}$. The **topology of a metric space** is determined by the requirement that the class of all spheres be a base. A metric space is completely regular. A closed set in a metric space is a G_δ. A metric space is separable if and only if it contains a countable dense set. If E is a subset of a metric space and $f(x) = d(E,x)$, then f is a continuous function and $\bar{E} = \{x: f(x) = 0\}$. If X is the real line, or the Cartesian product of a finite number of real lines, then X is a locally compact separable Hausdorff space; it is even a metric space if for $x = (x_1, \cdots, x_n)$ and $y = (y_1, \cdots, y_n)$ the distance $d(x,y)$ is defined to be $(\sum_{i=1}^{n} (x_i - y_i)^2)^{1/2}$. A closed interval in the real line is a compact set.

A transformation T from a topological space X into a topological space Y is **continuous** if the inverse image of every open set is open, or, equivalently, if the inverse image of every closed set is closed. The transformation T is **open** if the image of every open set is open. If $\mathbf{B}$ is a subbase in Y, then a necessary and sufficient condition that T be continuous is that $T^{-1}(B)$ be open for every B in $\mathbf{B}$. If a continuous transformation T maps X onto Y, and if X is compact, then Y is compact. A **homeomorphism** is a one to one, continuous transformation of X onto Y whose inverse is also continuous.

The sum of a uniformly convergent series of real valued, continuous functions is continuous. If f and g are real valued continuous functions, then $f \cup g$ and $f \cap g$ are continuous.

Topological Groups

A **group** is a non empty set X of elements for which an associative multiplication is defined so that, for any two elements a and b of X, the equations $ax = b$ and $ya = b$ are solvable. In every group X there is a unique **identity** element e, characterized by the fact that $ex = xe = x$ for every x in X. Each element x of X has a unique **inverse,** x^{-1}, characterized by the fact that $xx^{-1} = x^{-1}x = e$. A non empty subset Y of X is a **subgroup** if $x^{-1}y \in Y$ whenever x and y are in Y. If E is any subset of a group X, E^{-1} is the set of all elements of the form x^{-1}, where $x \in E$; if E and F are any two subsets of X, EF is the set of all elements of the form xy, where $x \in E$ and $y \in F$. A non empty subset Y of X is a subgroup if and only if $Y^{-1}Y \subset Y$. If $x \in X$, it is customary to write xE and Ex in place of $\{x\}E$ and $E\{x\}$ respectively; the set xE [or Ex] is called a **left translation** [or **right translation**] of E. If Y is a subgroup of X, the sets xY and Yx are called (left and right) **cosets** of Y. A subgroup Y of X is **invariant** if $xY = Yx$ for every x in X. If the product of two cosets Y_1 and Y_2 of an invariant subgroup Y is defined to be Y_1Y_2, then, with respect to this notion of multiplication, the class of all cosets is a group $\hat{X}$, called the **quotient group** of X modulo Y and denoted by X/Y. The identity element $\hat{e}$ of $\hat{X}$ is Y. If Y is an invariant subgroup of X, and if for every x in X, $\pi(x)$ is the coset of Y which contains x, then the transformation π is called the **projection** from X onto $\hat{X}$. A **homomorphism** is a transformation T from a group X into a group Y such that $T(xy) = T(x)T(y)$ for every two elements x and y of X. The projection from a group X onto a quotient group $\hat{X}$ is a homomorphism.

A **topological group** is a group X which is a Hausdorff space such that the transformation (from $X \times X$ onto X) which sends

(x,y) into $x^{-1}y$ is continuous. A class **N** of open sets containing e in a topological group is a **base at** e if (a) for every x different from e there exists a set U in **N** such that $x \,\varepsilon'\, U$, (b) for any two sets U and V in **N** there exists a set W in **N** such that $W \subset U \cap V$, (c) for any set U in **N** there exists a set V in **N** such that $V^{-1}V \subset U$, (d) for any set U in **N** and any element x in X, there exists a set V in **N** such that $V \subset xUx^{-1}$, and (e) for any set U in **N** and any element x in U there exists a set V in **N** such that $Vx \subset U$. The class of all neighborhoods of e is a base at e; conversely if, in any group X, **N** is a class of sets satisfying the conditions described above, and if the class of all translations of sets of **N** is taken for a base, then, with respect to the topology so defined, X becomes a topological group. A neighborhood V of e is **symmetric** if $V = V^{-1}$; the class of all symmetric neighborhoods of e is a base at e. If **N** is a base at e and if F is any closed set in X, then $F = \bigcap \{UF\colon U \,\varepsilon\, \mathbf{N}\}$.

The closure of a subgroup [or of an invariant subgroup] of a topological group X is a subgroup [or an invariant subgroup] of X. If Y is a closed invariant subgroup of X, and if a subset of $\hat{X} = X/Y$ is called open if and only if its inverse image (under the projection π) is open in X, then $\hat{X}$ is a topological group and the transformation π from X onto $\hat{X}$ is open and continuous.

If C is a compact set and U is an open set in a topological group X, and if $C \subset U$, then there exists a neighborhood V of e such that $VCV \subset U$. If C and D are two disjoint compact sets, then there exists a neighborhood U of e such that UCU and UDU are disjoint. If C and D are any two compact sets, then C^{-1} and CD are also compact.

A subset E of a topological group X is **bounded** if, for every neighborhood U of e, there exists a finite set $\{x_1, \cdots, x_n\}$ (which, in case $E \neq 0$, may be assumed to be a subset of E) such that $E \subset \bigcup_{i=1}^{n} x_iU$; if X is locally compact, then this definition of boundedness agrees with the one applicable in any locally compact space (i.e. the one which requires that the closure of E be compact). If a continuous, real valued function f on X is such that the set $N(f) = \{x\colon f(x) \neq 0\}$ is bounded, then f is **uniformly continuous** in the sense that to every positive number ϵ there corresponds a neighborhood U of e such that $|f(x_1) - f(x_2)| < \epsilon$ whenever $x_1x_2^{-1} \varepsilon U$.

A topological group is **locally bounded** if there exists in it a bounded neighborhood of e. To every locally bounded topological group X, there corresponds a locally compact topological group X^*, called the **completion** of X (uniquely determined to within an isomorphism), such that X is a dense subgroup of X^*. Every closed subgroup and every quotient group of a locally compact group is a locally compact group.

Chapter 1

SETS AND CLASSES

§ 1. SET INCLUSION

Throughout this book, whenever the word **set** is used, it will be interpreted to mean a subset of a given set, which, unless it is assigned a different symbol in a special context, will be denoted by X. The elements of X will be called **points**; the set X will be referred to as the **space**, or the **whole** or **entire** space, under consideration. The purpose of this introductory chapter is to define the basic concepts of the theory of sets, and to state the principal results which will be used constantly in what follows.

If x is a point of X and E is a subset of X, the notation

$$x \varepsilon E$$

means that x belongs to E (i.e. that one of the points of E is x); the negation of this assertion, i.e. the statement that x does not belong to E, will be denoted by

$$x \varepsilon' E.$$

Thus, for example, for every point x of X, we have

$$x \varepsilon X,$$

and for no point x of X do we have

$$x \varepsilon' X.$$

If E and F are subsets of X, the notation

$$E \subset F \quad \text{or} \quad F \supset E$$

means that E is a subset of F, i.e. that every point of E belongs to F. In particular therefore

$$E \subset E$$

for every set E. Two sets E and F are called **equal** if and only if they contain exactly the same points, or, equivalently, if and only if

$$E \subset F \quad \text{and} \quad F \subset E.$$

This seemingly innocuous definition has as a consequence the important principle that the only way to prove that two sets are equal is to show, in two steps, that every point of either set belongs also to the other.

It makes for tremendous simplification in language and notation to admit into the class of sets a set containing no points, which we shall call the **empty set** and denote by 0. For every set E we have

$$0 \subset E \subset X;$$

for every point x we have

$$x \,\varepsilon'\, 0.$$

In addition to sets of points we shall have frequent occasion to consider also sets of sets. If, for instance, X is the real line, then an interval is a set, i.e. a subset of X, but the set of all intervals is a set of sets. To help keep the notions clear, we shall always use the word **class** for a set of sets. The same notations and terminology will be used for classes as for sets. Thus, for instance, if E is a set and $\mathbf{E}$ is a class of sets, then

$$E \,\varepsilon\, \mathbf{E}$$

means that the set E belongs to (is a member of, is an element of) the class $\mathbf{E}$; if $\mathbf{E}$ and $\mathbf{F}$ are classes, then

$$\mathbf{E} \subset \mathbf{F}$$

means that every set of $\mathbf{E}$ belongs also to $\mathbf{F}$, i.e. that $\mathbf{E}$ is a subclass of $\mathbf{F}$.

On very rare occasions we shall also have to consider sets of classes, for which we shall always use the word **collection.** If, for instance, X is the Euclidean plane and $\mathbf{E}_y$ is the class of all intervals on the horizontal line at distance y from the origin, then each $\mathbf{E}_y$ is a class and the set of all these classes is a collection.

(1) The relation $\subset$ between sets is always reflexive and transitive; it is symmetric if and only if X is empty.

(2) Let $\mathbf{X}$ be the class of all subsets of X, including of course the empty set 0 and the whole space X; let x be a point of X, let E be a subset of X (i.e. a member of $\mathbf{X}$), and let $\mathbf{E}$ be a class of subsets of X (i.e. a subclass of $\mathbf{X}$). If u and v vary independently over the five symbols x, E, X, $\mathbf{E}$, $\mathbf{X}$, then some of the fifty relations of the forms

$$u \in v \quad \text{or} \quad u \subset v$$

are necessarily true, some are possibly true, some are necessarily false, and some are meaningless. In particular $u \in v$ is meaningless unless the right term is a subset of a space of which the left term is a point, and $u \subset v$ is meaningless unless u and v are both subsets of the same space.

§ 2. UNIONS AND INTERSECTIONS

If $\mathbf{E}$ is any class of subsets of X, the set of all those points of X which belong to at least one set of the class $\mathbf{E}$ is called the **union** of the sets of $\mathbf{E}$; it will be denoted by

$$\bigcup \mathbf{E} \quad \text{or} \quad \bigcup \{E: E \in \mathbf{E}\}.$$

The last written symbol is an application of an important and frequently used principle of notation. If we are given any set of objects denoted by the generic symbol x, and if, for each x, $\pi(x)$ is a proposition concerning x, then the symbol

$$\{x: \pi(x)\}$$

denotes the set of those points x for which the proposition $\pi(x)$ is true. If $\{\pi_n(x)\}$ is a sequence of propositions concerning x, the symbol

$$\{x: \pi_1(x), \pi_2(x), \cdots\}$$

denotes the set of those points x for which $\pi_n(x)$ is true for every $n = 1, 2, \cdots$. If, more generally, to every element γ of a certain index set Γ there corresponds a proposition $\pi_\gamma(x)$ concerning x, then we shall denote the set of all those points x for which the proposition $\pi_\gamma(x)$ is true for every γ in Γ by

$$\{x: \pi_\gamma(x), \quad \gamma \in \Gamma\}.$$

Thus, for instance,

$$\{x: x \varepsilon E\} = E$$

and

$$\{E: E \varepsilon \mathbf{E}\} = \mathbf{E}.$$

For more illuminating examples we consider the sets

$$\{t: 0 \leqq t \leqq 1\}$$

(= the closed unit interval),

$$\{(x,y): x^2 + y^2 = 1\}$$

(= the circumference of the unit circle in the plane), and

$$\{n^2: n = 1, 2, \cdots\}$$

(= the set of those positive integers which are squares). In accordance with this notation, the upper and lower bounds (supremum and infimum) of a set E of real numbers are denoted by

$$\sup \{x: x \varepsilon E\} \quad \text{and} \quad \inf \{x: x \varepsilon E\}$$

respectively.

In general the brace $\{\cdots\}$ notation will be reserved for the formation of sets. Thus, for instance, if x and y are points, then $\{x,y\}$ denotes the set whose only elements are x and y. It is important logically to distinguish between the point x and the set $\{x\}$ whose only element is x, and similarly to distinguish between the set E and the class $\{E\}$ whose only element is E. The empty set 0, for example, contains no points, but the class $\{0\}$ contains exactly one set, namely the empty set.

For the union of special classes of sets various special notations are used. If, for instance,

$$\mathbf{E} = \{E_1, E_2\}$$

then

$$\bigcup \mathbf{E} = \bigcup \{E_i: i = 1, 2\}$$

is denoted by

$$E_1 \cup E_2;$$

if, more generally,

$$\mathbf{E} = \{E_1, \cdots, E_n\}$$

is a finite class of sets, then

$$\bigcup \mathbf{E} = \bigcup \{E_i: i = 1, \cdots, n\}$$

is denoted by

$$E_1 \cup \cdots \cup E_n \quad \text{or} \quad \bigcup_{i=1}^n E_i.$$

If, similarly, $\{E_n\}$ is an infinite sequence of sets, then the union of the terms of this sequence is denoted by

$$E_1 \cup E_2 \cup \cdots \quad \text{or} \quad \bigcup_{i=1}^\infty E_i.$$

More generally, if to every element γ of a certain index set Γ there corresponds a set E_γ, then the union of the class of sets

$$\{E_\gamma: \gamma \in \Gamma\}$$

is denoted by

$$\bigcup_{\gamma \in \Gamma} E_\gamma \quad \text{or} \quad \bigcup_\gamma E_\gamma.$$

If the index set Γ is empty, we shall make the convention that

$$\bigcup_\gamma E_\gamma = 0.$$

The relations of the empty set 0 and the whole space X to the formation of unions are given by the identities

$$E \cup 0 = E \quad \text{and} \quad E \cup X = X.$$

More generally it is true that

$$E \subset F$$

if and only if

$$E \cup F = F.$$

If $\mathbf{E}$ is any class of subsets of X, the set of all those points of X which belong to every set of the class $\mathbf{E}$ is called the **intersection** of the sets of $\mathbf{E}$; it will be denoted by

$$\bigcap \mathbf{E} \quad \text{or} \quad \bigcap \{E: E \in \mathbf{E}\}.$$

Symbols similar to those used for unions are used, but with the symbol $\cup$ replaced by $\cap$, for the intersections of two sets, of a finite or countably infinite sequence of sets, or of the terms of any indexed class of sets. If the index set Γ is empty, we shall make the somewhat startling convention that

$$\bigcap_{\gamma \in \Gamma} E_\gamma = X.$$

There are several heuristic motivations for this convention. One of them is that if Γ_1 and Γ_2 are two (non empty) index sets for which $\Gamma_1 \subset \Gamma_2$, then clearly

$$\bigcap_{\gamma \varepsilon \Gamma_1} E_\gamma \supset \bigcap_{\gamma \varepsilon \Gamma_2} E_\gamma,$$

and that therefore to the smallest possible Γ, i.e. the empty one, we should make correspond the largest possible intersection. Another motivation is the identity

$$\bigcap_{\gamma \varepsilon \Gamma_1 \cup \Gamma_2} E_\gamma = \bigcap_{\gamma \varepsilon \Gamma_1} E_\gamma \cap \bigcap_{\gamma \varepsilon \Gamma_2} E_\gamma,$$

valid for all non empty index sets Γ_1 and Γ_2. If we insist that this identity remain valid for arbitrary Γ_1 and Γ_2, then we are committed to believing that, for every Γ,

$$\bigcap_{\gamma \varepsilon \Gamma} E_\gamma = \bigcap_{\gamma \varepsilon \Gamma \cup 0} E_\gamma = \bigcap_{\gamma \varepsilon \Gamma} E_\gamma \cap \bigcap_{\gamma \varepsilon 0} E_\gamma;$$

writing $E_\gamma = X$ for every γ in Γ, we conclude that

$$\bigcap_{\gamma \varepsilon 0} E_\gamma = X.$$

Union and intersection are sometimes called **join** and **meet**, respectively. As a mnemonic device for distinguishing between $\cup$ and $\cap$ (which, by the way, are usually read as **cup** and **cap**, respectively), it may be remarked that the symbol $\cup$ is similar to the initial letter of the word "union" and the symbol $\cap$ is similar to the initial letter of the word "meet."

The relations of 0 and X to the formation of intersections are given by the identities

$$E \cap 0 = 0 \quad \text{and} \quad E \cap X = E.$$

More generally it is true that

$$E \subset F$$

if and only if

$$E \cap F = E.$$

Two sets E and F are called **disjoint** if they have no points in common, i.e. if

$$E \cap F = 0.$$

A **disjoint class** is a class **E** of sets such that every two distinct sets of **E** are disjoint; in this case we shall refer to the union of the sets of **E** as a **disjoint union.**

We conclude this section with the introduction of the useful concept of characteristic function. If E is any subset of X, the function χ_E, defined for all x in X by the relations

$$\chi_E(x) = \begin{cases} 1 & \text{if } x \varepsilon E, \\ 0 & \text{if } x \varepsilon' E, \end{cases}$$

is called the **characteristic function** of the set E. The correspondence between sets and their characteristic functions is one to one, and all properties of sets and set operations may be expressed by means of characteristic functions. As one more relevant illustration of the brace notation, we mention

$$E = \{x: \chi_E(x) = 1\}.$$

(1) The formation of unions is commutative and associative, i.e.

$$E \cup F = F \cup E \quad \text{and} \quad E \cup (F \cup G) = (E \cup F) \cup G;$$

the same is true for the formation of intersections.

(2) Each of the two operations, the formation of unions and the formation of intersections, is distributive with respect to the other, i.e.

$$E \cap (F \cup G) = (E \cap F) \cup (E \cap G)$$

and

$$E \cup (F \cap G) = (E \cup F) \cap (E \cup G).$$

More generally the following extended distributive laws are valid:

$$F \cap \bigcup \{E: E \varepsilon \mathbf{E}\} = \bigcup \{E \cap F: E \varepsilon \mathbf{E}\}$$

and

$$F \cup \bigcap \{E: E \varepsilon \mathbf{E}\} = \bigcap \{E \cup F: E \varepsilon \mathbf{E}\}.$$

(3) Does the class of all subsets of X form a group with respect to either of the operations $\cup$ and $\cap$?

(4) $\chi_0(x) \equiv 0$, $\chi_X(x) \equiv 1$. The relation

$$\chi_E(x) \leqq \chi_F(x)$$

is valid for all x in X if and only if $E \subset F$. If $E \cap F = A$ and $E \cup F = B$, then

$$\chi_A = \chi_E\chi_F = \chi_E \cap \chi_F \quad \text{and} \quad \chi_B = \chi_E + \chi_F - \chi_A = \chi_E \cup \chi_F.$$

(5) Do the identities in (4), expressing χ_A and χ_B in terms of χ_E and χ_F, have generalizations to finite, countably infinite, and arbitrary unions and intersections?

§ 3. LIMITS, COMPLEMENTS, AND DIFFERENCES

If $\{E_n\}$ is a sequence of subsets of X, the set E^* of all those points of X which belong to E_n for infinitely many values of n is called the **superior limit** of the sequence and is denoted by

$$E^* = \limsup_n E_n.$$

The set E_* of all those points of X which belong to E_n for all but a finite number of values of n is called the **inferior limit** of the sequence and is denoted by

$$E_* = \liminf_n E_n.$$

If it so happens that

$$E_* = E^*,$$

we shall use the notation

$$\lim_n E_n$$

for this set. If the sequence is such that

$$E_n \subset E_{n+1}, \quad \text{for} \quad n = 1, 2, \cdots,$$

it is called **increasing**; if

$$E_n \supset E_{n+1}, \quad \text{for} \quad n = 1, 2, \cdots,$$

it is called **decreasing.** Both increasing and decreasing sequences will be referred to as **monotone.** It is easy to verify that if $\{E_n\}$ is a monotone sequence, then $\lim_n E_n$ exists and is equal to

$$\bigcup_n E_n \quad \text{or} \quad \bigcap_n E_n$$

according as the sequence is increasing or decreasing.

The **complement** of a subset E of X is the set of all those points of X which do not belong to E; it will be denoted by E'. The operation of forming complements satisfies the following algebraic identities:

$$E \cap E' = 0, \quad E \cup E' = X,$$

$$0' = X, \quad (E')' = E, \quad X' = 0, \quad \text{and}$$

$$\text{if} \quad E \subset F, \quad \text{then} \quad E' \supset F'.$$

The formation of complements also bears an interesting and very important relation to unions and intersections, expressed by the identities

$$(\bigcup \{E: E \in \mathbf{E}\})' = \bigcap \{E': E \in \mathbf{E}\},$$

$$(\bigcap \{E: E \in \mathbf{E}\})' = \bigcup \{E': E \in \mathbf{E}\}.$$

In words: the complement of the union of a class of sets is the intersection of their complements, and the complement of their intersection is the union of their complements. This fact, together with the elementary formulas relating to complements, proves the important **principle of duality:**

> any valid identity among sets, obtained by forming unions, intersections, and complements, remains valid if in it the symbols
>
> $$\bigcap, \subset, \text{ and } 0$$
>
> are interchanged with
>
> $$\bigcup, \supset, \text{ and } X$$
>
> respectively (and equality and complementation are left unchanged).

If E and F are subsets of X, the **difference** between E and F, in symbols

$$E - F,$$

is the set of all those points of E which do not belong to F. Since

$$X - F = F',$$

and, more generally,

$$E - F = E \cap F',$$

the difference $E - F$ is frequently called the **relative complement** of F in E. The operation of forming differences, similarly to the operation of forming complements, interchanges $\bigcup$ with $\bigcap$ and $\subset$ with $\supset$, so that, for instance,

$$E - (F \cup G) = (E - F) \cap (E - G).$$

The difference $E - F$ is called **proper** if $E \supset F$.

As the final and frequently very important operation on sets we introduce the **symmetric difference** of two sets E and F, denoted by

$$E \Delta F,$$

and defined by

$$E \Delta F = (E - F) \cup (F - E) = (E \cap F') \cup (E' \cap F).$$

The formation of limits, complements, and differences of sets requires a bit of practice for ease in manipulation. The reader is accordingly advised to carry through the proofs of the most important properties of these processes, listed in the exercises that follow.

(1) Another heuristic motivation of the convention

$$\bigcap_{\gamma \in 0} E_\gamma = X$$

is the desire to have the identity

$$\bigcap_{\gamma \in \Gamma} E_\gamma = (\bigcup_{\gamma \in \Gamma} E_\gamma')',$$

which is valid for all non empty index sets Γ, remain valid for $\Gamma = 0$.

(2) If $E_* = \liminf_n E_n$ and $E^* = \limsup_n E_n$, then

$$E_* = \bigcup_{n=1}^{\infty} \bigcap_{m=n}^{\infty} E_m \subset \bigcap_{n=1}^{\infty} \bigcup_{m=n}^{\infty} E_m = E^*.$$

(3) The superior limit, inferior limit, and limit (if it exists) of a sequence of sets are unaltered if a finite number of the terms of the sequence are changed.

(4) If $E_n = A$ or B according as n is even or odd, then

$$\liminf_n E_n = A \cap B \quad \text{and} \quad \limsup_n E_n = A \cup B.$$

(5) If $\{E_n\}$ is a disjoint sequence, then

$$\lim_n E_n = 0.$$

(6) If $E_* = \liminf_n E_n$ and $E^* = \limsup_n E_n$, then

$$(E_*)' = \limsup_n E_n' \quad \text{and} \quad (E^*)' = \liminf_n E_n'.$$

More generally,

$$F - E_* = \limsup_n (F - E_n) \quad \text{and} \quad F - E^* = \liminf_n (F - E_n).$$

(7) $E - F = E - (E \cap F) = (E \cup F) - F,$

$$E \cap (F - G) = (E \cap F) - (E \cap G), \quad (E \cup F) - G = (E - G) \cup (F - G).$$

(8) $(E - G) \cap (F - G) = (E \cap F) - G,$

$$(E - F) - G = E - (F \cup G), \quad E - (F - G) = (E - F) \cup (E \cap G),$$

$$(E - F) \cap (G - H) = (E \cap G) - (F \cup H).$$

(9) $E \Delta F = F \Delta E, E \Delta (F \Delta G) = (E \Delta F) \Delta G,$

$$E \cap (F \Delta G) = (E \cap F) \Delta (E \cap G),$$

$$E \Delta 0 = E, \quad E \Delta X = E',$$

$$E \Delta E = 0, \quad E \Delta E' = X,$$

$$E \Delta F = (E \cup F) - (E \cap F).$$

(10) Does the class of all subsets of X form a group with respect to the operation Δ?

(11) If $E_* = \liminf_n E_n$ and $E^* = \limsup_n E_n$, then

$$\chi_{E_*}(x) = \liminf_n \chi_{E_n}(x) \quad \text{and} \quad \chi_{E^*}(x) = \limsup_n \chi_{E_n}(x).$$

(The expressions on the right sides of these equations refer, of course, to the usual numerical concepts of superior limit and inferior limit.)

(12) $\chi_{E'} = 1 - \chi_E, \chi_{E-F} = \chi_E(1 - \chi_F),$

$$\chi_{E \Delta F} = |\chi_E - \chi_F| \equiv \chi_E + \chi_F \pmod 2.$$

(13) If $\{E_n\}$ is a sequence of sets, write

$$D_1 = E_1, \quad D_2 = D_1 \Delta E_2, \quad D_3 = D_2 \Delta E_3,$$

and, in general,

$$D_{n+1} = D_n \Delta E_{n+1} \quad \text{for} \quad n = 1, 2, \cdots.$$

The limit of the sequence $\{D_n\}$ exists if and only if $\lim_n E_n = 0$. If the operation Δ is thought of as addition (cf. (12)), then this result has the following verbal phrasing: an infinite series of sets converges if and only if its terms approach zero.

§ 4. RINGS AND ALGEBRAS

A **ring** (or **Boolean ring**) of sets is a non empty class **R** of sets such that if

$$E \in \mathbf{R} \quad \text{and} \quad F \in \mathbf{R},$$

then

$$E \cup F \in \mathbf{R} \quad \text{and} \quad E - F \in \mathbf{R}.$$

In other words a ring is a non empty class of sets which is closed under the formation of unions and differences.

The empty set belongs to every ring **R**, for if

$$E \in \mathbf{R},$$

then

$$0 = E - E \in \mathbf{R}.$$

Since

$$E - F = (E \cup F) - F,$$

it follows that a non empty class of sets closed under the formation of unions and proper differences is a ring. Since

$$E \Delta F = (E - F) \cup (F - E)$$

and

$$E \cap F = (E \cup F) - (E \Delta F),$$

it follows that a ring is closed under the formation of symmetric differences and intersections. An application of mathematical induction and the associative laws for unions and intersections shows that if $\mathbf{R}$ is a ring and

$$E_i \in \mathbf{R}, \quad i = 1, \cdots, n,$$

then

$$\bigcup_{i=1}^n E_i \in \mathbf{R} \quad \text{and} \quad \bigcap_{i=1}^n E_i \in \mathbf{R}.$$

Two extreme but useful examples of rings are the class $\{0\}$ containing the empty set only, and the class of all subsets of X. Another example, for an arbitrary set X, is the class of all finite sets. A more illuminating example is the following. Let

$$X = \{x: -\infty < x < +\infty\}$$

be the real line, and let $\mathbf{R}$ be the class of all finite unions of bounded, left closed, and right open intervals, i.e. the class of all sets of the form

$$\bigcup_{i=1}^n \{x: -\infty < a_i \leqq x < b_i < +\infty\}.$$

Union and intersection are treated unsymmetrically in the definition of rings. While, for instance, it is true that a ring is closed under the formation of intersections, it is not true that a class of sets closed under the formation of intersections and differences is necessarily closed also under the formation of unions. If, however, a non empty class of sets is closed under the formation of intersections, proper differences, and disjoint unions, then it is a ring. (Proof:

$$E \cup F = [E - (E \cap F)] \cup [F - (E \cap F)] \cup (E \cap F).)$$

It is easily possible to give a definition of rings which is more nearly symmetric in its treatment of union and intersection: a ring may be defined as a non empty class of sets closed under the formation of intersections and symmetric differences. The proof of this statement is in the identities:

$$E \cup F = (E \Delta F) \Delta (E \cap F), \quad E - F = E \Delta (E \cap F).$$

If in this form of the definition we replace intersection by union we obtain a true statement: a non empty class of sets closed under the formation of unions and symmetric differences is a ring.

An **algebra** (or **Boolean algebra**) of sets is a non empty class $\mathbf{R}$ of sets such that

(a) if $E \varepsilon \mathbf{R}$ and $F \varepsilon \mathbf{R}$, then $E \cup F \varepsilon \mathbf{R}$, and
(b) if $E \varepsilon \mathbf{R}$, then $E' \varepsilon \mathbf{R}$.

Since

$$E - F = E \cap F' = (E' \cup F)',$$

it follows that every algebra is a ring. The relation between the general concept of ring and the more special concept of algebra is simple: an algebra may be characterized as a ring containing X. Since

$$E' = X - E,$$

it is clear that every such ring is an algebra; if, conversely, $\mathbf{R}$ is an algebra and

$$E \varepsilon \mathbf{R}$$

(we recall that $\mathbf{R}$ is non empty), then

$$X = E \cup E' \varepsilon \mathbf{R}.$$

(1) The following classes of sets are examples of rings and algebras.

(1a) X is n-dimensional Euclidean space; $\mathbf{E}$ is the class of all finite unions of semiclosed "intervals" of the form

$$\{(x_1, \cdots, x_n): -\infty < a_i \leqq x_i < b_i < \infty, \quad i = 1, \cdots, n\}.$$

(1b) X is an uncountable set; $\mathbf{E}$ is the class of all countable subsets of X.

(1c) X is an uncountable set; $\mathbf{E}$ is the class of all sets which either are countable or have countable complements.

(2) Which topological spaces have the property that the class $\mathbf{E}$ of open sets is a ring?

(3) The intersection of any collection of rings or algebras is again a ring or an algebra, respectively.

(4) If $\mathbf{R}$ is a ring of sets and if we define, for E and F in $\mathbf{R}$,

$$E \odot F = E \cap F \quad \text{and} \quad E \oplus F = E \Delta F,$$

then, with respect to the operations of "addition" ($\oplus$) and "multiplication" ($\odot$), the system $\mathbf{R}$ is a ring in the algebraic sense of the word. Algebraic rings, such as this one, in which every element is idempotent (i.e. $E \odot E = E$ for every E in $\mathbf{R}$) are also called Boolean rings. The existence of a very close relation between Boolean rings of sets and Boolean rings in general is the main justification of the ring terminology in the set theoretic case.

(5) If $\mathbf{R}$ is a ring of sets and if $\mathbf{A}$ is the class of all those sets E for which

$$\text{either} \quad E \in \mathbf{R} \quad \text{or else} \quad E' \in \mathbf{R},$$

then $\mathbf{A}$ is an algebra.

(6) A **semiring** is a non empty class $\mathbf{P}$ of sets such that

(6a) if $E \in \mathbf{P}$ and $F \in \mathbf{P}$, then $E \cap F \in \mathbf{P}$, and

(6b) if $E \in \mathbf{P}$ and $F \in \mathbf{P}$ and $E \subset F$, then there is a finite class $\{C_0, C_1, \cdots, C_n\}$ of sets in $\mathbf{P}$ such that $E = C_0 \subset C_1 \subset \cdots \subset C_n = F$ and $D_i = C_i - C_{i-1} \in \mathbf{P}$ for $i = 1, \cdots, n$.

The empty set belongs to every semiring. If X is any set, then the class $\mathbf{P}$ consisting of the empty set and all one-point sets (i.e. sets of the form $\{x\}$ with $x \in X$) is a semiring. If X is the real line, the class of all bounded, left closed, and right open intervals is a semiring.

§ 5. GENERATED RINGS AND σ-RINGS

Theorem A. *If* $\mathbf{E}$ *is any class of sets, then there exists a unique ring* $\mathbf{R}_0$ *such that* $\mathbf{R}_0 \supset \mathbf{E}$ *and such that if* $\mathbf{R}$ *is any other ring containing* $\mathbf{E}$ *then* $\mathbf{R}_0 \subset \mathbf{R}$.

The ring $\mathbf{R}_0$, the smallest ring containing $\mathbf{E}$, is called the ring **generated** by $\mathbf{E}$; it will be denoted by $\mathbf{R}(\mathbf{E})$.

Proof. Since the class of all subsets of X is a ring, it is clear that at least one ring containing $\mathbf{E}$ always exists. Since moreover (cf. 4.3) the intersection of any collection of rings is also a ring, the intersection of all rings containing $\mathbf{E}$ is easily seen to be the desired ring $\mathbf{R}_0$. ∎

Theorem B. *If* $\mathbf{E}$ *is any class of sets, then every set in* $\mathbf{R}(\mathbf{E})$ *may be covered by a finite union of sets in* $\mathbf{E}$.

Proof. The class of all sets which may be covered by a finite union of sets in **E** is a ring; since this ring contains **E**, it also contains **R**(**E**). ∎

Theorem C. *If* **E** *is a countable class of sets, then* **R**(**E**) *is countable.*

Proof. For any class **C** of sets, we write **C*** for the class of all finite unions of differences of sets of **C**. It is clear that if **C** is countable, then so is **C***, and if

$$0 \in \mathbf{C},$$

then

$$\mathbf{C} \subset \mathbf{C}^*.$$

To prove the theorem we assume, as we may without any loss of generality, that

$$0 \in \mathbf{E},$$

and we write

$$\mathbf{E}_0 = \mathbf{E}, \quad \mathbf{E}_n = \mathbf{E}_{n-1}^*, \quad n = 1, 2, \cdots.$$

It is clear that

$$\mathbf{E} \subset \bigcup_{n=0}^{\infty} \mathbf{E}_n \subset \mathbf{R}(\mathbf{E}),$$

and that the class

$$\bigcup_{n=0}^{\infty} \mathbf{E}_n$$

is countable; we shall complete the proof by showing that $\bigcup_{n=0}^{\infty} \mathbf{E}_n$ is a ring. Since

$$\mathbf{E} = \mathbf{E}_0 \subset \mathbf{E}_1 \subset \mathbf{E}_2 \subset \cdots,$$

it follows that if A and B are any two sets in $\bigcup_{n=0}^{\infty} \mathbf{E}_n$, then there exists a positive integer n such that both A and B belong to $\mathbf{E}_n$. We have

$$A - B \in \mathbf{E}_{n+1},$$

and, since

$$0 \in \mathbf{E}_0 \subset \mathbf{E}_n,$$

it follows also that

$$A \cup B = (A - 0) \cup (B - 0) \in \mathbf{E}_{n+1}.$$

We have proved therefore that both $A - B$ and $A \cup B$ belong to $\bigcup_{n=0}^{\infty} \mathbf{E}_n$, i.e. that $\bigcup_{n=0}^{\infty} \mathbf{E}_n$ is indeed closed under the formation of unions and differences. ▮

A **σ-ring** is a non empty class $\mathbf{S}$ of sets such that

(a) if $E \in \mathbf{S}$ and $F \in \mathbf{S}$, then $E - F \in \mathbf{S}$, and
(b) if $E_i \in \mathbf{S}$, $i = 1, 2, \cdots$, then $\bigcup_{i=1}^{\infty} E_i \in \mathbf{S}$.

Equivalently a σ-ring is a ring closed under the formation of countable unions. If $\mathbf{S}$ is a σ-ring and if

$$E_i \in \mathbf{S}, \quad i = 1, 2, \cdots, \quad \text{and} \quad E = \bigcup_{i=1}^{\infty} E_i,$$

then the identity

$$\bigcap_{i=1}^{\infty} E_i = E - \bigcup_{i=1}^{\infty} (E - E_i)$$

shows that

$$\bigcap_{i=1}^{\infty} E_i \in \mathbf{S},$$

i.e. that a σ-ring is closed under the formation of countable intersections. It follows also (cf. 3.2) that if $\mathbf{S}$ is a σ-ring and

$$E_i \in \mathbf{S}, \quad i = 1, 2, \cdots,$$

then both $\liminf_i E_i$ and $\limsup_i E_i$ belong to $\mathbf{S}$.

Since the truth and proof of Theorem A remain unaltered if we replace "ring" by "σ-ring" throughout, we may define the σ-ring $\mathbf{S}(\mathbf{E})$ **generated** by any class $\mathbf{E}$ of sets as the smallest σ-ring containing $\mathbf{E}$.

Theorem D. *If* $\mathbf{E}$ *is any class of sets and* E *is any set in* $\mathbf{S} = \mathbf{S}(\mathbf{E})$, *then there exists a countable subclass* $\mathbf{D}$ *of* $\mathbf{E}$ *such that* $E \in \mathbf{S}(\mathbf{D})$.

Proof. The union of all those σ-subrings of $\mathbf{S}$ which are generated by some countable subclass of $\mathbf{E}$ is a σ-ring containing $\mathbf{E}$ and contained in $\mathbf{S}$; it is therefore identical with $\mathbf{S}$. ▮

For every class $\mathbf{E}$ of subsets of X and every fixed subset A of X, we shall denote by

$$\mathbf{E} \cap A$$

the class of all sets of the form $E \cap A$ with E in $\mathbf{E}$.

Theorem E. *If* $\mathbf{E}$ *is any class of sets and if* A *is any subset of* X, *then*

$$\mathbf{S}(\mathbf{E}) \cap A = \mathbf{S}(\mathbf{E} \cap A).$$

Proof. Denote by $\mathbf{C}$ the class of all sets of the form $B \cup (C - A)$, where

$$B \varepsilon \mathbf{S}(\mathbf{E} \cap A) \quad \text{and} \quad C \varepsilon \mathbf{S}(\mathbf{E});$$

it is easy to verify that $\mathbf{C}$ is a σ-ring. If $E \varepsilon \mathbf{E}$, then the relation

$$E = (E \cap A) \cup (E - A),$$

together with

$$E \cap A \varepsilon \mathbf{E} \cap A \subset \mathbf{S}(\mathbf{E} \cap A),$$

shows that $E \varepsilon \mathbf{C}$, and therefore that

$$\mathbf{E} \subset \mathbf{C}.$$

It follows that

$$\mathbf{S}(\mathbf{E}) \subset \mathbf{C}$$

and therefore that

$$\mathbf{S}(\mathbf{E}) \cap A \subset \mathbf{C} \cap A.$$

Since, however, it is obvious that

$$\mathbf{C} \cap A = \mathbf{S}(\mathbf{E} \cap A),$$

it follows that

$$\mathbf{S}(\mathbf{E}) \cap A \subset \mathbf{S}(\mathbf{E} \cap A).$$

The reverse inequality,

$$\mathbf{S}(\mathbf{E} \cap A) \subset \mathbf{S}(\mathbf{E}) \cap A,$$

follows from the facts that $\mathbf{S}(\mathbf{E}) \cap A$ is a σ-ring and

$$\mathbf{E} \cap A \subset \mathbf{S}(\mathbf{E}) \cap A. \quad \blacksquare$$

(1) For each of the following examples, what is the ring generated by the class $\mathbf{E}$ of sets there described?

(1a) For a fixed subset E of X, $\mathbf{E} = \{E\}$ is the class containing E only.

(1b) For a fixed subset E of X, $\mathbf{E}$ is the class of all sets of which E is a subset, i.e. $\mathbf{E} = \{F: E \subset F\}$.

(1c) $\mathbf{E}$ is the class of all sets which contain exactly two points.

(2) A **lattice** (of sets) is a class $\mathbf{L}$ such that $0 \varepsilon \mathbf{L}$ and such that if $E \varepsilon \mathbf{L}$ and $F \varepsilon \mathbf{L}$, then $E \cup F \varepsilon \mathbf{L}$ and $E \cap F \varepsilon \mathbf{L}$. Let $\mathbf{P} = \mathbf{P}(\mathbf{L})$ be the class of all sets

of the form $F - E$, where $E \varepsilon \mathbf{L}$, $F \varepsilon \mathbf{L}$, and $E \subset F$; then $\mathbf{P}$ is a semiring; (cf. 4.6). (Hint: if

$$D_i = F_i - E_i, \quad i = 1, 2$$

are representations of two sets of $\mathbf{P}$ as proper differences of sets of $\mathbf{L}$, and if $D_1 \supset D_2$, then for

$$C = (F_1 \cap F_2) - (E_1 \cap F_2),$$

or, alternatively, for

$$C = F_1 - [E_1 \cup (F_1 \cap E_2)],$$

we have $F_2 - E_2 \subset C \subset F_1 - E_1$.) Is $\mathbf{P}$ a ring?

(3) Let $\mathbf{P}$ be a semiring and let $\mathbf{R}$ be the class of all sets of the form $\bigcup_{i=1}^n E_i$, where $\{E_1, \cdots, E_n\}$ is an arbitrary finite, disjoint class of sets in $\mathbf{P}$.

(3a) $\mathbf{R}$ is closed under the formation of finite intersections and disjoint unions.

(3b) If $E \varepsilon \mathbf{P}$, $F \varepsilon \mathbf{P}$, and $E \subset F$, then $F - E \varepsilon \mathbf{R}$.

(3c) If $E \varepsilon \mathbf{P}$, $F \varepsilon \mathbf{R}$, and $E \subset F$, then $F - E \varepsilon \mathbf{R}$.

(3d) If $E \varepsilon \mathbf{R}$, $F \varepsilon \mathbf{R}$, and $E \subset F$, then $F - E \varepsilon \mathbf{R}$.

(3e) $\mathbf{R} = \mathbf{R}(\mathbf{P})$. It follows in particular that a semiring which is closed under the formation of unions is a ring.

(4) The fact that the analog of Theorem A for algebras is true may be seen either by replacing "ring" by "algebra" in its proof or by using 4.5.

(5) If $\mathbf{P}$ is a semiring and $\mathbf{R} = \mathbf{R}(\mathbf{P})$, then $\mathbf{S}(\mathbf{R}) = \mathbf{S}(\mathbf{P})$.

(6) Is it true that if a non empty class of sets is closed under the formation of symmetric differences and countable intersections, then it is a σ-ring?

(7) If $\mathbf{E}$ is a non empty class of sets, then every set in $\mathbf{S}(\mathbf{E})$ may be covered by a countable union of sets in $\mathbf{E}$; (cf. Theorem B).

(8) If $\mathbf{E}$ is an infinite class of sets, then $\mathbf{E}$ and $\mathbf{R}(\mathbf{E})$ have the same cardinal number; (cf. Theorem C).

(9) The following procedure yields an analog of Theorem C for σ-rings; (cf. also (8)). If $\mathbf{E}$ is any class of sets containing 0, write $\mathbf{E}_0 = \mathbf{E}$, and, for any ordinal $\alpha > 0$, write, inductively,

$$\mathbf{E}_\alpha = (\bigcup \{\mathbf{E}_\beta : \beta < \alpha\})^*,$$

where $\mathbf{C}^*$ denotes the class of all countable unions of differences of sets of $\mathbf{C}$.

(9a) If $0 < \alpha < \beta$, then $\mathbf{E} \subset \mathbf{E}_\alpha \subset \mathbf{E}_\beta \subset \mathbf{S}(\mathbf{E})$.

(9b) If Ω is the first uncountable ordinal, then $\mathbf{S}(\mathbf{E}) = \bigcup\{\mathbf{E}_\alpha : \alpha < \Omega\}$.

(9c) If the cardinal number of $\mathbf{E}$ is not greater than that of the continuum, then the same is true of the cardinal number of $\mathbf{S}(\mathbf{E})$.

(10) What are the analogs of Theorems D and E for rings instead of σ-rings?

§ 6. MONOTONE CLASSES

It is impossible to give a constructive process for obtaining the σ-ring generated by a class of sets. By studying, however, another type of class, less restricted than a σ-ring, it is possible to obtain a technically very helpful theorem concerning the structure of generated σ-rings.

A non empty class **M** of sets is **monotone** if, for every monotone sequence $\{E_n\}$ of sets in **M**, we have

$$\lim_n E_n \in \mathbf{M}.$$

Since it is true for monotone classes (just as for rings and σ-rings) that the class of all subsets of X is a monotone class, and that the intersection of any collection of monotone classes is a monotone class, we may define the monotone class **M(E)** **generated** by any class **E** of sets as the smallest monotone class containing **E**.

Theorem A. *A σ-ring is a monotone class; a monotone ring is a σ-ring.*

Proof. The first assertion is obvious. To prove the second assertion we must show that a monotone ring is closed under the formation of countable unions. If **M** is a monotone ring and if

$$E_i \in \mathbf{M}, \quad i = 1, 2, \cdots,$$

then, since **M** is a ring,

$$\bigcup_{i=1}^{n} E_i \in \mathbf{M}, \quad n = 1, 2, \cdots.$$

Since $\{\bigcup_{i=1}^{n} E_i\}$ is an increasing sequence of sets whose union is $\bigcup_{i=1}^{\infty} E_i$, the fact that **M** is a monotone class implies that

$$\bigcup_{i=1}^{\infty} E_i \in \mathbf{M}. \quad \blacksquare$$

Theorem B. *If* **R** *is a ring, then* **M(R)** = **S(R)**. *Hence if a monotone class contains a ring* **R**, *then it contains* **S(R)**.

Proof. Since a σ-ring is a monotone class and since **S(R)** $\supset$ **R**, it follows that

$$\mathbf{S(R)} \supset \mathbf{M} = \mathbf{M(R)}.$$

The proof will be completed by showing that **M** is a σ-ring; it will then follow, since **M(R)** $\supset$ **R**, that **M(R)** $\supset$ **S(R)**.

For any set F let $\mathbf{K}(F)$ be the class of all those sets E for which $E - F$, $F - E$, and $E \cup F$ are all in **M**. We observe that, because of the symmetric roles of E and F in the definition of $\mathbf{K}(F)$, the relations

$$E \in \mathbf{K}(F) \quad \text{and} \quad F \in \mathbf{K}(E)$$

are equivalent. If $\{E_n\}$ is a monotone sequence of sets in $\mathbf{K}(F)$, then

$$\lim_n E_n - F = \lim_n (E_n - F) \in \mathbf{M},$$

$$F - \lim_n E_n = \lim_n (F - E_n) \in \mathbf{M},$$

$$F \cup \lim_n E_n = \lim_n (F \cup E_n) \in \mathbf{M},$$

so that if $\mathbf{K}(F)$ is not empty, then it is a monotone class.

If $E \in \mathbf{R}$ and $F \in \mathbf{R}$, then, by the definition of a ring, $E \in \mathbf{K}(F)$. Since this is true for every E in $\mathbf{R}$, it follows that $\mathbf{R} \subset \mathbf{K}(F)$, and therefore, since $\mathbf{M}$ is the smallest monotone class containing $\mathbf{R}$, that

$$\mathbf{M} \subset \mathbf{K}(F).$$

Hence if $E \in \mathbf{M}$ and $F \in \mathbf{R}$, then $E \in \mathbf{K}(F)$, and therefore $F \in \mathbf{K}(E)$. Since this is true for every F in $\mathbf{R}$, it follows as before that

$$\mathbf{M} \subset \mathbf{K}(E).$$

The validity of this relation for every E in $\mathbf{M}$ is equivalent to the assertion that $\mathbf{M}$ is a ring; the desired conclusion follows from Theorem A. ▌

This theorem does not tell us, given a ring $\mathbf{R}$, how to construct the generated σ-ring. It does tell us that, instead of studying the σ-ring generated by $\mathbf{R}$, it is sufficient to study the monotone class generated by $\mathbf{R}$. In many applications that is quite easy to do.

(1) Is Theorem B true for semirings instead of rings?

(2) A class $\mathbf{N}$ of sets is **normal** if it is closed under the formation of countable decreasing intersections and countable disjoint unions. A σ-ring is a normal class; a normal ring is a σ-ring.

(3) If the smallest normal class containing a class $\mathbf{E}$ is denoted by $\mathbf{N}(\mathbf{E})$, then, for every semiring $\mathbf{P}$, $\mathbf{N}(\mathbf{P}) = \mathbf{S}(\mathbf{P})$.

(4) If a **σ-algebra** of sets is defined as a non empty class of sets closed under the formation of complements and countable unions, then a σ-algebra is a σ-ring containing X. If $\mathbf{R}$ is an algebra, then $\mathbf{M}(\mathbf{R})$ coincides with the smallest σ-algebra containing $\mathbf{R}$. Is this result true if $\mathbf{R}$ is a ring?

(5) For each of the following examples what is the σ-algebra, the σ-ring, and the monotone class generated by the class **E** of sets there described?

(5a) Let X be any set and let P be any permutation of the points of X, i.e. P is a one to one transformation of X onto itself. A subset E of X is **invariant** under P if, whenever $x \varepsilon E$, then $P(x) \varepsilon E$ and $P^{-1}(x) \varepsilon E$. Let **E** be the class of all invariant sets.

(5b) Let X and Y be any two sets and let T be any (not necessarily one to one) transformation defined on X and taking X into Y. For every subset E of Y denote by $T^{-1}(E)$ the set of all points x in X for which $T(x) \varepsilon E$. Let **E** be the class of all sets of the form $T^{-1}(E)$, where E varies over all subsets of Y.

(5c) X is a topological space; **E** is the class of all sets of the first category.

(5d) X is three dimensional Euclidean space. Let a subset E of X be called a **cylinder** if whenever $(x,y,z) \varepsilon E$, then $(x,y,\hat{z}) \varepsilon E$ for every real number $\hat{z}$. Let **E** be the class of all cylinders.

(5e) X is the Euclidean plane; **E** is the class of all sets which may be covered by countably many horizontal lines.

Chapter II

MEASURES AND OUTER MEASURES

§ 7. MEASURE ON RINGS

A **set function** is a function whose domain of definition is a class of sets. An extended real valued set function μ defined on a class $\mathbf{E}$ of sets is **additive** if, whenever

$$E \in \mathbf{E}, \quad F \in \mathbf{E}, \quad E \cup F \in \mathbf{E}, \quad \text{and} \quad E \cap F = 0,$$

then

$$\mu(E \cup F) = \mu(E) + \mu(F).$$

An extended real valued set function μ defined on a class $\mathbf{E}$ is **finitely additive** if, for every finite, disjoint class $\{E_1, \cdots, E_n\}$ of sets in $\mathbf{E}$ whose union is also in $\mathbf{E}$, we have

$$\mu(\textstyle\bigcup_{i=1}^n E_i) = \sum_{i=1}^n \mu(E_i).$$

An extended real valued set function μ defined on a class $\mathbf{E}$ is **countably additive** if, for every disjoint sequence $\{E_n\}$ of sets in $\mathbf{E}$ whose union is also in $\mathbf{E}$, we have

$$\mu(\textstyle\bigcup_{n=1}^\infty E_n) = \sum_{n=1}^\infty \mu(E_n).$$

A **measure** is an extended real valued, non negative, and countably additive set function μ, defined on a ring $\mathbf{R}$, and such that $\mu(0) = 0$.

We observe that, in view of the identity,

$$\textstyle\bigcup_{i=1}^n E_i = E_1 \cup \cdots \cup E_n \cup 0 \cup 0 \cup \cdots,$$

a measure is always finitely additive. A rather trivial example of a measure may be obtained as follows. Let f be an extended real valued, non negative function of the points of a set X. Let the ring **R** consist of all finite subsets of X; define μ by

$$\mu(\{x_1, \cdots, x_n\}) = \sum_{i=1}^n f(x_i) \quad \text{and} \quad \mu(0) = 0.$$

Less trivial examples will be presented in the following sections.

If μ is a measure on a ring **R**, a set E in **R** is said to have **finite measure** if $\mu(E) < \infty$; the measure of E is **σ-finite** if there exists a sequence $\{E_n\}$ of sets in **R** such that

$$E \subset \bigcup_{n=1}^\infty E_n \quad \text{and} \quad \mu(E_n) < \infty, \quad n = 1, 2, \cdots.$$

If the measure of every set E in **R** is finite [or σ-finite], the measure μ is called **finite** [or **σ-finite**] on **R**. If $X \in \mathbf{R}$ (i.e. if **R** is an algebra) and $\mu(X)$ is finite or σ-finite, then μ is called **totally finite** or **totally σ-finite** respectively. The measure μ is called **complete** if the conditions

$$E \in \mathbf{R}, \quad F \subset E, \quad \text{and} \quad \mu(E) = 0$$

imply that $F \in \mathbf{R}$.

(1) If μ is an extended real valued, non negative, and additive set function defined on a ring **R** and such that $\mu(E) < \infty$ for at least one E in **R**, then $\mu(0) = 0$.

(2) If **E** is a non empty class of sets and μ is a measure on $\mathbf{R}(\mathbf{E})$ such that $\mu(E) < \infty$ whenever $E \in \mathbf{E}$, then μ is finite on $\mathbf{R}(\mathbf{E})$; cf. 5.B.

(3) If μ is a measure on a σ-ring, then the class of all sets of finite measure is a ring and the class of all sets of σ-finite measure is a σ-ring. If, in addition, μ is σ-finite, then a necessary and sufficient condition that the class of all sets of finite measure be a σ-ring is that μ be finite. Is the latter statement true if μ is not σ-finite?

(4) Suppose that μ is a measure on a σ-ring **S** and that E is a set in **S** of σ-finite measure. If **D** is a disjoint class of sets in **S**, then $\mu(E \cap D) \neq 0$ for at most countably many sets D in **D**. (Hint: assume first that $\mu(E) < \infty$; for each positive integer n consider the class

$$\left\{D: D \in \mathbf{D}, \quad \mu(E \cap D) \geq \frac{1}{n}\right\}.)$$

(5) If μ is an extended real valued, non negative, and additive set function defined on a ring **R** and such that $\mu(0) = 0$, then μ is finitely additive. The proof of the same statement for a semiring **P** is not trivial; it may be achieved by the following considerations. A finite, disjoint class $\{E_1, \cdots, E_n\}$ of sets in **P** whose union, E, is also in **P** is called a **partition** of E. The partition $\{E_i\}$ is called a **μ-partition** if, for every F in **P**,

$$\mu(E \cap F) = \sum_{i=1}^n \mu(E_i \cap F).$$

If $\{E_i\}$ and $\{F_j\}$ are partitions of E, then $\{E_i\}$ is called a **subpartition** of $\{F_j\}$ if each set E_i is contained in one of the sets F_j.

(5a) If $\{E_i\}$ and $\{F_j\}$ are partitions of E, then so is their **product**, consisting of all sets of the form $E_i \cap F_j$.

(5b) If a subpartition of a partition $\{E_i\}$ is a μ-partition, then $\{E_i\}$ is a μ-partition.

(5c) The product of two μ-partitions is a μ-partition.

(5d) If $E = C_0 \subset C_1 \subset \cdots \subset C_n = F$, where $C_i \,\varepsilon\, \mathbf{P}$, $i = 0, 1, \cdots, n$, and if

$$D_i = C_i - C_{i-1} \,\varepsilon\, \mathbf{P}, \quad i = 1, \cdots, n,$$

then $\{E, D_1, \cdots, D_n\}$ is a μ-partition of F.

(5e) Every partition of a set E in $\mathbf{P}$ is a μ-partition.

§ 8. MEASURE ON INTERVALS

In order to motivate and illustrate the elementary notions of measure theory, we now propose to discuss an important and classical special case. Throughout this section the underlying space X is to be the real line. We shall denote by $\mathbf{P}$ the class of all bounded, left closed, and right open intervals, i.e. the class of all sets of the form

$$\{x: -\infty < a \leqq x < b < \infty\};$$

we shall denote by $\mathbf{R}$ the class of all finite, disjoint unions of sets of $\mathbf{P}$, i.e. the class of all sets of the form

$$\bigcup_{i=1}^{n} \{x: -\infty < a_i \leqq x < b_i < \infty\}.$$

(It is easy to verify that any union of this form may be written as a disjoint union of the same form.)

For simplicity of language we shall always use the expression "semiclosed interval" instead of "bounded, left closed, and right open interval." The consideration of semiclosed intervals, instead of open intervals or closed intervals, is a technical device. If, for instance, a, b, c, and d are real numbers, $-\infty < a < b < c < d < \infty$, then the difference between the open intervals

$$\{x: a < x < d\} \quad \text{and} \quad \{x: b < x < c\}$$

is neither an open interval nor a finite union of open intervals, and the same negative statement holds for the corresponding closed intervals. The fact that semiclosed intervals are better behaved in this respect is what makes them desirable.

We shall, as usual, write $[a,b]$ for a closed interval,

$$[a,b] = \{x: a \leqq x \leqq b\},$$

$[a,b)$ for a semiclosed interval,

$$[a,b) = \{x: a \leqq x < b\},$$

and (a,b) for an open interval,

$$(a,b) = \{x: a < x < b\}.$$

In writing any of these symbols it shall always be understood that $a \leqq b$.

On the class $\mathbf{P}$ of semiclosed intervals we define a set function μ by

$$\mu([a,b)) = b - a.$$

We observe that when $a = b$, the interval reduces to the empty set, so that

$$\mu(0) = 0.$$

We proceed to investigate the relation of the set function μ to some set theoretic notions in $\mathbf{P}$.

Theorem A. *If $\{E_1, \cdots, E_n\}$ is a finite, disjoint class of sets in $\mathbf{P}$, each contained in a given set E_0 in $\mathbf{P}$, then*

$$\sum_{i=1}^{n} \mu(E_i) \leqq \mu(E_0).$$

Proof. We write $E_i = [a_i,b_i)$, $i = 0, 1, \cdots, n$, and, without any loss of generality, we assume that

$$a_1 \leqq \cdots \leqq a_n.$$

It follows from the assumed properties of $\{E_1, \cdots, E_n\}$ that

$$a_0 \leqq a_1 \leqq b_1 \leqq \cdots \leqq a_n \leqq b_n \leqq b_0,$$

and therefore

$$\begin{aligned}\sum_{i=1}^{n} \mu(E_i) &= \sum_{i=1}^{n} (b_i - a_i) \leqq \\ &\leqq \sum_{i=1}^{n} (b_i - a_i) + \sum_{i=1}^{n-1} (a_{i+1} - b_i) = \\ &= b_n - a_1 \leqq b_0 - a_0 = \mu(E_0). \quad \blacksquare\end{aligned}$$

Theorem B. *If a closed interval F_0, $F_0 = [a_0,b_0]$, is contained in the union of a finite number of bounded, open intervals, $U_1, \cdots, U_n$, $U_i = (a_i,b_i)$, $i = 1, \cdots, n$, then*

$$b_0 - a_0 < \sum_{i=1}^n (b_i - a_i).$$

Proof. Let k_1 be such that $a_0 \in U_{k_1}$. If $b_{k_1} \leqq b_0$, then let k_2 be such that $b_{k_1} \in U_{k_2}$; if $b_{k_2} \leqq b_0$, then let k_3 be such that $b_{k_2} \in U_{k_3}$, and so on by induction. The process stops with k_m if $b_{k_m} > b_0$. There is no loss of generality in assuming that $m = n$ and $U_{k_i} = U_i$ for $i = 1, \cdots, n$, because this state of affairs may be achieved merely by omitting superfluous U_i's and changing the notation. In other words we may (and do) assume that

$$a_1 < a_0 < b_1, \quad a_n < b_0 < b_n,$$

and, in case $n > 1$,

$$a_{i+1} < b_i < b_{i+1} \quad \text{for} \quad i = 1, \cdots, n - 1;$$

it follows that

$$b_0 - a_0 < b_n - a_1 = (b_1 - a_1) + \sum_{1 \leqq i \leqq n-1} (b_{i+1} - b_i) \leqq$$
$$\leqq \sum_{i=1}^n (b_i - a_i). \quad \blacksquare$$

Theorem C. *If $\{E_0, E_1, E_2, \cdots\}$ is a sequence of sets in* **P**, *such that*

$$E_0 \subset \bigcup_{i=1}^\infty E_i,$$

then

$$\mu(E_0) \leqq \sum_{i=1}^\infty \mu(E_i).$$

Proof. We write $E_i = [a_i,b_i)$, $i = 0, 1, 2, \cdots$. If $a_0 = b_0$, the theorem is trivial; otherwise let ϵ be a positive number such that $\epsilon < b_0 - a_0$. If we write, for any positive number δ,

$$F_0 = [a_0, b_0 - \epsilon] \quad \text{and} \quad U_i = \left(a_i - \frac{\delta}{2^i}, b_i\right), \quad i = 1, 2, \cdots,$$

then

$$F_0 \subset \bigcup_{i=1}^\infty U_i$$

and therefore, by the Heine–Borel theorem, there is a positive integer n such that $F_0 \subset \bigcup_{i=1}^{n} U_i$. From Theorem B we obtain

$$\mu(E_0) - \epsilon = (b_0 - a_0) - \epsilon < \sum_{i=1}^{n} \left(b_i - a_i + \frac{\delta}{2^i}\right) \leqq$$
$$\leqq \sum_{i=1}^{\infty} \mu(E_i) + \delta.$$

Since ϵ and δ are arbitrary, the conclusion of the theorem follows. ∎

Theorem D. *The set function μ is countably additive on* **P**.

Proof. If $\{E_i\}$ is a disjoint sequence of sets in **P** whose union, E, is also in **P**, then from Theorem A we have

$$\sum_{i=1}^{n} \mu(E_i) \leqq \mu(E) \quad \text{for} \quad n = 1, 2, \cdots.$$

It follows that

$$\sum_{i=1}^{\infty} \mu(E_i) \leqq \mu(E);$$

an application of Theorem C completes the proof. ∎

Theorem E. *There exists a unique, finite measure $\bar{\mu}$ on the ring* **R** *such that, whenever $E \,\varepsilon\, \mathbf{P}$, $\bar{\mu}(E) = \mu(E)$.*

Proof. We know that every set E in **R** may be represented as a finite, disjoint union of sets in **P**. Suppose that

$$E = \bigcup_{i=1}^{n} E_i \quad \text{and} \quad E = \bigcup_{j=1}^{m} F_j$$

are two such representations of the same set E. Then, for each $i = 1, \cdots, n$,

$$E_i = \bigcup_{j=1}^{m} (E_i \cap F_j)$$

is a representation of the set E_i in **P** as a finite, disjoint union of sets in **P**, and therefore, since μ is finitely additive,

$$\sum_{i=1}^{n} \mu(E_i) = \sum_{i=1}^{n} \sum_{j=1}^{m} \mu(E_i \cap F_j).$$

Similarly, of course, we have

$$\sum_{j=1}^{m} \mu(F_j) = \sum_{j=1}^{m} \sum_{i=1}^{n} \mu(E_i \cap F_j).$$

It follows that, for every E in **R**, the function $\bar{\mu}$ is unambiguously defined by the equation

$$\bar{\mu}(E) = \sum_{i=1}^{n} \mu(E_i),$$

where $\{E_1, \cdots, E_n\}$ is a finite, disjoint class of sets in **P** whose union is E.

It is clear from its definition that the function $\bar{\mu}$ thus defined coincides with μ on **P** and is finitely additive. Since any function satisfying these conditions must in particular be finitely additive when the terms of the union to be formed are in **P**, it is also clear that $\bar{\mu}$ is unique. The only non trivial thing left to prove is that $\bar{\mu}$ is countably additive.

Let $\{E_i\}$ be a disjoint sequence of sets in **R** whose union, E, is also in **R**; then each E_i is a finite, disjoint union of sets in **P**,

$$E_i = \bigcup_j E_{ij} \quad \text{and} \quad \bar{\mu}(E_i) = \sum_j \mu(E_{ij}).$$

If $E \in \mathbf{P}$, then, since the class of all E_{ij} is countable and disjoint, it follows from the countable additivity of μ that

$$\bar{\mu}(E) = \mu(E) = \sum_i \sum_j \mu(E_{ij}) = \sum_i \bar{\mu}(E_i).$$

In the general case E is a finite, disjoint union of sets in **P**,

$$E = \bigcup_k F_k,$$

and, using the result just obtained, we have

$$\bar{\mu}(E) = \sum_k \bar{\mu}(F_k) = \sum_k \sum_i \bar{\mu}(E_i \cap F_k) =$$
$$= \sum_i \sum_k \bar{\mu}(E_i \cap F_k) = \sum_i \bar{\mu}(E_i). \quad \blacksquare$$

In view of Theorem E we shall, as we may without any possibility of confusion, write $\mu(E)$ instead of $\bar{\mu}(E)$ even for sets E which are in **R** but not in **P**.

(1) In the notation of the proof of Theorem D, let E_{n_1} be that term of the sequence $\{E_i\}$ whose left end point is the left end point of E; let E_{n_2} be the term whose left end point is the right end point of E_{n_1}, and so on. It may be shown, without using Theorems A, B, and C, that

$$\bigcup_{i=1}^{\infty} E_{n_i} \in \mathbf{P} \quad \text{and} \quad \mu(\bigcup_{i=1}^{\infty} E_{n_i}) = \sum_{i=1}^{\infty} \mu(E_{n_i}).$$

(2) An alternative proof of Theorem D (which does not use Theorems A, B, and C) proceeds by arranging the terms of the sequence $\{E_i\}$ in the order of magnitude of their left end points and then applying transfinite induction; cf. (1).

(3) Let g be a finite, increasing, and continuous function of a real variable, and write

$$\mu_g([a,b)) = g(b) - g(a).$$

Theorems D and E remain true if μ is replaced by μ_g.

(4) Theorems D and E may be generalized to n-dimensional Euclidean space by considering "intervals" of the form

$$E = \{(x_1, \cdots, x_n): a_i \leq x_i < b_i, \quad i = 1, \cdots, n\},$$

and defining μ by

$$\mu(E) = \prod_{i=1}^n (b_i - a_i).$$

(5) If μ is a countably additive and non negative set function on a semiring **P**, such that $\mu(0) = 0$, then there is a unique measure $\bar{\mu}$ on the ring **R**(**P**) such that, whenever $E \varepsilon \mathbf{P}$, $\bar{\mu}(E) = \mu(E)$. If μ is [totally] finite or σ-finite, then so is $\bar{\mu}$; (cf. 5.3 and the proof of Theorem E).

§ 9. PROPERTIES OF MEASURES

An extended real valued set function μ on a class **E** is **monotone** if, whenever $E \varepsilon \mathbf{E}$, $F \varepsilon \mathbf{E}$, and $E \subset F$, then $\mu(E) \leq \mu(F)$. An extended real valued set function μ on a class **E** is **subtractive** if, whenever $E \varepsilon \mathbf{E}$, $F \varepsilon \mathbf{E}$, $E \subset F$, $F - E \varepsilon \mathbf{E}$, and $|\mu(E)| < \infty$, then

$$\mu(F - E) = \mu(F) - \mu(E).$$

Theorem A. *If μ is a measure on a ring* **R**, *then μ is monotone and subtractive.*

Proof. If $E \varepsilon \mathbf{R}$, $F \varepsilon \mathbf{R}$, and $E \subset F$, then $F - E \varepsilon \mathbf{R}$ and $\mu(F) = \mu(E) + \mu(F - E)$. The fact that μ is monotone follows now from the fact that it is non negative; the fact that it is subtractive follows from the fact that $\mu(E)$, if it is finite, may be subtracted from both sides of the last written equation. ▮

Theorem B. *If μ is a measure on a ring* **R**, *if $E \varepsilon \mathbf{R}$, and if $\{E_i\}$ is a finite or infinite sequence of sets in* **R** *such that $E \subset \bigcup_i E_i$,* then

$$\mu(E) \leq \sum_i \mu(E_i).$$

Proof. We make use of the elementary but important fact that if $\{F_i\}$ is any sequence of sets in a ring **R**, then there exists a disjoint sequence $\{G_i\}$ of sets in **R** such that

$$G_i \subset F_i \quad \text{and} \quad \bigcup_i G_i = \bigcup_i F_i.$$

(Write $G_i = F_i - \bigcup \{F_j: 1 \leq j < i\}$.) Applying this result to the sequence $\{E \cap E_i\}$, the desired result follows from the countable additivity and monotoneness of μ. ▮

Theorem C. *If μ is a measure on a ring* $\mathbf{R}$, *if* $E \in \mathbf{R}$, *and if* $\{E_i\}$ *is a finite or infinite disjoint sequence of sets in* $\mathbf{R}$ *such that* $\bigcup_i E_i \subset E$, *then*

$$\sum_i \mu(E_i) \leqq \mu(E).$$

Proof. If the sequence $\{E_i\}$ is finite, then $\bigcup_i E_i \in \mathbf{R}$, and it follows that

$$\sum_i \mu(E_i) = \mu(\bigcup_i E_i) \leqq \mu(E).$$

The validity of the inequality for an infinite sequence of sets is a consequence of its validity for every finite subsequence. ∎

Theorem D. *If μ is a measure on a ring* $\mathbf{R}$ *and if* $\{E_n\}$ *is an increasing sequence of sets in* $\mathbf{R}$ *for which* $\lim_n E_n \in \mathbf{R}$, *then* $\mu(\lim_n E_n) = \lim_n \mu(E_n)$.

Proof. If we write $E_0 = 0$, then

$$\mu(\lim_n E_n) = \mu(\bigcup_{i=1}^{\infty} E_i) = \mu(\bigcup_{i=1}^{\infty} (E_i - E_{i-1})) =$$
$$= \sum_{i=1}^{\infty} \mu(E_i - E_{i-1}) = \lim_n \sum_{i=1}^{n} \mu(E_i - E_{i-1}) =$$
$$= \lim_n \mu(\bigcup_{i=1}^{n} (E_i - E_{i-1})) = \lim_n \mu(E_n).$$ ∎

Theorem E. *If μ is a measure on a ring* $\mathbf{R}$, *and if* $\{E_n\}$ *is a decreasing sequence of sets in* $\mathbf{R}$ *of which at least one has finite measure and for which* $\lim_n E_n \in \mathbf{R}$, *then* $\mu(\lim_n E_n) = \lim_n \mu(E_n)$.

Proof. If $\mu(E_m) < \infty$, then $\mu(E_n) \leqq \mu(E_m) < \infty$ for $n \geqq m$, and therefore $\mu(\lim_n E_n) < \infty$. It follows from Theorems A and D, and the fact that $\{E_m - E_n\}$ is an increasing sequence, that

$$\mu(E_m) - \mu(\lim_n E_n) = \mu(E_m - \lim_n E_n) =$$
$$= \mu(\lim_n (E_m - E_n)) = \lim_n \mu(E_m - E_n) =$$
$$= \lim_n (\mu(E_m) - \mu(E_n)) =$$
$$= \mu(E_m) - \lim_n \mu(E_n).$$

Since $\mu(E_m) < \infty$, the proof of the theorem is complete. ∎

We shall say that an extended real valued set function μ defined on a class **E** is **continuous from below** at a set E (in **E**) if, for every increasing sequence $\{E_n\}$ of sets in **E** for which $\lim_n E_n = E$, we have $\lim_n \mu(E_n) = \mu(E)$. Similarly μ is **continuous from above** at E if, for every decreasing sequence $\{E_n\}$ of sets in **E** for which $|\mu(E_m)| < \infty$ for at least one value of m and for which $\lim_n E_n = E$, we have $\lim_n \mu(E_n) = \mu(E)$. Theorems D and E assert that if μ is a measure, then μ is continuous from above and from below (at every set in the ring of definition of μ); the following result goes in the converse direction.

Theorem F. *Let μ be a finite, non negative, and additive set function on a ring* **R**. *If μ is either continuous from below at every E in* **R**, *or continuous from above at* 0, *then μ is a measure on* **R**.

Proof. We observe first that the additivity of μ, together with the fact that **R** is a ring, implies, by mathematical induction, that μ is finitely additive. Let $\{E_n\}$ be a disjoint sequence of sets in **R**, whose union, E, is also in **R** and write

$$F_n = \bigcup_{i=1}^n E_i, \quad G_n = E - F_n.$$

If μ is continuous from below, then, since $\{F_n\}$ is an increasing sequence of sets in **R** with $\lim_n F_n = E$, we have

$$\mu(E) = \lim_n \mu(F_n) = \lim_n \sum_{i=1}^n \mu(E_i) = \sum_{i=1}^\infty \mu(E_i).$$

If μ is continuous from above at 0, then, since $\{G_n\}$ is a decreasing sequence of sets in **R** with $\lim_n G_n = 0$, and since μ is finite, we have

$$\mu(E) = \left(\sum_{i=1}^n \mu(E_i)\right) + \mu(G_n) =$$

$$= \lim_n \sum_{i=1}^n \mu(E_i) + \lim_n \mu(G_n) = \sum_{i=1}^\infty \mu(E_i). \quad \blacksquare$$

(1) Theorems A, B, C, D, and E are true for semirings in place of rings. The proofs may be carried out directly or they may be reduced to the corresponding results for rings by means of 8.5.

(2) If μ is a measure on a ring **R**, and if E and F are any two sets in **R**, then

$$\mu(E) + \mu(F) = \mu(E \cup F) + \mu(E \cap F).$$

If E, F, and G are any three sets in $\mathbf{R}$, then

$$\mu(E) + \mu(F) + \mu(G) + \mu(E \cap F \cap G) =$$
$$= \mu(E \cup F \cup G) + \mu(E \cap F) + \mu(F \cap G) + \mu(G \cap E).$$

These statements may be generalized to any finite union.

(3) If μ is a measure on a ring $\mathbf{R}$, and E and F are sets in $\mathbf{R}$, we write $E \sim F$ whenever $\mu(E \Delta F) = 0$. The relation "$\sim$" is reflexive, symmetric, and transitive; if $E \sim F$, then $\mu(E) = \mu(F) = \mu(E \cap F)$. Is the class of all those sets E in $\mathbf{R}$ for which $E \sim 0$ a ring?

(4) Continuing in the notation of (3), we write $\rho(E,F) = \mu(E \Delta F)$. Then $\rho(E,F) \geqq 0$, $\rho(E,F) = \rho(F,E)$, and $\rho(E,F) \leqq \rho(E,G) + \rho(G,F)$. If $E_1 \sim E_2$ and $F_1 \sim F_2$, then $\rho(E_1,F_1) = \rho(E_2,F_2)$.

(5) The following generalizations of Theorems D and E are valid. If μ is a measure on a ring $\mathbf{R}$ and if $\{E_n\}$ is a sequence of sets in $\mathbf{R}$ for which

$$\bigcap_{i=n}^{\infty} E_i \,\varepsilon\, \mathbf{R}, \quad n = 1, 2, \cdots \quad \text{and} \quad \liminf_n E_n = \bigcup_{n=1}^{\infty} \bigcap_{i=n}^{\infty} E_i \,\varepsilon\, \mathbf{R},$$

then $\mu(\liminf_n E_n) \leqq \liminf_n \mu(E_n)$. If, similarly,

$$\bigcup_{i=n}^{\infty} E_i \,\varepsilon\, \mathbf{R}, \quad n = 1, 2, \cdots \quad \text{and} \quad \limsup_n E_n = \bigcap_{n=1}^{\infty} \bigcup_{i=n}^{\infty} E_i \,\varepsilon\, \mathbf{R},$$

and if $\mu(\bigcup_{i=n}^{\infty} E_i) < \infty$ for at least one value of n, then $\mu(\limsup_n E_n) \geqq \limsup_n \mu(E_n)$.

(6) Under the hypotheses of the second part of (5), if $\sum_{n=1}^{\infty} \mu(E_n) < \infty$, then $\mu(\limsup_n E_n) = 0$.

(7) Let X be the set of all rational numbers x for which $0 \leqq x \leqq 1$, and let $\mathbf{P}$ be the class of all "semiclosed intervals" of the form $\{x\colon x \,\varepsilon\, X, a \leqq x < b\}$, where $0 \leqq a \leqq b \leqq 1$, and a and b are rational. Define μ on $\mathbf{P}$ by

$$\mu(\{x\colon a \leqq x < b\}) = b - a.$$

The set function μ is finitely additive and continuous from above and below but it is not countably additive, so that Theorem F is not true for semirings in place of rings.

(8) Let X be the set of all positive integers and let $\mathbf{R}$ be the class of all finite sets and their complements. For E in $\mathbf{R}$ write $\mu(E) = 0$ or $\mu(E) = \infty$ according as E is finite or infinite. The set function μ is continuous from above at 0 but it is not countably additive, so that the second half of Theorem F is not true if infinite values are admitted.

(9) Is Theorem E true without the finiteness condition described in its statement?

(10) If μ is a measure on the Borel sets of a separable, complete, metric space X such that $\mu(X) = 1$, then there exists a subset E of X such that E is a countable union of compact sets and such that $\mu(E) = 1$. (Hint: let $\{x_n\}$ be a sequence of points dense in X and write U_n^k for the closed sphere of radius $\frac{1}{k}$ with center at x_n. If $0 < \epsilon < 1$ and $F_m^k = \bigcup_{n=1}^{m} U_n^k$, let m_k be defined inductively as the smallest positive integer for which

$$\mu(\bigcap_{i=1}^{k} F_{m_i}^i) > 1 - \epsilon.$$

If $C = \bigcap_{i=1}^{\infty} F_{m_i}^i$, then C is compact and $\mu(C) \geqq 1 - \epsilon$.)

§ 10. OUTER MEASURES

A non empty class **E** of sets is **hereditary** if, whenever $E \varepsilon \mathbf{E}$ and $F \subset E$, then $F \varepsilon \mathbf{E}$.

A typical example of a hereditary class is the class of all subsets of some subset E of X. The part of the algebraic theory of hereditary classes that we shall need is very easy and it is similar in every detail to the theories of rings, σ-rings, and other classes of sets we have considered. It is, in particular, true that the intersection of every collection of hereditary classes is again a hereditary class, and that, therefore, corresponding to any class of sets, there is a smallest hereditary class containing it. We shall be especially interested in hereditary classes which are σ-rings; it is easy to see that a hereditary class is a σ-ring if and only if it is closed under the formation of countable unions. If **E** is any class of sets, we shall denote the hereditary σ-ring **generated** by **E**, i.e. the smallest hereditary σ-ring containing **E**, by **H**(**E**). The hereditary σ-ring generated by **E** is, in fact, the class of all sets which can be covered by countably many sets in **E**; if **E** is a non empty class closed under the formation of countable unions (for instance if **E** is a σ-ring), then **H**(**E**) is the class of all sets which are subsets of some set in **E**.

An extended real valued set function μ^* defined on a class **E** of sets is **subadditive** if, whenever $E \varepsilon \mathbf{E}$, $F \varepsilon \mathbf{E}$, and $E \cup F \varepsilon \mathbf{E}$, then

$$\mu^*(E \cup F) \leqq \mu^*(E) + \mu^*(F).$$

An extended real valued set function μ^* on **E** is **finitely subadditive** if, for every finite class $\{E_1, \cdots, E_n\}$ of sets in **E** whose union is also in **E**, we have

$$\mu^*(\bigcup_{i=1}^n E_i) \leqq \sum_{i=1}^n \mu^*(E_i).$$

An extended real valued set function μ^* on **E** is **countably subadditive** if, for every sequence $\{E_i\}$ of sets in **E** whose union is also in **E**, we have

$$\mu^*(\bigcup_{i=1}^\infty E_i) \leqq \sum_{i=1}^\infty \mu^*(E_i).$$

An **outer measure** is an extended real valued, non negative, monotone, and countably subadditive set function μ^*, defined on a hereditary σ-ring $\mathbf{H}$, and such that $\mu^*(0) = 0$. We observe that an outer measure is necessarily finitely subadditive. The same terminology concerning [total] finiteness and σ-finiteness is used for outer measures as for measures.

Outer measures arise naturally in the attempt to extend measures from rings to larger classes of sets. The first precise formulation of some of the details is contained in the following statement.

Theorem A. *If μ is a measure on a ring $\mathbf{R}$ and if, for every E in $\mathbf{H}(\mathbf{R})$,*

$$\mu^*(E) = \inf\{\textstyle\sum_{n=1}^{\infty} \mu(E_n): E_n \in \mathbf{R}, \quad n = 1, 2, \cdots;$$
$$E \subset \textstyle\bigcup_{n=1}^{\infty} E_n\},$$

then μ^* is an extension of μ to an outer measure on $\mathbf{H}(\mathbf{R})$; if μ is [totally] σ-finite, then so is μ^*.

Verbally $\mu^*(E)$ may be described as the lower bound of sums of the type $\sum_{n=1}^{\infty} \mu(E_n)$, where $\{E_n\}$ is a sequence of sets in $\mathbf{R}$ whose union contains E. The outer measure μ^* is called the outer measure **induced** by the measure μ.

Proof. If $E \in \mathbf{R}$, then $E \subset E \cup 0 \cup 0 \cup \cdots$ and therefore $\mu^*(E) \leq \mu(E) + \mu(0) + \mu(0) + \cdots = \mu(E)$. On the other hand if $E \in \mathbf{R}$, $E_n \in \mathbf{R}$, $n = 1, 2, \cdots$, and $E \subset \bigcup_{n=1}^{\infty} E_n$, then, by 9.B, $\mu(E) \leq \sum_{n=1}^{\infty} \mu(E_n)$, so that $\mu(E) \leq \mu^*(E)$. This proves that μ^* is indeed an extension of μ, i.e. that if $E \in \mathbf{R}$, then $\mu^*(E) = \mu(E)$; it follows in particular that $\mu^*(0) = 0$.

If $E \in \mathbf{H}(\mathbf{R})$, $F \in \mathbf{H}(\mathbf{R})$, $E \subset F$, and $\{E_n\}$ is a sequence of sets in $\mathbf{R}$ which covers F, then $\{E_n\}$ also covers E, and therefore $\mu^*(E) \leq \mu^*(F)$.

To prove that μ^* is countably subadditive, suppose that E and E_i are sets in $\mathbf{H}(\mathbf{R})$ such that $E \subset \bigcup_{i=1}^{\infty} E_i$; let ϵ be an arbitrary positive number, and choose, for each $i = 1, 2, \cdots$, a sequence $\{E_{ij}\}$ of sets in $\mathbf{R}$ such that

$$E_i \subset \textstyle\bigcup_{j=1}^{\infty} E_{ij} \quad \text{and} \quad \sum_{j=1}^{\infty} \mu(E_{ij}) \leq \mu^*(E_i) + \frac{\epsilon}{2^i}.$$

(The possibility of such a choice follows from the definition of $\mu^*(E_i)$.) Then, since the sets E_{ij} form a countable class of sets in **R** which covers E,

$$\mu^*(E) \leqq \sum_{i=1}^{\infty} \sum_{j=1}^{\infty} \mu(E_{ij}) \leqq \sum_{i=1}^{\infty} \mu^*(E_i) + \epsilon.$$

The arbitrariness of ϵ implies that

$$\mu^*(E) \leqq \sum_{i=1}^{\infty} \mu^*(E_i).$$

Suppose, finally, that μ is σ-finite and let E be any set in **H**(**R**). Then, by the definition of **H**(**R**), there exists a sequence $\{E_i\}$ of sets in **R** such that $E \subset \bigcup_{i=1}^{\infty} E_i$. Since μ is σ-finite, there exists, for each $i = 1, 2, \cdots$, a sequence $\{E_{ij}\}$ of sets in **R** such that

$$E_i \subset \bigcup_{j=1}^{\infty} E_{ij} \quad \text{and} \quad \mu(E_{ij}) < \infty.$$

Consequently

$$E \subset \bigcup_{i=1}^{\infty} \bigcup_{j=1}^{\infty} E_{ij} \quad \text{and} \quad \mu^*(E_{ij}) = \mu(E_{ij}) < \infty. \quad \blacksquare$$

(1) Is it necessarily true, under the hypotheses of Theorem A, that if μ is finite, then so is μ^*?

(2) For any class **E** of sets we denote by **J**(**E**) the smallest hereditary ring containing **E**. If μ is a real valued, finite, non negative, and finitely additive set function defined on a ring **R**, and if, for every E in **J**(**R**),

$$\mu^*(E) = \inf \{\mu(F): E \subset F \epsilon \mathbf{R}\},$$

then μ^* is a real valued, finite, non negative, and finitely subadditive set function on **J**(**R**). Is it true that, for E in **R**, $\mu^*(E) = \mu(E)$?

(3) A necessary and sufficient condition that a class **H** of subsets of a set X be an ideal in the Boolean ring of all subsets of X is that **H** be a hereditary ring; cf. 4.4.

(4) The following are some examples of set functions defined on hereditary σ-rings; some of them are outer measures, while others violate exactly one of the defining conditions of outer measures.

(4a) X is arbitrary, **H** is the class of all subsets of X. For any fixed point x_0 in X, write $\mu^*(E) = \chi_E(x_0)$.

(4b) X and **H** are as in (4a); $\mu^*(E) = 1$ for every E in **H**.

(4c) $X = \{x,y\}$ is a set consisting of exactly two distinct points x and y, **H** is the class of all subsets of X, and μ^* is defined by the relations

$$\mu^*(0) = 0, \quad \mu^*(\{x\}) = \mu^*(\{y\}) = 10, \quad \mu^*(X) = 1.$$

(4d) X is a set of 100 points arranged in a square array of 10 columns each with 10 points; **H** is the class of all subsets of X; $\mu^*(E)$ is the number of columns which contain at least one point of E.

(4e) X is the set of all positive integers, $\mathbf{H}$ is the class of all subsets of X. For every finite subset E of X, $\nu(E)$ is the number of points in E;

$$\mu^*(E) = \limsup_n \frac{1}{n} \nu(E \cap \{1, \cdots, n\}).$$

(4f) X is arbitrary, $\mathbf{H}$ is the class of all countable subsets of X, $\mu^*(E)$ is the number of points in E, ($= \infty$ if E is infinite).

(5) If μ^* is an outer measure on a hereditary σ-ring $\mathbf{H}$ and E_0 is any set in $\mathbf{H}$, then the set function μ_0^*, defined by $\mu_0^*(E) = \mu^*(E \cap E_0)$, is an outer measure on $\mathbf{H}$.

(6) If λ^* and μ^* are outer measures on a hereditary σ-ring $\mathbf{H}$, then the set function ν^*, defined by $\nu^*(E) = \lambda^*(E) \cup \mu^*(E)$, is an outer measure on $\mathbf{H}$.

(7) If $\{\mu_n^*\}$ is a sequence of outer measures on a hereditary σ-ring $\mathbf{H}$ and $\{a_n\}$ is a sequence of positive numbers, then the set function μ^* defined by $\mu^*(E) = \sum_{n=1}^{\infty} a_n\mu_n^*(E)$, is an outer measure on $\mathbf{H}$.

§ 11. MEASURABLE SETS

Let μ^* be an outer measure on a hereditary σ-ring $\mathbf{H}$. A set E in $\mathbf{H}$ is **μ^*-measurable** if, for every set A in $\mathbf{H}$,

$$\mu^*(A) = \mu^*(A \cap E) + \mu^*(A \cap E').$$

The concept of μ^*-measurability is the most important one in the theory of outer measures. It is rather difficult to get an intuitive understanding of the meaning of μ^*-measurability except through familiarity with its implications, which we propose to develop below. The following comment may, however, be helpful. An outer measure is not necessarily a countably, nor even finitely, additive set function (cf. 10.4d). In an attempt to satisfy the reasonable requirement of additivity, we single out those sets which split every other set additively—the definition of μ^*-measurability is the precise formulation of this rather loose description. The greatest justification of this apparently complicated concept is, however, its possibly surprising but absolutely complete success as a tool in proving the important and useful extension theorem of § 13.

Theorem A. *If μ^* is an outer measure on a hereditary σ-ring* $\mathbf{H}$ *and if* $\overline{\mathbf{S}}$ *is the class of all μ^*-measurable sets, then* $\overline{\mathbf{S}}$ *is a ring.*

Proof. If E and F are in $\bar{\mathbf{S}}$ and $A \varepsilon \mathbf{H}$, then

(a) $$\mu^*(A) = \mu^*(A \cap E) + \mu^*(A \cap E'),$$

(b) $$\mu^*(A \cap E) = \mu^*(A \cap E \cap F) + \mu^*(A \cap E \cap F'),$$

(c) $$\mu^*(A \cap E') = \mu^*(A \cap E' \cap F) + \mu^*(A \cap E' \cap F').$$

Substituting (b) and (c) into (a) we obtain

(d) $$\mu^*(A) = \mu^*(A \cap E \cap F) + \mu^*(A \cap E \cap F') + \mu^*(A \cap E' \cap F) + \mu^*(A \cap E' \cap F').$$

If in equation (d) we replace A by $A \cap (E \cup F)$, the first three terms of the right hand side remain unaltered and the last term drops out; we get

(e) $$\mu^*(A \cap (E \cup F)) = \mu^*(A \cap E \cap F) + \mu^*(A \cap E \cap F') + \mu^*(A \cap E' \cap F).$$

Since $E' \cap F' = (E \cup F)'$, substituting (e) into (d) yields

(f) $$\mu^*(A) = \mu^*(A \cap (E \cup F)) + \mu^*(A \cap (E \cup F)'),$$

which proves that $E \cup F \varepsilon \bar{\mathbf{S}}$.

If, similarly, we replace A in equation (d) by $A \cap (E - F)' = A \cap (E' \cup F)$, we get

(g) $$\mu^*(A \cap (E - F)') = \mu^*(A \cap E \cap F) + \mu^*(A \cap E' \cap F) + \mu^*(A \cap E' \cap F').$$

Since $E \cap F' = E - F$, substituting (g) into (d) yields

(h) $$\mu^*(A) = \mu^*(A \cap (E - F)) + \mu^*(A \cap (E - F)'),$$

which proves that $E - F \varepsilon \bar{\mathbf{S}}$. Since it is clear that $E = 0$ satisfies (a), it follows that $\bar{\mathbf{S}}$ is a ring. ▮

Before proceeding with the study of the deeper properties of μ^*-measurability, we remark on the following elementary but frequently useful fact.

If μ^* is an outer measure on a hereditary σ-ring $\mathbf{H}$ and if a set E in $\mathbf{H}$ is such that, for every A in $\mathbf{H}$,

$$\mu^*(A) \geqq \mu^*(A \cap E) + \mu^*(A \cap E'),$$

then E is μ^*-measurable.

The proof of this remark is achieved simply by recalling that the reverse inequality, $\mu^*(A) \leqq \mu^*(A \cap E) + \mu^*(A \cap E')$, is an automatic consequence of the subadditivity of μ^*.

Theorem B. *If μ^* is an outer measure on a hereditary σ-ring $\mathbf{H}$ and if $\overline{\mathbf{S}}$ is the class of all μ^*-measurable sets, then $\overline{\mathbf{S}}$ is a σ-ring. If $A \in \mathbf{H}$ and if $\{E_n\}$ is a disjoint sequence of sets in $\overline{\mathbf{S}}$ with $\bigcup_{n=1}^{\infty} E_n = E$, then*

$$\mu^*(A \cap E) = \sum_{n=1}^{\infty} \mu^*(A \cap E_n).$$

Proof. Replacing E and F in (e) by E_1 and E_2 respectively, we see that

$$\mu^*(A \cap (E_1 \cup E_2)) = \mu^*(A \cap E_1) + \mu^*(A \cap E_2).$$

It follows by mathematical induction that

$$\mu^*(A \cap \bigcup_{i=1}^{n} E_i) = \sum_{i=1}^{n} \mu^*(A \cap E_i)$$

for every positive integer n. If we write

$$F_n = \bigcup_{i=1}^{n} E_i, \quad n = 1, 2, \cdots,$$

then it follows from Theorem A that

$$\mu^*(A) = \mu^*(A \cap F_n) + \mu^*(A \cap F_n') \geqq$$
$$\geqq \sum_{i=1}^{n} \mu^*(A \cap E_i) + \mu^*(A \cap E').$$

Since this is true for every n, we obtain

$$\text{(i)} \qquad \mu^*(A) \geqq \sum_{i=1}^{\infty} \mu^*(A \cap E_i) + \mu^*(A \cap E') \geqq$$
$$\geqq \mu^*(A \cap E) + \mu^*(A \cap E').$$

It follows that $E \in \overline{\mathbf{S}}$ (so that, by the way, $\overline{\mathbf{S}}$ is closed under the formation of disjoint countable unions), and therefore that

$$\text{(j)} \quad \sum_{i=1}^{\infty} \mu^*(A \cap E_i) + \mu^*(A \cap E') =$$
$$= \mu^*(A \cap E) + \mu^*(A \cap E').$$

Replacing A by $A \cap E$ in (j), we obtain the second assertion of the theorem. (Since $\mu^*(A \cap E')$ may be infinite, it is not permissible simply to subtract it from both sides of (j).) Since every countable union of sets in a ring may be written as a disjoint countable union of sets in the ring, we see also that $\overline{\mathbf{S}}$ is a σ-ring. ∎

Theorem C. *If μ^* is an outer measure on a hereditary σ-ring* **H** *and if* $\bar{\mathbf{S}}$ *is the class of all μ^*-measurable sets, then every set of outer measure zero belongs to* $\bar{\mathbf{S}}$ *and the set function $\bar{\mu}$, defined for E in* $\bar{\mathbf{S}}$ *by $\bar{\mu}(E) = \mu^*(E)$, is a complete measure on* $\bar{\mathbf{S}}$.

The measure $\bar{\mu}$ is called the measure **induced** by the outer measure μ^*.

Proof. If $E \in \mathbf{H}$ and $\mu^*(E) = 0$, then, for every A in **H**, we have

$$\mu^*(A) = \mu^*(E) + \mu^*(A) \geq \mu^*(A \cap E) + \mu^*(A \cap E'),$$

so that indeed $E \in \bar{\mathbf{S}}$. The fact that $\bar{\mu}$ is countably additive on $\bar{\mathbf{S}}$ follows from (j) upon replacing A by E. If

$$E \in \bar{\mathbf{S}}, \quad F \subset E, \quad \text{and} \quad \bar{\mu}(E) = \mu^*(E) = 0,$$

then $\mu^*(F) = 0$, so that $F \in \bar{\mathbf{S}}$, which proves that $\bar{\mu}$ is complete. ▌

(1) For the example 10.4d, a set E is μ^*-measurable if and only if with every point x in E the entire column which includes x is contained in E. Which sets are μ^*-measurable in 10.4f?

(2) If μ^* is an outer measure on a hereditary σ-ring **H**, under what additional conditions is the class of μ^*-measurable sets an algebra?

(3) In the notation of Theorem A, replacing A in equation (d) by $A \cap (E' \cup F')$ may be used to give a direct proof of the fact that $\bar{\mathbf{S}}$ is closed under the formation of intersections. What would the same technique prove if A were replaced by $A \cap (F - E)' = A \cap (E \cup F')$?

(4) Let μ^* be a finite, non negative, monotone, and finitely subadditive set function with $\mu^*(0) = 0$ on a hereditary ring **J**; cf. 10.2. The class of all μ^*-measurable sets is a ring, and the set function μ^* is additive on this ring.

(5) Suppose that μ^* is an outer measure on a hereditary σ-ring **H** and that $\bar{\mathbf{S}}$ is the class of all μ^*-measurable sets. If $A \in \mathbf{H}$ and $\{E_n\}$ is an increasing sequence of sets in $\bar{\mathbf{S}}$, then $\mu^*(\lim_n (A \cap E_n)) = \lim_n \mu^*(A \cap E_n)$. Similarly, if $\{E_n\}$ is a decreasing sequence of sets in $\bar{\mathbf{S}}$, and if a set A in **H** is such that $\mu^*(A \cap E_m) < \infty$ for at least one value of m, then $\mu^*(\lim_n (A \cap E_n)) = \lim_n \mu^*(A \cap E_n)$.

(6) If μ^* is an outer measure on a hereditary σ-ring **H** and if E and F are two sets in **H** of which at least one is μ^*-measurable, then (cf. 9.2)

$$\mu^*(E) + \mu^*(F) = \mu^*(E \cup F) + \mu^*(E \cap F).$$

(7) The results of this section could also have been obtained by means of partitions (cf. 7.5). A **partition** is a finite or infinite disjoint sequence $\{E_i\}$ of

sets such that $\bigcup_i E_i = X$. If μ^* is an outer measure on a hereditary σ-ring **H**, the partition $\{E_i\}$ is called a **μ^*-partition** if

$$\mu^*(A) = \sum_i \mu^*(A \cap E_i)$$

for every A in **H**; a set E is a **μ^*-set** if the partition $\{E,E'\}$ is a μ^*-partition. If $\{E_i\}$ and $\{F_j\}$ are partitions, then $\{E_i\}$ is called a **subpartition** of $\{F_j\}$ if each E_i is contained in one of the sets F_j. The **product** of two partitions, $\{E_i\}$ and $\{F_j\}$, is the partition consisting of all sets of the form $E_i \cap F_j$. We note that a set in **H** is a μ^*-set if and only if it is μ^*-measurable in the sense of this section.

(7a) The product of two μ^*-partitions is a μ^*-partition.

(7b) If a subpartition of a partition $\{E_i\}$ is a μ^*-partition, then $\{E_i\}$ is a μ^*-partition.

(7c) A partition $\{E_i\}$ is a μ^*-partition if and only if each E_i is a μ^*-set.

(7d) The class of all μ^*-sets is a σ-ring. (Hint: the class of all μ^*-sets is a ring closed under the formation of countable disjoint unions.)

(8a) An outer measure μ^* on the class **H** of all subsets of a metric space X is a **metric** outer measure if

$$\mu^*(E \cup F) = \mu^*(E) + \mu^*(F)$$

whenever $\rho(E,F) > 0$, where ρ is the metric on X. If μ^* is a metric outer measure, if E is a subset of an open set U in X, and if $E_n = E \cap \left\{x\colon \rho(x,U') \geqq \frac{1}{n}\right\}$, $n = 1, 2, \cdots$, then $\lim_n \mu^*(E_n) = \mu^*(E)$. (Hint: observe that $\{E_n\}$ is an increasing sequence of sets whose union is E. If $E_0 = 0$, $D_n = E_{n+1} - E_n$, and if neither D_{n+1} nor E_n is empty, then $\rho(D_{n+1},E_n) > 0$, and it follows that

$$\mu^*(E_{2n+1}) \geqq \sum_{i=1}^{n} \mu^*(D_{2i}) \quad \text{and} \quad \mu^*(E_{2n}) \geqq \sum_{i=1}^{n} \mu^*(D_{2i-1}).$$

The desired conclusion is trivial if either of the two series,

$$\sum_{i=1}^{\infty} \mu^*(D_{2i}) \quad \text{and} \quad \sum_{i=1}^{\infty} \mu^*(D_{2i-1}),$$

diverges; if they both converge, then it follows from the relation

$$\mu^*(E) \leqq \mu^*(E_{2n}) + \sum_{i=n}^{\infty} \mu^*(D_{2i}) + \sum_{i=n+1}^{\infty} \mu^*(D_{2i-1}).)$$

(8b) If μ^* is a metric outer measure, then every open set (and therefore every Borel set) is μ^*-measurable. (Hint: if U is an open set and A is an arbitrary subset of X, apply (8a) to $E = A \cap U$. Since $\rho(E_n, A \cap U') > 0$, it follows that

$$\mu^*(A) \geqq \mu^*(E_n \cup (A \cap U')) = \mu^*(E_n) + \mu^*(A \cap U').)$$

(8c) If μ^* is an outer measure on the class of all subsets of a metric space X such that every open set is μ^*-measurable, then μ^* is a metric outer measure. (Hint: if $\rho(E,F) > 0$, let U be an open set such that $E \subset U$ and $F \cap U = 0$, and test the μ^*-measurability of U with $A = E \cup F$.)

Chapter III

EXTENSION OF MEASURES

§ 12. PROPERTIES OF INDUCED MEASURES

We have seen that a measure induces an outer measure and that an outer measure induces a measure, both in a certain natural way. If we start with a measure μ, form the induced outer measure μ^*, and then form the measure $\bar{\mu}$ induced by μ^*, what is the relation between μ and $\bar{\mu}$? The main purpose of the present section is to answer this question. Throughout this section we shall assume that

> μ is a measure on a ring $\mathbf{R}$, μ^* is the induced outer measure on $\mathbf{H}(\mathbf{R})$, and $\bar{\mu}$ is the measure induced by μ^* on the σ-ring $\bar{\mathbf{S}}$ of all μ^*-measurable sets.

Theorem A. *Every set in* $\mathbf{S}(\mathbf{R})$ *is* μ^**-measurable.*

Proof. If $E \in \mathbf{R}$, $A \in \mathbf{H}(\mathbf{R})$, and $\epsilon > 0$, then, by the definition of μ^*, there exists a sequence $\{E_n\}$ of sets in $\mathbf{R}$ such that $A \subset \bigcup_{n=1}^{\infty} E_n$ and

$$\mu^*(A) + \epsilon \geqq \sum_{n=1}^{\infty} \mu(E_n) = \sum_{n=1}^{\infty} (\mu(E_n \cap E) + \mu(E_n \cap E')) \geqq$$

$$\geqq \mu^*(A \cap E) + \mu^*(A \cap E').$$

Since this is true for every ϵ, it follows that E is μ^*-measurable. In other words, we have proved that $\mathbf{R} \subset \bar{\mathbf{S}}$; it follows from the fact that $\bar{\mathbf{S}}$ is a σ-ring that $\mathbf{S}(\mathbf{R}) \subset \bar{\mathbf{S}}$. ∎

Theorem B. *If* $E \in \mathbf{H}(\mathbf{R})$, *then*

$$\mu^*(E) = \inf\{\bar{\mu}(F): E \subset F \in \bar{\mathbf{S}}\} = $$
$$= \inf\{\bar{\mu}(F): E \subset F \in \mathbf{S}(\mathbf{R})\}.$$

Equivalent to the statement of Theorem B is the assertion that the outer measure induced by $\bar{\mu}$ on $\mathbf{S}(\mathbf{R})$ and the outer measure induced by $\bar{\mu}$ on $\bar{\mathbf{S}}$ both coincide with μ^*.

Proof. Since, for F in $\mathbf{R}$, $\mu(F) = \bar{\mu}(F)$ (by the definition of $\bar{\mu}$ and 10.A), it follows that

$$\mu^*(E) = \inf\{\textstyle\sum_{n=1}^{\infty} \mu(E_n): E \subset \bigcup_{n=1}^{\infty} E_n, \quad E_n \in \mathbf{R},$$
$$n = 1, 2, \cdots\} \geqq$$
$$\geqq \inf\{\textstyle\sum_{n=1}^{\infty} \bar{\mu}(E_n): E \subset \bigcup_{n=1}^{\infty} E_n, \quad E_n \in \mathbf{S}(\mathbf{R}),$$
$$n = 1, 2, \cdots\}.$$

Since every sequence $\{E_n\}$ of sets in $\mathbf{S}(\mathbf{R})$ for which

$$E \subset \textstyle\bigcup_{n=1}^{\infty} E_n = F$$

may be replaced by a disjoint sequence with the same property, without increasing the sum of the measures of the terms of the sequence, and since, by the definition of $\bar{\mu}$, $\bar{\mu}(F) = \mu^*(F)$ for F in $\bar{\mathbf{S}}$, it follows that

$$\mu^*(E) \geqq \inf\{\bar{\mu}(F): E \subset F \in \mathbf{S}(\mathbf{R})\} \geqq$$
$$\geqq \inf\{\bar{\mu}(F): E \subset F \in \bar{\mathbf{S}}\} \geqq \mu^*(E). \quad ▮$$

If $E \in \mathbf{H}(\mathbf{R})$ and $F \in \mathbf{S}(\mathbf{R})$, we shall say that F is a **measurable cover** of E if $E \subset F$ and if, for every set G in $\mathbf{S}(\mathbf{R})$ for which $G \subset F - E$, we have $\bar{\mu}(G) = 0$. Loosely speaking, a measurable cover of a set E in $\mathbf{H}(\mathbf{R})$ is a minimal set in $\mathbf{S}(\mathbf{R})$ which covers E.

Theorem C. *If a set E in* $\mathbf{H}(\mathbf{R})$ *is of σ-finite outer measure, then there exists a set F in* $\mathbf{S}(\mathbf{R})$ *such that $\mu^*(E) = \bar{\mu}(F)$ and such that F is a measurable cover of E.*

Proof. If $\mu^*(E) = \infty$, and $E \subset F \in \mathbf{S}(\mathbf{R})$, then clearly $\bar{\mu}(F) = \infty$, so that it is sufficient to prove the assertion $\mu^*(E) = \bar{\mu}(F)$ only in the case in which $\mu^*(E) < \infty$. Since a set of σ-finite outer measure is a countable disjoint union of sets of finite outer measure, it is sufficient to prove the entire theorem under the added assumption that $\mu^*(E) < \infty$.

It follows from Theorem B that, for every $n = 1, 2, \cdots$, there exists a set F_n in $\mathbf{S}(\mathbf{R})$ such that

$$E \subset F_n \quad \text{and} \quad \bar{\mu}(F_n) \leqq \mu^*(E) + \frac{1}{n}.$$

If we write $F = \bigcap_{n=1}^{\infty} F_n$, then

$$E \subset F \,\varepsilon\, \mathbf{S}(\mathbf{R}) \quad \text{and} \quad \mu^*(E) \leqq \bar{\mu}(F) \leqq \bar{\mu}(F_n) \leqq \mu^*(E) + \frac{1}{n}.$$

Since n is arbitrary, it follows that $\mu^*(E) = \bar{\mu}(F)$. If $G \,\varepsilon\, \mathbf{S}(\mathbf{R})$ and $G \subset F - E$, then $E \subset F - G$ and therefore

$$\bar{\mu}(F) = \mu^*(E) \leqq \bar{\mu}(F - G) = \bar{\mu}(F) - \bar{\mu}(G) \leqq \bar{\mu}(F);$$

the fact that F is a measurable cover of E follows from the finiteness of $\bar{\mu}(F)$. ▮

Theorem D. *If $E \,\varepsilon\, \mathbf{H}(\mathbf{R})$ and F is a measurable cover of E, then $\mu^*(E) = \bar{\mu}(F)$; if both F_1 and F_2 are measurable covers of E, then $\bar{\mu}(F_1 \,\Delta\, F_2) = 0$.*

Proof. Since the relation $E \subset F_1 \cap F_2 \subset F_1$ implies that $F_1 - (F_1 \cap F_2) \subset F_1 - E$, it follows from the fact that F_1 is a measurable cover of E that

$$\bar{\mu}(F_1 - (F_1 \cap F_2)) = 0.$$

Since, similarly,

$$\bar{\mu}(F_2 - (F_1 \cap F_2)) = 0,$$

we have, indeed, $\bar{\mu}(F_1 \,\Delta\, F_2) = 0$.

If $\mu^*(E) = \infty$, then the relation $\mu^*(E) = \bar{\mu}(F)$ is trivial; if $\mu^*(E) < \infty$, then it follows from Theorem C that there exists a measurable cover F_0 of E with

$$\bar{\mu}(F_0) = \mu^*(E).$$

Since the result of the preceding paragraph implies that every two measurable covers have the same measure, the proof of the theorem is complete. ▮

Theorem E. *If μ on $\mathbf{R}$ is σ-finite, then so are the measures $\bar{\mu}$ on $\mathbf{S}(\mathbf{R})$ and $\bar{\mu}$ on $\bar{\mathbf{S}}$.*

Proof. According to 10.A, if μ is σ-finite, then so is μ^*. Hence for every E in $\bar{\mathbf{S}}$ there exists a sequence $\{E_i\}$ of sets in $\mathbf{H}(\mathbf{R})$ such that

$$E \subset \bigcup_{i=1}^{\infty} E_i, \quad \mu^*(E_i) < \infty, \quad i = 1, 2, \cdots.$$

An application of Theorem C to each set E_i concludes the proof of the theorem. ▮

The main question at the beginning of this section could have been asked in the other direction. Suppose that we start with an outer measure μ^*, form the induced measure $\bar{\mu}$, and then form the outer measure $\bar{\mu}^*$ induced by $\bar{\mu}$. What is the relation between μ^* and $\bar{\mu}^*$? In general these two set functions are not the same; if, however, the induced outer measure $\bar{\mu}^*$ does coincide with the original outer measure μ^*, then μ^* is called **regular.** The assertion of Theorem B is exactly that the outer measure induced by a measure on a ring is always regular. The converse of this last statement is also true: if μ^* is regular, then $\mu^* = \bar{\mu}^*$ is induced by a measure on a ring, namely by $\bar{\mu}$ on the class of μ^*-measurable sets. Thus the notions of induced outer measure and regular outer measure are coextensive.

(1) Theorem D asserts that a measurable cover is uniquely determined to within a set of measure zero, if it exists at all; Theorem C asserts that for sets of σ-finite outer measure a measurable cover does exist. The following considerations show that the hypothesis of σ-finiteness cannot be omitted from Theorem C.

If L is a line in the Euclidean plane X, and E is any subset of X, we shall say that E is **full** on L if $L - E$ is countable. Let $\mathbf{R}_0$ be the class of all those sets E which may be covered by countably many horizontal lines on each of which E is either full or countable; let $\mathbf{R}$ be the algebra generated by $\mathbf{R}_0$; (cf. 4.5). If, for every E in $\mathbf{R}$, $\mu(E) = 0$ or ∞ according as E is countable or not, then μ is a measure on $\mathbf{R}$; it is easy to verify that in this case $\mathbf{R} = \mathbf{S}(\mathbf{R})$ and $\bar{\mathbf{S}} = \mathbf{H}(\mathbf{R})$ is the class of all subsets of X. If E is the y-axis and $E \subset F \varepsilon \mathbf{S}(\mathbf{R})$, then there always exists a set G in $\mathbf{S}(\mathbf{R})$ such that $G \subset F - E$ and $\mu(G) \neq 0$.

(2) A subset E of the real line is said to have an **infinite condensation point** if there are uncountably many points of E outside every finite interval. Let X be the real line and define a set function μ^* on every subset E of X as follows: if E is finite or countably infinite, then $\mu^*(E) = 0$; if E is uncountable but does not have an infinite condensation point, then $\mu^*(E) = 1$; if E has an infinite condensation point, then $\mu^*(E) = \infty$. Then μ^* is a totally σ-finite outer measure, but, since the only μ^*-measurable sets are the countable sets and their complements, the induced measure $\bar{\mu}$ is not σ-finite. Is μ^* regular? What can

be said if, instead, $\mu^*(E)$ is defined to be 17 whenever E has an infinite condensation point?

(3) Let n be a fixed positive integer, and let $\aleph_0, \aleph_1, \cdots, \aleph_n$ be the first $n + 1$ infinite cardinal numbers in the well ordering of the cardinals according to magnitude. If X is a set of cardinal number $\aleph_n$, and E is a finite subset of X, write $\mu^*(E) = 0$; if the cardinal number of a subset E of X is the infinite cardinal $\aleph_k$, $0 \leq k \leq n$, write $\mu^*(E) = k$. The set function μ^* is an outer measure; is it regular?

(4) If μ^* is a regular outer measure on a hereditary σ-ring $\mathbf{H}$, and if $\{E_n\}$ is an increasing sequence of sets in $\mathbf{H}$ with $\lim_n E_n = E$, then $\mu^*(E) = \lim_n \mu^*(E_n)$. (Hint: if $\lim_n \mu^*(E_n) = \infty$, the result is clear. If not, then let F_n be a μ^*-measurable cover of E_n, $n = 1, 2, \cdots$, so that the sequence $\{F_n\}$ is increasing, and write $F = \lim_n F_n$. Since $\mu^*(F_n) = \mu^*(E_n) \leq \mu^*(E)$, we have $\lim_n \mu^*(F_n) = \mu^*(F) \leq \mu^*(E)$; since $E \subset F$, $\mu^*(E) \leq \mu^*(F)$. Hence F is a measurable cover of E.) This result is not true for non regular outer measures; a counter example may be constructed on the basis of (2) above.

(5) For every subset E of an arbitrary set X write $\mu^*(E) = 0$ or 1 according as E is empty or not; the set function μ^* is a regular outer measure on the class of all subsets of X. If $\{E_n\}$ is a decreasing sequence of non empty sets with an empty intersection (such a sequence exists whenever X is infinite), then

$$\lim_n \mu^*(E_n) = 1 \quad \text{and} \quad \mu^*(\lim_n E_n) = 0;$$

in other words the analog of (4) for decreasing sequences is false even for totally finite, regular outer measures.

(6) Let μ_1^* and μ_2^* be two finite outer measures on the class of all subsets of a set X, and let $\bar{\mathbf{S}}_i$, $i = 1, 2$, be the class of μ_i^*-measurable sets. If, for all subsets E of X,

$$\mu^*(E) = \mu_1^*(E) + \mu_2^*(E),$$

then the class $\bar{\mathbf{S}}$ of all μ^*-measurable sets is the intersection of $\bar{\mathbf{S}}_1$ and $\bar{\mathbf{S}}_2$. (Hint: if $\mu^*(A \cap E) + \mu^*(A \cap E') = \mu^*(A)$, then both the inequalities, $\mu_i^*(A \cap E) + \mu_i^*(A \cap E') \geq \mu_i^*(A)$, $i = 1, 2$, must become equalities.) What can be said if μ_1^* and μ_2^* are not necessarily finite?

(7) Let μ_1^* be any finite, regular outer measure on the class of all subsets of a set X, and write $\mu_2^*(E) = 0$ or 1 according as E is empty or not. Then μ_2^* is also a finite, regular outer measure, but, if μ_1^* assumes more than two values, then $\mu_1^* + \mu_2^*$ is not regular.

(8) If X is a metric space, p is a positive real number, and E is a subset of X, then the p-dimensional **Hausdorff** (outer) **measure** of E is defined to be the number

$$\mu_p^*(E) = \sup_{\epsilon>0} \inf \{\textstyle\sum_{i=1}^\infty (\delta(E_i))^p : E = \bigcup_{i=1}^\infty E_i, \quad \delta(E_i) < \epsilon, \quad i = 1, 2, \cdots\},$$

where $\delta(E)$ denotes the diameter of E.

(8a) The set function μ_p^* is a metric outer measure; cf. 11.8a.

(8b) The outer measure μ_p^* is regular; in fact, for every subset E of X, there exists a decreasing sequence $\{U_n\}$ of open sets containing E such that

$$\mu_p^*(E) = \mu_p^*(\textstyle\bigcap_{n=1}^\infty U_n).$$

§ 13. EXTENSION, COMPLETION, AND APPROXIMATION

Can we always extend a measure on a ring to the generated σ-ring? The answer to this question is essentially contained in the results of the preceding sections; it is formally summarized in the following theorem.

Theorem A. *If* μ *is a* σ*-finite measure on a ring* **R**, *then there is a unique measure* $\bar{\mu}$ *on the* σ*-ring* **S**(**R**) *such that, for* E *in* **R**, $\bar{\mu}(E) = \mu(E)$; *the measure* $\bar{\mu}$ *is* σ*-finite.*

The measure $\bar{\mu}$ is called the **extension** of μ; except when it is likely to lead to confusion, we shall write $\mu(E)$ instead of $\bar{\mu}(E)$ even for sets E in **S**(**R**).

Proof. The existence of $\bar{\mu}$ (even without the restriction of σ-finiteness) is proved by 11.C and 12.A. To prove uniqueness, suppose that μ_1 and μ_2 are two measures on **S**(**R**) such that $\mu_1(E) = \mu_2(E)$ whenever $E \in \mathbf{R}$, and let **M** be the class of all sets E in **S**(**R**) for which $\mu_1(E) = \mu_2(E)$. If one of the two measures is finite, and if $\{E_n\}$ is a monotone sequence of sets in **M**, then, since

$$\mu_i(\lim_n E_n) = \lim_n \mu_i(E_n), \quad i = 1, 2,$$

we have $\lim_n E_n \in \mathbf{M}$. (The full justification of this step in the reasoning makes use of the fact that one of the two numbers $\mu_1(E_n)$ and $\mu_2(E_n)$, and therefore also the other one, is finite for every $n = 1, 2, \cdots$; cf. 9.D and 9.E.) Since this means that **M** is a monotone class, and since **M** contains **R**, it follows from 6.B that **M** contains **S**(**R**).

In the general, not necessarily finite, case we proceed as follows. Let A be any fixed set in **R**, of finite measure with respect to one of the two measures μ_1 and μ_2. Since $\mathbf{R} \cap A$ is a ring and $\mathbf{S}(\mathbf{R}) \cap A$ is the σ-ring it generates (cf. 5.E), it follows that the reasoning of the preceding paragraph applies to $\mathbf{R} \cap A$ and $\mathbf{S}(\mathbf{R}) \cap A$, and proves that if $E \in \mathbf{S}(\mathbf{R}) \cap A$, then $\mu_1(E) = \mu_2(E)$. Since every E in **S**(**R**) may be covered by a countable, disjoint union of sets of finite measure in **R** (with respect to either of the measures μ_1 and μ_2), the proof of the theorem is complete. ▌

The extension procedure employed in the proofs of § 12 yields slightly more than Theorem A states; the given measure μ can actually be extended to a class (the class of all μ^*-measurable sets) which is in general larger than the generated σ-ring. The following theorems show that it is not necessary to make use of the theory of outer measures in order to obtain this slight enlargement of the domain of μ.

Theorem B. *If μ is a measure on a σ-ring $\mathbf{S}$, then the class $\overline{\mathbf{S}}$ of all sets of the form $E \Delta N$, where $E \in \mathbf{S}$ and N is a subset of a set of measure zero in $\mathbf{S}$, is a σ-ring, and the set function $\bar{\mu}$ defined by $\bar{\mu}(E \Delta N) = \mu(E)$ is a complete measure on $\overline{\mathbf{S}}$.*

The measure $\bar{\mu}$ is called the **completion** of μ.

Proof. If $E \in \mathbf{S}$, $N \subset A \in \mathbf{S}$, and $\mu(A) = 0$, then the relations

$$E \cup N = (E - A) \Delta [A \cap (E \cup N)]$$

and

$$E \Delta N = (E - A) \cup [A \cap (E \Delta N)]$$

show that the class $\overline{\mathbf{S}}$ may also be described as the class of all sets of the form $E \cup N$, where $E \in \mathbf{S}$ and N is a subset of a set of measure zero in $\mathbf{S}$. Since this implies that the class $\overline{\mathbf{S}}$, which is obviously closed under the formation of symmetric differences, is closed also under the formation of countable unions, it follows that $\overline{\mathbf{S}}$ is a σ-ring. If

$$E_1 \Delta N_1 = E_2 \Delta N_2,$$

where $E_i \in \mathbf{S}$ and N_i is a subset of a set of measure zero in $\mathbf{S}$, $i = 1, 2$, then

$$E_1 \Delta E_2 = N_1 \Delta N_2,$$

and therefore $\mu(E_1 \Delta E_2) = 0$. It follows that $\mu(E_1) = \mu(E_2)$, and hence that $\bar{\mu}$ is indeed unambiguously defined by the relations

$$\bar{\mu}(E \Delta N) = \bar{\mu}(E \cup N) = \mu(E).$$

Using the union (instead of the symmetric difference) representation of sets in $\overline{\mathbf{S}}$, it is easy to verify that $\bar{\mu}$ is a measure; the completeness of $\bar{\mu}$ is an immediate consequence of the fact that $\overline{\mathbf{S}}$ contains all subsets of sets of measure zero in $\mathbf{S}$. ∎

The following theorem establishes the connection between the general concept of completion and the particular complete extension obtained by using outer measures.

Theorem C. *If μ is a σ-finite measure on a ring* **R**, *and if μ^* is the outer measure induced by μ, then the completion of the extension of μ to* **S(R)** *is identical with μ^* on the class of all μ^*-measurable sets.*

Proof. Let us denote the class of all μ^*-measurable sets by $\mathbf{S}^*$ and the domain of the completion $\bar{\mu}$ of μ by $\bar{\mathbf{S}}$. Since μ^* on $\mathbf{S}^*$ is a complete measure, it follows that $\bar{\mathbf{S}}$ is contained in $\mathbf{S}^*$ and that $\bar{\mu}$ and μ^* coincide on $\bar{\mathbf{S}}$. All that we have left to prove is that $\mathbf{S}^*$ is contained in $\bar{\mathbf{S}}$; in view of the σ-finiteness of μ^* on $\mathbf{S}^*$ (cf. 12.E) it is sufficient to prove that if $E \in \mathbf{S}^*$ and $\mu^*(E) < \infty$, then $E \in \bar{\mathbf{S}}$.

By 12.C, E has a measurable cover F. Since $\mu^*(F) = \mu(F) = \mu^*(E)$, it follows from the finiteness of $\mu^*(E)$, and the fact that μ^* is a measure on $\mathbf{S}^*$, that $\mu^*(F - E) = 0$. Since $F - E$ also has a measurable cover G, and since

$$\mu(G) = \mu^*(F - E) = 0,$$

the relation

$$E = (F - G) \cup (E \cap G)$$

exhibits E as a union of a set in **S(R)** and a set which is a subset of a set of measure zero in **S(R)**. This shows that $E \in \bar{\mathbf{S}}$, and thus completes the proof of Theorem C. ∎

Loosely speaking, Theorem C says that in the σ-finite case the σ-ring of all μ^*-measurable sets and the generated σ-ring **S(R)** are not very different; every μ^*-measurable set suitably modified by a set of measure zero belongs to **S(R)**.

We conclude this section with a very useful result concerning the relation between a measure on a ring and its extension to the generated σ-ring.

Theorem D. *If μ is a σ-finite measure on a ring* **R**, *then, for every set E of finite measure in* **S(R)** *and for every positive number ϵ, there exists a set E_0 in* **R** *such that $\mu(E \Delta E_0) \leqq \epsilon$.*

Proof. The results of §§ 10, 11, and 12, together with Theorem A, imply that

$$\mu(E) = \inf\{\sum_{i=1}^{\infty} \mu(E_i) \colon E \subset \bigcup_{i=1}^{\infty} E_i, \quad E_i \varepsilon \mathbf{R}, \quad i = 1, 2, \cdots\}.$$

Consequently there exists a sequence $\{E_i\}$ of sets in $\mathbf{R}$ such that

$$E \subset \bigcup_{i=1}^{\infty} E_i \quad \text{and} \quad \mu(\bigcup_{i=1}^{\infty} E_i) \leqq \mu(E) + \frac{\epsilon}{2}.$$

Since

$$\lim_n \mu(\bigcup_{i=1}^{n} E_i) = \mu(\bigcup_{i=1}^{\infty} E_i),$$

there exists a positive integer n such that if

$$E_0 = \bigcup_{i=1}^{n} E_i,$$

then

$$\mu(\bigcup_{i=1}^{\infty} E_i) \leqq \mu(E_0) + \frac{\epsilon}{2}.$$

Clearly $E_0 \varepsilon \mathbf{R}$; since

$$\mu(E - E_0) \leqq \mu(\bigcup_{i=1}^{\infty} E_i - E_0) = \mu(\bigcup_{i=1}^{\infty} E_i) - \mu(E_0) \leqq \frac{\epsilon}{2}$$

and

$$\mu(E_0 - E) \leqq \mu(\bigcup_{i=1}^{\infty} E_i - E) = \mu(\bigcup_{i=1}^{\infty} E_i) - \mu(E) \leqq \frac{\epsilon}{2},$$

the proof of the theorem is complete. ∎

(1) Let μ be a finite, non negative, and finitely additive set function defined on a ring $\mathbf{R}$. The function μ^* defined by the procedure of § 10 is still an outer measure, and, therefore, the $\bar{\mu}$ of 11.C may still be formed, but it is no longer necessarily true that $\bar{\mu}$ is an extension of μ; (cf. 10.2, 10.4e, and 11.4).

(2) If $\bar{\mu}$ is the extension of the measure μ on the ring $\mathbf{R}$ described in § 8, then, for any countable set E, $E \varepsilon \mathbf{S}(\mathbf{R})$ and $\bar{\mu}(E) = 0$.

(3) The uniqueness assertion of Theorem A is not true if the class $\mathbf{R}$ is not a ring. (Hint: let $X = \{a,b,c,d\}$ be a space of four points and define the measures μ_1 and μ_2 on the class of all subsets of X by

$$\mu_1(\{a\}) = \mu_1(\{d\}) = \mu_2(\{b\}) = \mu_2(\{c\}) = 1,$$
$$\mu_1(\{b\}) = \mu_1(\{c\}) = \mu_2(\{a\}) = \mu_2(\{d\}) = 2.)$$

(4) Is Theorem A true for semirings instead of rings?

(5) Let $\mathbf{R}$ be a ring of subsets of a countable set X, with the property that every non empty set in $\mathbf{R}$ is infinite and such that $\mathbf{S}(\mathbf{R})$ is the class of all subsets of X; (cf. 9.7). If, for every subset E of X, $\mu_1(E)$ is the number of points in E and $\mu_2(E) = 2\mu_1(E)$, then μ_2 and μ_1 agree on $\mathbf{R}$ but not on $\mathbf{S}(\mathbf{R})$. In other words,

the uniqueness assertion of Theorem A is not true without the restriction of σ–finiteness on $\mathbf{R}$, even for measures which are totally σ–finite on $\mathbf{S}(\mathbf{R})$.

(6) Suppose that μ is a measure on a σ–ring $\mathbf{S}$ and that $\bar{\mu}$ on $\bar{\mathbf{S}}$ is its completion. If A and B are in $\mathbf{S}$ and if $A \subset E \subset B$, and $\mu(B - A) = 0$, then $E \varepsilon \bar{\mathbf{S}}$.

(7) Let X be an uncountable set, let $\mathbf{S}$ be the class of all countable sets and their complements, and, for every E in $\mathbf{S}$, let $\mu(E)$ be the number of points in E. Then μ is a complete measure on $\mathbf{S}$, but every subset of X is μ^*–measurable; in other words, Theorem C is false without the assumption of σ–finiteness.

(8) If μ and ν are σ–finite measures on a ring $\mathbf{R}$, then, for every E in $\mathbf{S}(\mathbf{R})$ for which both $\mu(E)$ and $\nu(E)$ are finite and for every positive number ϵ, there exists a set E_0 in $\mathbf{R}$ such that

$$\mu(E \,\Delta\, E_0) \leqq \epsilon \quad \text{and} \quad \nu(E \,\Delta\, E_0) \leqq \epsilon.$$

§ 14. INNER MEASURES

We return now to the general study of measures, outer measures, and the relations among them, in order to describe an interesting and historically important part of the theory.

We have seen that if μ is a measure on a σ–ring $\mathbf{S}$, then the set function μ^* (defined for every E in the hereditary σ–ring $\mathbf{H}(\mathbf{S})$ by

$$\mu^*(E) = \inf \{\mu(F): E \subset F \varepsilon \mathbf{S}\})$$

is an outer measure; in the σ–finite case the induced measure $\bar{\mu}$ on the σ–ring $\bar{\mathbf{S}}$ of all μ^*–measurable sets is the completion of μ. Analogously we now define the **inner measure** μ_* **induced** by μ; for every E in $\mathbf{H}(\mathbf{S})$ we write

$$\mu_*(E) = \sup \{\mu(F): E \supset F \varepsilon \mathbf{S}\}.$$

In this section we shall study μ_* and its relation to μ^*; we shall show that the properties of μ_* are in a very legitimate sense the duals of those of μ^*. It is very easy to see that the set function μ_* is non negative, monotone, and such that $\mu_*(0) = 0$; in what follows we shall make use of these elementary facts without any reference. Throughout this section we shall assume that

> μ is a σ–finite measure on a σ–ring $\mathbf{S}$, μ^* and μ_* are the outer measure and the inner measure induced by μ, respectively, and $\bar{\mu}$ on $\bar{\mathbf{S}}$ is the completion of μ;

we recall that $\bar{\mu}$ on $\bar{\mathbf{S}}$ coincides with μ^* on the class of all μ^*–measurable sets (cf. 13.C).

Theorem A. *If* $E \varepsilon \mathbf{H}(\mathbf{S})$, *then*

$$\mu_*(E) = \sup \{\bar{\mu}(F): E \supset F \varepsilon \bar{\mathbf{S}}\}.$$

Proof. Since $\mathbf{S} \subset \bar{\mathbf{S}}$, it is clear from the definition of μ_* that

$$\mu_*(E) \leqq \sup \{\bar{\mu}(F): E \supset F \varepsilon \bar{\mathbf{S}}\}.$$

On the other hand 13.B implies that, for every F in $\bar{\mathbf{S}}$, there is a G in $\mathbf{S}$ with $G \subset F$ and $\bar{\mu}(F) = \mu(G)$. Since this means that every value of $\bar{\mu}$ on subsets of E in $\bar{\mathbf{S}}$ is also attained by μ on subsets of E in $\mathbf{S}$, the proof is complete. ∎

If $E \varepsilon \mathbf{H}(\mathbf{S})$ and $F \varepsilon \mathbf{S}$, we shall say that F is a **measurable kernel** of E if $F \subset E$ and if, for every set G in $\mathbf{S}$ for which $G \subset E - F$, we have $\mu(G) = 0$. Loosely speaking a measurable kernel of a set E in $\mathbf{H}(\mathbf{S})$ is a maximal set in $\mathbf{S}$ which is contained in E.

Theorem B. *Every set* E *in* $\mathbf{H}(\mathbf{S})$ *has a measurable kernel.*

Proof. Let $\hat{E}$ be a measurable cover of E, let N be a measurable cover of $\hat{E} - E$, and write $F = \hat{E} - N$. We have

$$F = \hat{E} - N \subset \hat{E} - (\hat{E} - E) = E,$$

and, if $G \subset E - F$, then

$$G \subset E - (\hat{E} - N) = E \cap N \subset N - (\hat{E} - E).$$

It follows (since N is a measurable cover of $\hat{E} - E$), that F is a measurable kernel of E. ∎

Theorem C. *If* $E \varepsilon \mathbf{H}(\mathbf{S})$ *and* F *is a measurable kernel of* E, *then* $\mu(F) = \mu_*(E)$; *if both* F_1 *and* F_2 *are measurable kernels of* E, *then* $\mu(F_1 \Delta F_2) = 0$.

Proof. Since $F \subset E$, it is clear that $\mu(F) \leqq \mu_*(E)$. If $\mu(F) < \mu_*(E)$, then $\mu(F)$ is finite and, by the definition of $\mu_*(E)$, there exists a set F_0 in $\mathbf{S}$ such that $F_0 \subset E$ and $\mu(F_0) > \mu(F)$. Since

$$F_0 - F \subset E - F \quad \text{and} \quad \mu(F_0 - F) \geqq \mu(F_0) - \mu(F) > 0,$$

this is a contradiction, and therefore indeed $\mu(F) = \mu_*(E)$.

Since the relation $F_1 \subset F_1 \cup F_2 \subset E$ implies that $(F_1 \cup F_2) -$

$F_1 \subset E - F_1$, it follows from the fact that F_1 is a measurable kernel of E that

$$\mu((F_1 \cup F_2) - F_1) = 0.$$

Since, similarly,

$$\mu((F_1 \cup F_2) - F_2) = 0,$$

we have $\mu(F_1 \Delta F_2) = 0$. ▌

Theorem D. *If $\{E_n\}$ is a disjoint sequence of sets in* $\mathbf{H}(\mathbf{S})$, *then*

$$\mu_*(\bigcup_{n=1}^{\infty} E_n) \geqq \sum_{n=1}^{\infty} \mu_*(E_n).$$

Proof. If F_n is a measurable kernel of E_n, $n = 1, 2, \cdots$ then the countable additivity of μ implies that

$$\sum_{n=1}^{\infty} \mu_*(E_n) = \sum_{n=1}^{\infty} \mu(F_n) = \mu(\bigcup_{n=1}^{\infty} F_n) \leqq \mu_*(\bigcup_{n=1}^{\infty} E_n). \quad ▌$$

Theorem E. *If $A \varepsilon \mathbf{H}(\mathbf{S})$ and if $\{E_n\}$ is a disjoint sequence of sets in $\bar{\mathbf{S}}$ with $\bigcup_{n=1}^{\infty} E_n = E$, then*

$$\mu_*(A \cap E) = \sum_{n=1}^{\infty} \mu_*(A \cap E_n).$$

Proof. If F is a measurable kernel of $A \cap E$, then

$$\mu_*(A \cap E) = \mu(F) = \sum_{n=1}^{\infty} \bar{\mu}(F \cap E_n) \leqq \sum_{n=1}^{\infty} \mu_*(A \cap E_n);$$

the desired result follows from Theorem D. ▌

Theorem F. *If $E \varepsilon \bar{\mathbf{S}}$, then*

$$\mu^*(E) = \mu_*(E) = \bar{\mu}(E),$$

and, conversely, if $E \varepsilon \mathbf{H}(\mathbf{S})$ and

$$\mu^*(E) = \mu_*(E) < \infty,$$

then $E \varepsilon \bar{\mathbf{S}}$.

Proof. If $E \varepsilon \bar{\mathbf{S}}$, then both the supremum in Theorem A and the infimum in 12.B are attained by $\bar{\mu}(E)$. To prove the converse, let A and B be a measurable kernel and a measurable cover of E, respectively. Since $\mu(A) = \mu_*(E) < \infty$, we have

$$\mu(B - A) = \mu(B) - \mu(A) = \mu^*(E) - \mu_*(E) = 0,$$

and the desired conclusion follows from the completeness of $\bar{\mu}$ on $\bar{\mathbf{S}}$; (cf. 11.C and 13.6). ▮

Theorem G. *If E and F are disjoint sets in* $\mathbf{H}(\mathbf{S})$, *then*

$$\mu_*(E \cup F) \leqq \mu_*(E) + \mu^*(F) \leqq \mu^*(E \cup F).$$

Proof. Let A be a measurable cover of F and let B be a measurable kernel of $E \cup F$. Since $B - A \subset E$, it follows that

$$\mu_*(E \cup F) = \mu(B) \leqq \mu(B - A) + \mu(A) \leqq \mu_*(E) + \mu^*(F).$$

Dually, let A be a measurable kernel of E and let B be a measurable cover of $E \cup F$. Since $B - A \supset F$, it follows that

$$\mu^*(E \cup F) = \mu(B) = \mu(A) + \mu(B - A) \geqq \mu_*(E) + \mu^*(F). \quad ▮$$

Theorem H. *If $E \in \bar{\mathbf{S}}$, then, for every subset A of X,*

$$\mu_*(A \cap E) + \mu^*(A' \cap E) = \bar{\mu}(E).$$

Proof. Applying Theorem G to $A \cap E$ and $A' \cap E$, we obtain

$$\mu_*(E) \leqq \mu_*(A \cap E) + \mu^*(A' \cap E) \leqq \mu^*(E).$$

Since $E \in \bar{\mathbf{S}}$, we have, by Theorem F, $\mu_*(E) = \mu^*(E) = \bar{\mu}(E)$. ▮

The results of this section enable us to sketch the steps of an alternative approach to the extension theorem, an approach that is frequently employed. If μ is a σ-finite measure on a ring $\mathbf{R}$, and if μ^* is the induced outer measure on $\mathbf{H}(\mathbf{R})$, then, for every set E in $\mathbf{R}$ with $\mu(E) < \infty$ and for every A in $\mathbf{H}(\mathbf{R})$, we have

$$\mu_*(A \cap E) = \mu(E) - \mu^*(A' \cap E).$$

If we prove now that whenever E and F are two sets of finite measure in $\mathbf{R}$ for which $A \cap E = A \cap F$, then it follows that $\mu(E) - \mu^*(A' \cap E) = \mu(F) - \mu^*(A' \cap F)$, then we may use the equation for $\mu_*(A \cap E)$ as a definition of inner measure, and we may define a set E in $\mathbf{H}(\mathbf{R})$ of finite outer measure to be μ^*-measurable if and only if $\mu_*(E) = \mu^*(E)$. The details of this procedure may be easily carried out by the interested reader, using the techniques we have introduced in our development of the extension theory.

(1) Do the results of 12.4 remain true if μ^* is replaced by μ_*?

(2) With suitable finiteness restrictions the dual of 12.4 is true for inner measures, but the unaltered result of 12.4 is not; (cf. 12.5).

(3) If E is a set of finite measure in $\bar{\mathbf{S}}$, if $F \subset E$, and if $\bar{\mu}(E) = \mu^*(F) + \mu^*(E - F)$, then $F \in \bar{\mathbf{S}}$. In other words the μ^*-measurability of F may be tested by employing a fixed set E (containing F) in $\bar{\mathbf{S}}$ instead of an arbitrary A in $\mathbf{H}(\mathbf{S})$. (Hint: use Theorem H.)

(4) Is an analog of 11.6 true for inner measures?

(5) If $E \in \mathbf{H}(\mathbf{S})$ and F is a measurable cover of E, then, for every μ^*-measurable set M, $\bar{\mu}(F \cap M) = \mu^*(E \cap M)$. (Hint: apply Theorem H to $E = F \cap M$ and $A = E'$.) Conversely, any set F with this property and such that $E \subset F \in \mathbf{S}$ is a measurable cover of E. Similarly, F is a measurable kernel of E if and only if $E \supset F \in \mathbf{S}$ and $\bar{\mu}(F \cap M) = \mu_*(E \cap M)$ for every M in $\bar{\mathbf{S}}$.

§ 15. LEBESGUE MEASURE

The purpose of this section is to apply the general extension theory to the special measure discussed in § 8, and to introduce some of the classical results and terminology pertinent to this special case. Throughout this section we shall assume that

> X is the real line, $\mathbf{P}$ is the class of all bounded, semiclosed intervals of the form $[a,b)$, $\mathbf{S}$ is the σ-ring generated by $\mathbf{P}$, and μ is the set function on $\mathbf{P}$ defined by $\mu([a,b)) = b - a$.

The sets of the σ-ring $\mathbf{S}$ are called the **Borel sets** of the line; according to the extension theorems 8.E and 13.A, we may assume that μ is defined for all Borel sets. If $\bar{\mu}$ on $\bar{\mathbf{S}}$ is the completion of μ on $\mathbf{S}$, the sets of $\bar{\mathbf{S}}$ are the **Lebesgue measurable sets** of the line; the measure $\bar{\mu}$ is **Lebesgue measure.** (The incomplete measure μ on the class $\mathbf{S}$ of all Borel sets is usually called Lebesgue measure also.)

Since the entire line X is the union of countably many sets in $\mathbf{P}$, we see that $X \in \mathbf{S}$, so that the σ-rings $\mathbf{S}$ and $\bar{\mathbf{S}}$ are even σ-algebras. Since clearly $\mu(X) = \infty$, μ is not finite on $\mathbf{S}$, but, since μ is finite on $\mathbf{P}$, both μ on $\mathbf{S}$ and $\bar{\mu}$ on $\bar{\mathbf{S}}$ are totally σ-finite. Some of the other interesting properties of μ and $\bar{\mu}$ are contained in the following theorems.

Theorem A. *Every countable set is a Borel set of measure zero.*

Proof. For any a, $-\infty < a < \infty$, we have

$$\{a\} = \{x: x = a\} = \bigcap_{n=1}^{\infty} \left\{x: a \leqq x < a + \frac{1}{n}\right\},$$

and therefore

$$\mu(\{a\}) = \lim_n \mu\left(\left[a, a + \frac{1}{n}\right)\right) = \lim_n \frac{1}{n} = 0,$$

so that every one-point set is a Borel set of measure zero. Since the Borel sets form a σ-ring and since μ is countably additive, the theorem follows. ▮

Theorem B. *The class* **S** *of all Borel sets coincides with the σ-ring generated by the class* **U** *of all open sets.*

Proof. Since, for every real number a, the set $\{a\}$ is a Borel set, it follows from the relation $(a,b) = [a,b) - \{a\}$, that every bounded open interval is a Borel set. Since every open set on the line is a countable union of bounded open intervals, it follows that $\mathbf{S} \supset \mathbf{U}$ and consequently that $\mathbf{S} \supset \mathbf{S}(\mathbf{U})$. To prove the reverse inequality, we observe that, for every real number a,

$$\{a\} = \bigcap_{n=1}^{\infty} \left(a - \frac{1}{n}, a + \frac{1}{n}\right),$$

so that $\{a\} \in \mathbf{S}(\mathbf{U})$. It follows from the relation $[a,b) = (a,b) \cup \{a\}$ that $\mathbf{P} \subset \mathbf{S}(\mathbf{U})$ and consequently that

$$\mathbf{S} = \mathbf{S}(\mathbf{P}) \subset \mathbf{S}(\mathbf{U}). \quad ▮$$

Theorem C. *If* **U** *is the class of all open sets, then, for every E in X,*

$$\mu^*(E) = \inf\{\mu(U): E \subset U \in \mathbf{U}\}.$$

Proof. Since $\mu^*(E) = \inf\{\mu(F): E \subset F \in \mathbf{S}\}$, it follows from the fact that $\mathbf{U} \subset \mathbf{S}$ that

$$\mu^*(E) \leqq \inf\{\mu(U): E \subset U \in \mathbf{U}\}.$$

If, on the other hand, ϵ is any positive number, then it follows from the definition of μ^* that there exists a sequence $\{[a_n,b_n)\}$ of sets in $\mathbf{P}$ such that

$$E \subset \bigcup_{n=1}^{\infty} [a_n,b_n) \quad \text{and} \quad \sum_{n=1}^{\infty} (b_n - a_n) \leqq \mu^*(E) + \frac{\epsilon}{2}.$$

Consequently

$$E \subset \bigcup_{n=1}^{\infty} \left(a_n - \frac{\epsilon}{2^{n+1}}, b_n\right) = U \,\varepsilon\, \mathbf{U},$$

and

$$\mu(U) \leqq \sum_{n=1}^{\infty} (b_n - a_n) + \frac{\epsilon}{2} \leqq \mu^*(E) + \epsilon.$$

The desired result follows from the arbitrariness of ϵ. ▮

Theorem D. *Let T be the one to one transformation of the entire real line onto itself, defined by $T(x) = \alpha x + \beta$, where α and β are real numbers and $\alpha \neq 0$. If, for every subset E of X, $T(E)$ denotes the set of all points of the form $T(x)$ with x in E, i.e. $T(E) = \{\alpha x + \beta : x \,\varepsilon\, E\}$, then*

$$\mu^*(T(E)) = |\alpha| \mu^*(E) \quad \text{and} \quad \mu_*(T(E)) = |\alpha| \mu_*(E).$$

The set $T(E)$ is a Borel set or a Lebesgue measurable set if and only if E is a Borel set or a Lebesgue measurable set, respectively.

Proof. It is sufficient to prove the theorem for $\alpha > 0$. For, if $\alpha < 0$, then the transformation T is the result of the iteration of two transformations T_1 and T_2, $T(x) = T_1(T_2(x))$, where $T_1(x) = |\alpha| x + \beta$ and $T_2(x) = -x$. We leave to the reader the verification of the fact that the transformation T_2 sends Borel sets and Lebesgue measurable sets into Borel sets and Lebesgue measurable sets, respectively, and that it preserves the inner and outer measures of every set.

Suppose then that $\alpha > 0$, and let $T(\mathbf{S})$ be the class of all sets of the form $T(E)$ with E in $\mathbf{S}$. It is clear that $T(\mathbf{S})$ is a σ-ring; we are to prove that $T(\mathbf{S}) = \mathbf{S}$. If $E = [a,b) \,\varepsilon\, \mathbf{P}$, then $E = T(F)$, where

$$F = \left[\frac{a - \beta}{\alpha}, \frac{b - \beta}{\alpha}\right) \varepsilon\, \mathbf{P},$$

so that $E \,\varepsilon\, T(\mathbf{S})$ and therefore $\mathbf{S} \subset T(\mathbf{S})$. By the same reasoning applied to the transformation T^{-1}, $T^{-1}(x) = \dfrac{x - \beta}{\alpha}$, we may conclude that $\mathbf{S} \subset T^{-1}(\mathbf{S})$, and, applying the transformation T to both sides of the last written relation, we obtain, $T(\mathbf{S}) \subset \mathbf{S}$ and therefore $T(\mathbf{S}) = \mathbf{S}$.

If, for every Borel set E, we write

$$\mu_1(E) = \mu(T(E)) \quad \text{and} \quad \mu_2(E) = \alpha\mu(E),$$

then both μ_1 and μ_2 are measures on $\mathbf{S}$. If $E = [a,b) \in \mathbf{P}$, then $T(E) = [\alpha a + \beta, \alpha b + \beta)$, and

$$\mu_1(E) = \mu(T(E)) = (\alpha b + \beta) - (\alpha a + \beta) = \alpha(b - a) =$$
$$= \alpha\mu(E) = \mu_2(E),$$

so that, by 8.E and 13.A, $\mu(T(E)) = \alpha\mu(E)$ for every E in $\mathbf{S}$.

Applying the results of the preceding two paragraphs to the transformation T^{-1}, we obtain the relations

$$\begin{aligned}\mu^*(T(E)) &= \inf\{\mu(F)\colon T(E) \subset F \in \mathbf{S}\} = \\ &= \inf\{\alpha\mu(T^{-1}(F))\colon E \subset T^{-1}(F) \in \mathbf{S}\} = \\ &= \alpha \inf\{\mu(G)\colon E \subset G \in \mathbf{S}\} = \\ &= \alpha\mu^*(E),\end{aligned}$$

and, replacing inf by sup, μ^* by μ_*, and $\subset$ by $\supset$ throughout,

$$\mu_*(T(E)) = \alpha\mu_*(E),$$

for every set E.

If E is a Lebesgue measurable set and A is any set, then

$$\begin{aligned}&\mu^*(A \cap T(E)) + \mu^*(A \cap (T(E))') = \\ &\qquad = \mu^*(T(T^{-1}(A) \cap E)) + \mu^*(T(T^{-1}(A) \cap E')) = \\ &\qquad = \alpha[\mu^*(T^{-1}(A) \cap E) + \mu^*(T^{-1}(A) \cap E')] = \\ &\qquad = \alpha\mu^*(T^{-1}(A)) = \\ &\qquad = \mu^*(A),\end{aligned}$$

so that $T(E)$ is Lebesgue measurable. This result applied to T^{-1} proves its own converse and completes the proof of the theorem. ▌

(1) The class of all Borel sets is the σ-ring generated by the class $\mathbf{C}$ of all closed sets, and, for every set E,

$$\mu_*(E) = \sup \{\mu(C): E \supset C \,\varepsilon\, \mathbf{C}\}.$$

(2) To every Lebesgue measurable set E there correspond two Borel sets A and B such that

$$A \subset E \subset B, \quad \mu(B - A) = 0,$$

and such that A is an F_σ and B is a G_δ.

(3) A bounded set has finite outer measure. Is the converse of this statement true?

(4) Let $\{x_1, x_2, \cdots\}$ be an enumeration of the set M of rational numbers in the closed unit interval X. For every $\epsilon > 0$ and $i = 1, 2, \cdots$, let $F_i(\epsilon)$ be the open interval of length $\frac{\epsilon}{2^i}$ whose center is at x_i, and write

$$F(\epsilon) = \bigcup_{i=1}^{\infty} F_i(\epsilon), \quad F = \bigcap_{n=1}^{\infty} F\left(\frac{1}{n}\right).$$

The following statements are true.

(4a) There exists an $\epsilon > 0$ and a point x in X such that $x \,\varepsilon'\, F(\epsilon)$.

(4b) The set $F(\epsilon)$ is open and $\mu(F(\epsilon)) \leqq \epsilon$.

(4c) The set $X - F(\epsilon)$ is nowhere dense.

(4d) The set $X - F$ is of the first category and therefore, since X is a complete metric space, F is uncountable. (Hence, in particular, $F \neq M$.)

(4e) The measure of F is zero.

Since $F \supset M$, the statement (4e) yields a new proof of the fact that the set M of rational numbers (as every countable set) has measure zero. More interesting than this, however, is the implied existence of an uncountable set of measure zero; cf. (5).

(5) Expand every number x in the closed unit interval X in the ternary system, i.e. write

$$x = \sum_{n=1}^{\infty} \frac{\alpha_n}{3^n}, \quad \alpha_n = 0, 1, 2, \quad n = 1, 2, \cdots,$$

and let C be the set of all those numbers x in whose expansion the digit 1 is not needed. (Observe that if, motivated by the customary decimal notation, we write $.\alpha_1\alpha_2\cdots$ for $\sum_{n=1}^{\infty} \alpha_n/3^n$, then for instance $\frac{1}{3} = .1000\cdots = .0222\cdots$, and therefore $\frac{1}{3} \,\varepsilon\, C$, but since $\frac{1}{2} = .111\cdots$ and since this is the only ternary expansion of $\frac{1}{2}$, therefore $\frac{1}{2} \,\varepsilon'\, C$.) Let X_1 be the open middle third of X, $X_1 = (\frac{1}{3}, \frac{2}{3})$; let X_2 and X_3 be the open middle thirds of the two closed intervals which make up $X - X_1$, i.e. $X_2 = (\frac{1}{9}, \frac{2}{9})$ and $X_3 = (\frac{7}{9}, \frac{8}{9})$; let X_4, X_5, X_6, and X_7 be the the open middle thirds of the four closed intervals which make up

$$X - (X_1 \cup X_2 \cup X_3),$$

and so on *ad infinitum*. The following statements are true.

(5a) $C = X - \bigcup_{n=1}^{\infty} X_n$. (Hint: for every x in X write $x = .\alpha_1\alpha_2\cdots$, $\alpha_n = 0, 1, 2, n = 1, 2, \cdots$, in such a way that if $x \,\varepsilon\, C$, then $\alpha_n = 0$ or 2, $n = 1, 2, \cdots$. Then the expansion of x is unique and (i) $x \,\varepsilon\, X_1$ if and only if $\alpha_1 = 1$,

(ii) if $\alpha_1 \neq 1$, then $x \in X_2 \cup X_3$ if and only if $\alpha_2 = 1$; (iii) if $\alpha_1 \neq 1$ and $\alpha_2 \neq 1$, then $x \in X_4 \cup X_5 \cup X_6 \cup X_7$ if and only if $\alpha_3 = 1$; $\cdots$.)

(5b) $\mu(C) = 0$.

(5c) C is nowhere dense. (Hint: assume that X contains an open subinterval whose intersection with $\bigcup_{n=1}^{\infty} X_n$ is empty.)

(5d) C is perfect. (Hint: no two of the intervals $X_1, X_2, \cdots$ have a common point.)

(5e) C has the cardinal number of the continuum. (Hint: consider the correspondence which associates with every x in C, $x = .\alpha_1\alpha_2\cdots$, $\alpha_n = 0$ or 2, $n = 1, 2, \cdots$, the number y whose *binary* expansion is $y = .\beta_1\beta_2\cdots$, $\beta_n = \alpha_n/2$, $n = 1, 2, \cdots$, or, equivalently, $y = \sum_{n=1}^{\infty} \alpha_n/2^{n+1}$. This correspondence is not one to one between C and X, but it is one to one between the irrational numbers in C and the irrational numbers in X. Alternative hint: use (5d).)

The set C is called the **Cantor set.**

(6) Since the cardinal number of the class of all Borel sets is that of the continuum (cf. 5.9c), and since every subset of the Cantor set is Lebesgue measurable (cf. (5b)), there exists a Lebesgue measurable set which is not a Borel set.

(7) The set of those points in the closed unit interval in whose binary expansion all the digits in the even places are 0 is a Lebesgue measurable set of measure zero.

(8) Let X be the perimeter of a circle in the Euclidean plane. There exists a unique measure μ defined on the Borel sets of X such that $\mu(X) = 1$ and such that μ is invariant under all rotations of X. (A subset of a circle is a Borel set if it belongs to the σ-ring generated by the class of all open arcs.)

(9) If g is a finite, increasing, and continuous function of a real variable, then there exists a unique complete measure $\bar{\mu}_g$ defined on a σ-ring $\bar{\mathbf{S}}_g$ containing all Borel sets, such that $\bar{\mu}_g([a,b)) = g(b) - g(a)$ and such that for every E in $\bar{\mathbf{S}}_g$ there is a Borel set F with $\bar{\mu}_g(E \,\Delta\, F) = 0$; (cf. 8.3). The measure $\bar{\mu}_g$ is called the **Lebesgue-Stieltjes measure induced** by g.

§ 16. NON MEASURABLE SETS

The discussion in the preceding section is not delicate enough to reveal the complete structure of Lebesgue measurable sets on the real line. It is, for instance, a non trivial task to decide whether or not any non measurable sets exist. It is the purpose of this section to answer this question, as well as some related ones. Some of the techniques used in obtaining the answer are very different from any we have hitherto employed. Since, however, most of them have repeated applications in measure theory, usually in the construction of illuminating examples, we shall present them in considerable detail. Throughout this section we shall employ the same notation as in § 15.

If E is any subset of the real line and a is any real number, then $E + a$ denotes the set of all numbers of the form $x + a$, with x in E; more generally if E and F are both subsets of the real line, then $E + F$ denotes the set of all numbers of the form $x + y$ with x in E and y in F. The symbol $D(E)$ will be used to denote the **difference set** of E, i.e. the set of all numbers of the form $x - y$ with x in E and y in E.

Theorem A. *If E is a Lebesgue measurable set of positive, finite measure, and if $0 \leqq \alpha < 1$, then there exists an open interval U such that $\bar{\mu}(E \cap U) \geqq \alpha\mu(U)$.*

Proof. Let $\mathbf{U}$ be the class of all open sets. Since, by 15.C, $\bar{\mu}(E) = \inf\{\mu(U): E \subset U \in \mathbf{U}\}$, we can find an open set U_0 such that $E \subset U_0$ and $\alpha\mu(U_0) \leqq \bar{\mu}(E)$. If $\{U_n\}$ is the disjoint sequence of open intervals whose union is U_0, then it follows that

$$\alpha \sum_{n=1}^{\infty} \mu(U_n) \leqq \sum_{n=1}^{\infty} \bar{\mu}(E \cap U_n).$$

Consequently we must have $\alpha\mu(U_n) \leqq \bar{\mu}(E \cap U_n)$ for at least one value of n; the interval U_n may be chosen for U. ▮

Theorem B. *If E is a Lebesgue measurable set of positive measure, then there exists an open interval containing the origin and entirely contained in the difference set $D(E)$.*

Proof. If E is, or at least contains, an open interval, the result is trivial. In the general case we make use of Theorem A, which asserts essentially that a suitable subset of E is arbitrarily close to an interval, to find a bounded open interval U such that

$$\bar{\mu}(E \cap U) \geqq \tfrac{3}{4}\mu(U).$$

If $-\frac{1}{2}\mu(U) < x < \frac{1}{2}\mu(U)$, then the set

$$(E \cap U) \cup ((E \cap U) + x)$$

is contained in an interval (namely $U \cup (U + x)$) whose length is less than $\frac{3}{2}\mu(U)$. If $E \cap U$ and $(E \cap U) + x$ were disjoint, then, since they have the same measure, we should have

$$\bar{\mu}((E \cap U) \cup [(E \cap U) + x]) = 2\bar{\mu}(E \cap U) \geqq \tfrac{3}{2}\mu(U).$$

Hence at least one point of $E \cap U$ belongs also to $(E \cap U) + x$, which proves that $x \in D(E)$. In other words the interval $(-\frac{1}{2}\mu(U), \frac{1}{2}\mu(U))$ satisfies the conditions stated in the theorem. ▮

Theorem C. *If ξ is an irrational number, then the set A of all numbers of the form $n + m\xi$, where n and m are arbitrary integers, is everywhere dense on the line; the same is true of the subset B of all numbers of the form $n + m\xi$ with n even, and the subset C of numbers of the form $n + m\xi$ with n odd.*

Proof. For every positive integer i there exists a unique integer n_i (which may be positive, negative, or zero) such that $0 \leq n_i + i\xi < 1$; we write $x_i = n_i + i\xi$. If U is any open interval, then there is a positive integer k such that $\mu(U) > \frac{1}{k}\cdot$ Among the $k + 1$ numbers, $x_1, \cdots, x_{k+1}$, in the unit interval, there must be at least two, say x_i and x_j, such that $|x_i - x_j| < \frac{1}{k}\cdot$ It follows that some integral multiple of $x_i - x_j$, i.e. some element of A, belongs to the interval U, and this concludes the proof of the assertion concerning A. The proof for B is similar; we have merely to replace the unit interval by the interval $[0,2)$. The proof for C follows from the fact that $C = B + 1$. ▮

Theorem D. *There exists at least one set E_0 which is not Lebesgue measurable.*

Proof. For any two real numbers x and y we write (for the purposes of this proof only) $x \sim y$ if $x - y \in A$, (where A is the set described in Theorem C). It is easy to verify that the relation "$\sim$" is reflexive, symmetric, and transitive, and that, accordingly, the set of all real numbers is the union of a disjoint class of sets, each set consisting of all those numbers which are in the relation "$\sim$" with a given number. By the axiom of choice we may find a set E_0 containing exactly one point from each such set; we shall prove that E_0 is not measurable.

Suppose that F is a Borel set such that $F \subset E_0$. Since the difference set $D(F)$ cannot contain any non zero elements of the dense set A, it follows from Theorem B that F must have measure zero, so that $\mu_*(E_0) = 0$. In other words, if E_0 is Lebesgue measurable, then its measure must be zero.

Observe next that if a_1 and a_2 are two different elements of A, then the sets $E_0 + a_1$ and $E_0 + a_2$ are disjoint. (If $x_1 + a_1 = x_2 + a_2$, with x_1 in E_0 and x_2 in E_0, then $x_1 - x_2 = a_2 - a_1 \varepsilon A$.) Since moreover the countable class of sets of the form $E_0 + a$, where $a \varepsilon A$, covers the entire real line, i.e. $E_0 + A = X$, and since the Lebesgue measurability of E_0 would imply that each $E_0 + a$ is Lebesgue measurable and of the same measure as E_0, we see that the Lebesgue measurability of E_0 would imply the nonsensical result $\mu(X) = 0$. ∎

The construction in the proof of Theorem D is well known, but it is not strong enough to yield certain counter examples needed for later purposes. The following theorem is an improvement.

Theorem E. *There exists a subset M of the real line such that, for every Lebesgue measurable set E,*

$$\mu_*(M \cap E) = 0 \quad \textit{and} \quad \mu^*(M \cap E) = \bar{\mu}(E).$$

Proof. Write $A = B \cup C$, as in Theorem C, and, if E_0 is the set constructed in the proof of Theorem D, write

$$M = E_0 + B.$$

If F is a Borel set such that $F \subset M$, then the difference set $D(F)$ cannot contain any elements of the dense set C, and it follows from Theorem B that $\mu_*(M) = 0$. The relations

$$M' = E_0 + C = E_0 + (B + 1) = M + 1$$

imply that $\mu_*(M') = 0$; (cf. 15.D). If E is any Lebesgue measurable set, then the monotone character of μ_* implies that $\mu_*(M \cap E) = \mu_*(M' \cap E) = 0$, and therefore (14.H) $\mu^*(M \cap E) = \bar{\mu}(E)$. ∎

The proofs of this section imply among other things that it is impossible to extend Lebesgue measure to the class of all subsets of the real line so that the extended set function is still a measure and is invariant under translations.

(1) If E is a Lebesgue measurable set such that, for every number x in an everywhere dense set,

$$\bar{\mu}(E \,\Delta\, (E + x)) = 0,$$

then either $\bar{\mu}(E) = 0$ or else $\bar{\mu}(E') = 0$.

(2) Let μ be a σ–finite measure on a σ–ring $\mathbf{S}$ of subsets of a set X, and let μ^* and μ_* be the outer measure and the inner measure, respectively, induced by μ on $\mathbf{H}(\mathbf{S})$. Let M be any set in $\mathbf{H}(\mathbf{S})$, and let $\tilde{\mathbf{S}}$ be the σ–ring generated by the class of all sets in $\mathbf{S}$ together with M. The chain of assertions below is designed to lead up to a proof of the assertion that μ may be extended to a measure $\tilde{\mu}$ on $\tilde{\mathbf{S}}$.

(2a) The σ–ring $\tilde{\mathbf{S}}$ is the class of all sets of the form $(E \cap M) \,\Delta\, (F \cap M')$, where E and F are in $\mathbf{S}$. (Hint: it is sufficient to prove that the class of all sets of the indicated form is a σ–ring. Observe that

$$(E \cap M) \,\Delta\, (F \cap M') = (E \cap M) \cup (F \cap M').)$$

(2b) If $\mu^*(M) < \infty$, if G and H are a measurable kernel and a measurable cover of M respectively, and if $D = H - G$, then the intersection of any set in $\tilde{\mathbf{S}}$ with D' belongs to $\mathbf{S}$.

(2c) There exist two sets G and H in $\mathbf{S}$ such that $G \subset M \subset H$ and $\mu_*(M - G) = \mu_*(H - M) = 0$, and such that if $D = H - G$, then the intersection of any set in $\tilde{\mathbf{S}}$ with D' belongs to $\mathbf{S}$. (Hint: there exists a disjoint sequence $\{X_n\}$ of sets in $\mathbf{S}$ with $\mu(X_n) < \infty$ and $M = \bigcup_{n=1}^{\infty} (M \cap X_n)$.)

(2d) In the notation of (2c), $\mu_*(M \cap D) = \mu_*(M' \cap D) = 0$, and therefore $\mu^*(M \cap D) = \mu^*(M' \cap D) = \mu(D)$.

(2e) In the notation of (2c), if

$$[(E_1 \cap M) \,\Delta\, (F_1 \cap M')] \cap D = [(E_2 \cap M) \,\Delta\, (F_2 \cap M')] \cap D,$$

where E_1, F_1, E_2, and F_2 are in $\mathbf{S}$, then

$$\mu(E_1 \cap D) = \mu(E_2 \cap D) \quad \text{and} \quad \mu(F_1 \cap D) = \mu(F_2 \cap D).$$

(Hint: use the fact that the condition

$$[(E_1 \,\Delta\, E_2) \cap M \cap D] \,\Delta\, [(F_1 \,\Delta\, F_2) \cap M' \cap D] = 0$$

implies that

$$(E_1 \cap D) \,\Delta\, (E_2 \cap D) \subset M' \cap D \quad \text{and} \quad (F_1 \cap D) \,\Delta\, (F_2 \cap D) \subset M \cap D.)$$

(2f) Let α and β be non negative real numbers with $\alpha + \beta = 1$. In the notation of (2c) the set function $\tilde{\mu}$ on $\tilde{\mathbf{S}}$, defined by

$$\tilde{\mu}((E \cap M) \,\Delta\, (F \cap M')) =$$
$$= \mu([(E \cap M) \,\Delta\, (F \cap M')] \cap D') + \alpha\mu(E \cap D) + \beta\mu(F \cap D),$$

is a measure on $\tilde{\mathbf{S}}$ which is an extension of μ on $\mathbf{S}$.

(3) If μ is a σ–finite measure on a σ–ring $\mathbf{S}$ and if $\{M_1, \cdots, M_n\}$ is a finite class of sets in the hereditary σ–ring $\mathbf{H}(\mathbf{S})$, then $\{M_1, \cdots, M_n\}$ may be adjoined to $\mathbf{S}$ and a measure $\tilde{\mu}$ may be defined on the generated σ–ring $\tilde{\mathbf{S}}$ so that it is an

extension of μ on **S**. The analogous statement, for an infinite sequence $\{M_n\}$ of sets in **H**(**S**), is not known.

(4) The following example is useful for developing intuition about non measurable sets; virtually all general properties of non measurable sets may be illustrated by it. Let $X = \{(x,y): 0 \leqq x \leqq 1, 0 \leqq y \leqq 1\}$ be the unit square. For every subset E of the interval $[0,1]$, write

$$\hat{E} = \{(x,y): x \,\varepsilon\, E, \quad 0 \leqq y \leqq 1\} \subset X.$$

Let **S** be the class of all sets of the form $\hat{E}$, for Lebesgue measurable sets E; define $\mu(\hat{E})$ as the Lebesgue measure of E. A set M such as $M = \{(x,y): 0 \leqq x \leqq 1, y = \frac{1}{2}\}$ is non measurable; $\mu_*(M) = 0$ and $\mu^*(M) = 1$.

(5) Let μ^* be a regular outer measure on the class of all subsets of a set X such that $\mu^*(X) = 1$, and let M be a subset of X such that $\mu_*(M) = 0$ and $\mu^*(M) = 1$; (cf. Theorem E and (4)). If $\nu^*(E) = \mu^*(E) + \mu^*(E \cap M)$, then ν^* is an outer measure; (cf. 10.5 and 10.7).

(5a) A set E is ν^*–measurable if and only if it is μ^*–measurable; (cf. 12.6).

(5b) The infimum of the values of $\nu^*(E)$ over all ν^*–measurable sets E containing a given set A is $2\mu^*(A)$. (Hint: if E is ν^*–measurable, then $\mu^*(E \cap M) = \mu^*(E)$.)

(5c) The outer measure ν^* is not regular. (Hint: test regularity with M'.)

Chapter IV

MEASURABLE FUNCTIONS

§ 17. MEASURE SPACES

A **measurable space** is a set X and a σ-ring **S** of subsets of X with the property that $\bigcup \mathbf{S} = X$. Ordinarily it causes no confusion to denote a measurable space by the same symbol as the underlying set X; on the occasions when it is desirable to call attention to the particular σ-ring under consideration, we shall write $(X,\mathbf{S})$ for X. It is customary to call a subset E of X **measurable** if and only if it belongs to the σ-ring **S**. This terminology is not meant to indicate that **S** is the σ-ring of all μ^*-measurable sets with respect to some outer measure μ^*, nor even that a non trivial measure is or may be defined on **S**.

In the language of measurable sets, the condition in the definition of measurable spaces may be expressed by saying that the union of all measurable sets is the entire space, or, equivalently, that every point is contained in some measurable set. The purpose of this restriction is to eliminate certain obvious and not at all useful pathological considerations, by excluding from the space points (and sets of points) of no measure theoretic relevance.

A **measure space** is a measurable space $(X,\mathbf{S})$ and a measure μ on **S**; just as for measurable spaces we shall ordinarily allow ourselves to confuse a measure space whose underlying set is X with the set X. On the occasions when it is desirable to call attention to the particular σ-ring and measure under consideration, we shall write $(X,\mathbf{S},\mu)$ for X. The measure space X is called [totally] finite, σ-finite, or complete, according as the measure μ is [totally] finite, σ-finite, or complete. For measure spaces we may and shall make use, without any further explanation, of the outer measure μ^* and (in the σ-finite case) the inner measure μ_* induced by μ on the hereditary σ-ring $\mathbf{H}(\mathbf{S})$.

Most of the considerations of the preceding chapter show by deductions and examples how certain measurable spaces may be made into measure spaces. In this section we shall make a few general remarks on measurable spaces and measure spaces and then, in the remainder of this chapter and in the following chapters, turn to the discussion of functions on measure spaces, useful ways of making new measure spaces out of old ones, and the theory of some particularly important special cases.

We observe first that a measurable subset X_0 of a measure space $(X,\mathbf{S},\mu)$ may itself be considered as a measure space $(X_0,\mathbf{S}_0,\mu_0)$, where $\mathbf{S}_0$ is the class of all measurable subsets of X_0, and, for E in $\mathbf{S}_0$, $\mu_0(E) = \mu(E)$. Conversely, if a subset X_0 of a set X is a measure space $(X_0,\mathbf{S}_0,\mu_0)$, then X may be made into a measure space $(X,\mathbf{S},\mu)$, where $\mathbf{S}$ is the class of all those subsets of X whose intersection with X_0 is in $\mathbf{S}_0$, and, for E in $\mathbf{S}$, $\mu(E) = \mu_0(E \cap X_0)$. (Entirely similar remarks are valid, of course, for measurable spaces.) A modification of this last construction is frequently useful even if X is already a measure space. If X_0 is a measurable subset of X, a new measure μ_0 may be defined on the class of all measurable subsets E of X by the equation $\mu_0(E) = \mu(E \cap X_0)$; it is easy to verify that $(X,\mathbf{S},\mu_0)$ is indeed a measure space.

What happens to the considerations of the preceding paragraph if the subset X_0 is not measurable? In order to give the most useful answer to this question, we introduce a new concept. A subset X_0 of a measure space $(X,\mathbf{S},\mu)$ is **thick** if $\mu_*(E - X_0) = 0$ for every measurable set E. If X itself is measurable, then X_0 is thick if and only if $\mu_*(X - X_0) = 0$; if μ is totally finite, then X_0 is thick if and only if $\mu^*(X_0) = \mu(X)$. (For examples of thick sets cf. 16.E and 16.4.) Slightly deeper than any of the comments in the preceding paragraph is the following result, which asserts essentially that a thick subset of a measure space may itself be regarded as a measure space.

Theorem A. *If X_0 is a thick subset of a measure space $(X,\mathbf{S},\mu)$, if $\mathbf{S}_0 = \mathbf{S} \cap X_0$, and if, for E in $\mathbf{S}$, $\mu_0(E \cap X_0) = \mu(E)$, then $(X_0,\mathbf{S}_0,\mu_0)$ is a measure space.*

Proof. If two sets, E_1 and E_2, in $\mathbf{S}$ are such that $E_1 \cap X_0 = E_2 \cap X_0$, then $(E_1 \Delta E_2) \cap X_0 = 0$, so that $\mu(E_1 \Delta E_2) = 0$ and therefore $\mu(E_1) = \mu(E_2)$. In other words μ_0 is indeed unambiguously defined on $\mathbf{S}_0$.

Suppose next that $\{F_n\}$ is a disjoint sequence of sets in $\mathbf{S}_0$, and let E_n be a set in $\mathbf{S}$ such that

$$F_n = E_n \cap X_0, \quad n = 1, 2, \cdots.$$

If $\tilde{E}_n = E_n - \bigcup\{E_i: 1 \leq i < n\}$, $n = 1, 2, \cdots$, then

$$(\tilde{E}_n \Delta E_n) \cap X_0 = (F_n - \bigcup\{F_i: 1 \leq i < n\}) \Delta F_n = $$
$$= F_n \Delta F_n = 0,$$

so that $\mu(\tilde{E}_n \Delta E_n) = 0$, and therefore

$$\sum_{n=1}^{\infty} \mu_0(F_n) = \sum_{n=1}^{\infty} \mu(E_n) = \sum_{n=1}^{\infty} \mu(\tilde{E}_n) = \mu(\bigcup_{n=1}^{\infty} \tilde{E}_n) =$$
$$= \mu(\bigcup_{n=1}^{\infty} E_n) = \mu_0(\bigcup_{n=1}^{\infty} F_n).$$

In other words μ_0 is indeed a measure, and the proof of the theorem is complete. ▮

(1) The following converse of Theorem A is true. If $(X,\mathbf{S},\mu)$ is a measure space and if X_0 is a subset of X such that, for every two measurable sets E_1 and E_2, the condition $E_1 \cap X_0 = E_2 \cap X_0$ implies that $\mu(E_1) = \mu(E_2)$, then X_0 is thick. (Hint: if $F \subset E - X_0$, then

$$(E - F) \cap X_0 = E \cap X_0.)$$

(2) The extension theorem 16.2 may be used to give an alternative proof of Theorem A in the σ-finite case.

(3) The following proposition shows that the concept of a finite measure space is not very different from the apparently much more special concept of a totally finite measure space. If $(X,\mathbf{S},\mu)$ is a finite measure space, then there exists a thick measurable set X_0. (Hint: write $c = \sup\{\mu(E): E \in \mathbf{S}\}$. Let $\{E_n\}$ be a sequence of measurable sets such that $\lim_n \mu(E_n) = c$ and write $X_0 = \bigcup_{n=1}^{\infty} E_n$. Observe that $\mu(X_0) = c$.) This result enables us, in most applications, to assume that a finite measure space is totally finite, since we may replace X by X_0 without significant loss of generality. For an example of a finite measure space which is not totally finite, let X be the real line, let $\mathbf{S}$ be the class of all sets of the form $E \cup C$, where E is a Lebesgue measurable subset of $[0,1]$ and C is countable, and let μ on $\mathbf{S}$ be Lebesgue measure. The method suggested above

to show the existence of X_0 has frequent application in measure theory; it is called the method of **exhaustion.**

(4) If $(X,\mathbf{S},\mu)$ is a complete, σ-finite measure space, then every μ^*-measurable set is measurable. Hence for complete, σ-finite measure spaces the two concepts of measurability collapse into one.

§ 18. MEASURABLE FUNCTIONS

Suppose that f is a real valued function on a set X and let M be any subset of the real line. We shall write

$$f^{-1}(M) = \{x: f(x) \, \varepsilon \, M\},$$

i.e. $f^{-1}(M)$ is the set of all those points of X which are mapped into M by f. The set $f^{-1}(M)$ is called the **inverse image** (under f, or with respect to f) of the set M. If, for instance, f is the characteristic function of a set E in X, then $f^{-1}(\{1\}) = E$ and $f^{-1}(\{0\}) = E'$; more generally

$$f^{-1}(M) = 0, \quad E, \quad E', \quad \text{or} \quad X,$$

according as M contains neither 0 nor 1, 1 but not 0, 0 but not 1, or both 0 and 1.

It is easy to verify that, for every f,

$$f^{-1}(\textstyle\bigcup_{n=1}^{\infty} M_n) = \bigcup_{n=1}^{\infty} f^{-1}(M_n),$$

$$f^{-1}(M - N) = f^{-1}(M) - f^{-1}(N);$$

in other words the mapping f^{-1}, from subsets of the line to subsets of X, preserves all set operations. It follows in particular that if $\mathbf{E}$ is a class of subsets of the line (such as a ring or a σ-ring) with certain algebraic properties, then $f^{-1}(\mathbf{E})$ (= the class of all those subsets of X which have the form $f^{-1}(M)$ for some M in $\mathbf{E}$) is a class with the same algebraic properties. Of particular interest for later applications is the case in which $\mathbf{E}$ is the class of all Borel sets on the line.

Suppose now that in addition to the set X we are given also a σ-ring $\mathbf{S}$ of subsets of X so that $(X,\mathbf{S})$ is a measurable space. For every real valued (and also for every extended real valued) function f on X, we shall write

$$N(f) = \{x: f(x) \neq 0\};$$

if a real valued function f is such that, for every Borel subset M of the real line the set $N(f) \cap f^{-1}(M)$ is measurable, then f is called a **measurable function.**

Several comments are called for in connection with this definition of measurability. First of all, the special role played by the value 0 should be emphasized. The reason for singling out 0 lies in the fact that it is the identity element of the additive group of real numbers. In the next chapter we shall introduce the concept of integral, defined for certain measurable functions; the fact that integration (which is without doubt the most important concept in measure theory) may be viewed as generalized addition necessitates treating 0 differently from other real numbers.

If f is a measurable function on X and if we take for M the entire real line, then it follows that $N(f)$ is a measurable set. Hence if E is a measurable subset of X and if M is a Borel subset of the real line, then it follows from the identity

$$E \cap \{f^{-1}(M)\} =$$

$$= [E \cap N(f) \cap f^{-1}(M)] \cup [(E - N(f)) \cap f^{-1}(M)],$$

that $E \cap f^{-1}(M)$ is measurable. (Observe that the second term in the last written union is either empty or else equal to $E - N(f)$.) If, in other words, we say that a real valued function f defined on a measurable set E is to be called **measurable on** E whenever $E \cap f^{-1}(M)$ is measurable for every Borel set M, then we have proved that a measurable function is measurable on every measurable set. If, in particular, the entire space X happens to be measurable, then the requirement of measurability on f is simply that $f^{-1}(M)$ be measurable for every Borel subset M of the real line. In other words, in case X is measurable, a measurable function is one whose inverse maps the sets of one prescribed σ-ring (namely the Borel sets on the line) into the sets of another prescribed σ-ring (namely **S**).

It is clear that the concept of measurability for a function depends on the σ-ring **S** and therefore, on the rare occasions when we shall have more than one σ-ring under consideration at the same time, we shall say that a function is measurable with respect to **S**, or, more concisely, that it is measurable (**S**). If in particular

X is the real line, and $\mathbf{S}$ and $\bar{\mathbf{S}}$ are the class of Borel sets and the class of Lebesgue measurable sets respectively, then we shall call a function measurable with respect to $\mathbf{S}$ a **Borel measurable function,** and a function measurable with respect to $\bar{\mathbf{S}}$ a **Lebesgue measurable function.**

It is important to emphasize also that the concept of measurability for functions, just as the concept of measurability for sets, as used in § 17, does *not* depend on the numerical values of a prescribed measure μ, but merely on the prescribed σ-ring $\mathbf{S}$. A set or a function is, from this point of view, declared measurable by *fiat;* the concept is purely set theoretical and is quite independent of *measure* theory.

The situation is analogous to that in the modern theory of topological spaces, where certain sets are declared open and certain functions continuous, without reference to a numerical distance. The existence or non existence of a metric, in terms of which openness and continuity can be *defined*, is an interesting but usually quite irrelevant question. The analogy is deeper than it seems: the reader familiar with the theory of continuous functions on topological spaces will recall that a function f is continuous if and only if, for every open set M in the range (in our case the real line), the set $f^{-1}(M)$ belongs to the prescribed family of sets which are called open in the domain.

We shall need the concept of measurability for extended real functions also. We define this concept simply by making the convention that the one-point sets $\{\infty\}$ and $\{-\infty\}$ of the extended real line are to be regarded as Borel sets, and then repeating verbatim the definition for real valued functions. Accordingly a possibly infinite valued function f is measurable, if, for every Borel set M of real numbers, each of the three sets

$$f^{-1}(\{\infty\}),\quad f^{-1}(\{-\infty\}),\quad \text{and}\quad N(f)\cap f^{-1}(M)$$

is measurable. We observe that for the extended concept of Borel set it is no longer true that the class of Borel sets is the σ-ring generated by semiclosed intervals.

We shall study and attempt to make clear the structure of

measurable functions in great detail below. The following is a preliminary result of considerable use.

Theorem A. *A real function f on a measurable space $(X,\mathbf{S})$ is measurable if and only if, for every real number c, the set $N(f) \cap \{x: f(x) < c\}$ is measurable.*

Proof. If M is the open ray extending from c to $-\infty$ on the real line, i.e. $M = \{t: t < c\}$, then M is a Borel set and $f^{-1}(M) = \{x: f(x) < c\}$. It is clear therefore that the stated condition is indeed necessary for the measurability of f.

Suppose next that the condition is satisfied. If c_1 and c_2 are real numbers, $c_1 \leqq c_2$, then

$$\{x: f(x) < c_2\} - \{x: f(x) < c_1\} = \{x: c_1 \leqq f(x) < c_2\}.$$

In other words if M is any semiclosed interval, then $N(f) \cap f^{-1}(M)$ is the difference of two measurable sets and is therefore measurable. Let $\mathbf{E}$ be the class of all those subsets M of the extended real line for which $N(f) \cap f^{-1}(M)$ is measurable. Since $\mathbf{E}$ is a σ-ring, and since a σ-ring containing all semiclosed intervals contains also all Borel sets, the proof of the theorem is complete. ∎

(1) Theorem A remains true if $<$ is replaced by $\leqq$ or $>$ or $\geqq$. (Hint: if $-\infty < c < \infty$, then

$$\{x: f(x) \leqq c\} = \bigcap_{n=1}^{\infty} \left\{x: f(x) < c + \frac{1}{n}\right\}.)$$

(2) Theorem A remains true if c is restricted to belong to an everywhere dense set of real numbers.

(3) If f is a measurable function and c is a real number, then cf is measurable.

(4) If a set E is a measurable set, then its characteristic function is a measurable function. Is the converse of this statement true?

(5) A non zero constant function is measurable if and only if X is measurable.

(6) If X is the real line and f is an increasing function, then f is Borel measurable. Is every continuous function Borel measurable?

(7) Let X be the real line and let E be a set which is not Lebesgue measurable; write $f(x) = x$ or $-x$ according as $x \in E$ or $x \in' E$. Is f Lebesgue measurable?

(8) If f is measurable, then, for every real number c, the set $N(f) \cap \{x: f(x) = c\}$ is measurable. Is the converse of this statement true?

(9) A complex valued function is called measurable if both its real and imaginary parts are measurable. A complex valued function f is measurable if and only if, for every open set M in the complex plane, the set $N(f) \cap f^{-1}(M)$ is measurable.

(10) Suppose that f is a real valued function on a measurable space $(X,\mathbf{S})$, and, for every real number t, write $B(t) = \{x: f(x) \leqq t\}$. Then

(10a) $$s < t \text{ implies } B(s) \subset B(t),$$

(10b) $$\bigcup_t B(t) = X \text{ and } \bigcap_t B(t) = 0,$$

(10c) $$\bigcap_{s<t} B(t) = B(s).$$

Conversely, if $\{B(t)\}$ is a class of sets with the properties (10a), (10b), and (10c), then there exists a unique, finite, real valued function f such that $\{x: f(x) \leqq t\} = B(t)$. (Hint: write $f(x) = \inf \{t: x \in B(t)\}$.)

(11) If f is a measurable function on a totally finite measure space $(X,\mathbf{S},\mu)$ and if, for every Borel set M on the extended real line, we write $\nu(M) = \mu(f^{-1}(M))$, then ν is a measure on the class of all Borel sets. If f is finite valued, then the function g of a real variable, defined by $g(t) = \mu(\{x: f(x) < t\})$, is monotone increasing, continuous on the left, and such that $g(-\infty) = 0$ and $g(\infty) = \mu(X)$; g is called the **distribution function** of f. If g is continuous, then the Lebesgue–Stieltjes measure μ_g, induced by g according to the procedure of 15.9, is the completion of ν. If f is the characteristic function of a measurable set E, then $\nu(M) = \chi_M(1)\mu(E) + \chi_M(0)\mu(E')$.

§ 19. COMBINATIONS OF MEASURABLE FUNCTIONS

Theorem A. *If f and g are extended real valued measurable functions on a measurable space $(X,\mathbf{S})$, and if c is any real number, then each of the three sets*

$$A = \{x: f(x) < g(x) + c\},$$

$$B = \{x: f(x) \leqq g(x) + c\},$$

$$C = \{x: f(x) = g(x) + c\},$$

has a measurable intersection with every measurable set.

Proof. Let M be the set of rational numbers on the line. Since

$$A = \bigcup_{r \in M} [\{x: f(x) < r\} \cap \{x: r - c < g(x)\}],$$

it follows that A has the desired property. The conclusions for B and C are consequences of the relations

$$B = X - \{x: g(x) < f(x) - c\} \quad \text{and} \quad C = B - A$$

respectively. ∎

Theorem B. *If ϕ is an extended real valued Borel measurable function on the extended real line such that $\phi(0) = 0$, and if f is an extended real valued measurable function on a measurable space X, then the function $\tilde{f}$, defined by $\tilde{f}(x) = \phi(f(x))$, is a measurable function on X.*

Proof. It is convenient to use here the definition of measurability (instead of the necessary and sufficient condition of § 18). If M is any Borel set on the extended real line, then

$$N(\tilde{f}) \cap \tilde{f}^{-1}(M) = \{x: \phi(f(x)) \varepsilon M - \{0\}\} = \\ = \{x: f(x) \varepsilon \phi^{-1}(M - \{0\})\}.$$

Since $\phi(0) = 0$, we have

$$\phi^{-1}(M - \{0\}) = \phi^{-1}(M - \{0\}) - \{0\}.$$

Since ϕ is Borel measurable, $\phi^{-1}(M - \{0\})$ is a Borel set and the measurability of the set

$$N(\tilde{f}) \cap \tilde{f}^{-1}(M) = N(f) \cap f^{-1}(\phi^{-1}(M - \{0\}))$$

follows from the measurability of f. ▮

Since it is easy to verify that, for any positive real number α, the function ϕ, defined for every real number t by $\phi(t) = |t|^\alpha$, is Borel measurable, it follows that the measurability of a function f implies the measurability of $|f|^\alpha$. Similarly any positive integral power of a measurable function is again a measurable function, and it follows similarly, by an even simpler argument, that a constant (real) multiple of a measurable function is also measurable. By considering Borel measurable functions ϕ of two or more real variables a similar argument may be used to prove such statements as that the sum and product of two measurable functions are measurable. Since, however, we have not yet defined and proved any properties of Borel measurability for functions of several variables, we postpone these considerations and turn now to a direct proof of the measurability of sums and products.

Theorem C. *If f and g are extended real valued measurable functions on a measurable space X, then so also are $f + g$ and fg.*

Proof. Since the behavior of $f+g$ and fg at those points x at which at least one of the two numbers, $f(x)$ and $g(x)$, is infinite is easily understood, after the examination of a small number of cases, we restrict our attention to finite valued functions. (We recall incidentally that if $f(x) = \pm\infty$ and $g(x) = \mp\infty$, then $f(x) + g(x)$ is not defined.)

Since if f and g are finite and if c is a real number, then

$$\{x: f(x) + g(x) < c\} = \{x: f(x) < c - g(x)\},$$

the measurability of $f+g$ follows from Theorem A (with $-g$ in place of g). The measurability of fg is a consequence of the identity

$$fg = \tfrac{1}{4}[(f+g)^2 - (f-g)^2]. \quad \blacksquare$$

Since if f and g are finite we have

$$f \cup g = \tfrac{1}{2}(f + g + |f - g|)$$

and

$$f \cap g = \tfrac{1}{2}(f + g - |f - g|),$$

Theorems B and C show that the measurability of f and g implies that of $f \cup g$ and $f \cap g$. If for every extended real valued function f we write

$$f^+ = f \cup 0 \quad \text{and} \quad f^- = -(f \cap 0),$$

then

$$f = f^+ - f^- \quad \text{and} \quad |f| = f^+ + f^-.$$

(The functions f^+ and f^- are called the **positive part** and the **negative part** of f, respectively.) The comment at the beginning of this paragraph implies that the positive and negative parts of a measurable function are both measurable; conversely, a function with measurable positive and negative parts is itself measurable.

(1) If f is such that $|f|$ is measurable, does f have to be measurable?

(2) If X is measurable, then Theorem B is true even without the assumption that $\phi(0) = 0$; in other words, in this case a Borel measurable function of a measurable function is a measurable function.

(3) It is not true, even if X is measurable, that a Lebesgue measurable function of a measurable function is a measurable function. The purpose of the sequence of statements below is to indicate the proof of this negative statement by the construction of a suitable example. The construction will yield a Lebesgue measurable function ϕ of a real variable y, and a continuous and strictly increasing function f of a real variable x, $0 \leqq x \leqq 1$, such that if $\tilde{f}(x) = \phi(f(x))$, then $\tilde{f}$ is not Lebesgue measurable.

For every x in X (where $X = [0,1]$ is the closed unit interval), write

$$x = \sum_{i=1}^{\infty} \alpha_i/3^i = .\alpha_1\alpha_2\alpha_3\cdots,$$

where $\alpha_i = 0, 1,$ or 2, $i = 1, 2, \cdots$, so that if $x \varepsilon C$, then $\alpha_i = 0$ or 2, $i = 1, 2, \cdots$. (The set C is the Cantor set, defined in 15.5.) Let $n = n(x)$ be the first index for which $\alpha_n = 1$. (If there is no such n, i.e. if $x \varepsilon C$, write $n(x) = \infty$.) Define the function ψ by the equation

$$\psi(x) = \sum_{1 \leqq i < n} \alpha_i/2^{i+1} + \frac{1}{2^n}.$$

(The function ψ is sometimes called the **Cantor function.**)

(3a) If $0 \leqq x \leqq y \leqq 1$, then

$$0 = \psi(0) \leqq \psi(x) \leqq \psi(y) \leqq \psi(1) = 1.$$

(Hint: if $x = .\alpha_1\alpha_2\alpha_3\cdots \leqq y = .\beta_1\beta_2\beta_3\cdots$, and if $\alpha_i = \beta_i$ for $1 \leqq i < j$, then $\alpha_j \leqq \beta_j$.)

(3b) The function ψ is continuous. (Hint: if $x = .\alpha_1\alpha_2\alpha_3\cdots$, $y = .\beta_1\beta_2\beta_3\cdots$, and $\alpha_i = \beta_i$ for $1 \leqq i < j$, then

$$|\psi(x) - \psi(y)| \leqq \frac{1}{2^{j-1}}.)$$

(3c) For every x in X there is one and only one number y, $0 \leqq y \leqq 1$, such that $x = \frac{1}{2}(y + \psi(y))$, and therefore the equation $y = f(x)$ defines a strictly increasing, continuous function f on X. (Hint: $\frac{1}{2}(y + \psi(y))$ is strictly increasing and continuous.)

(3d) The set $f^{-1}(C)$ is Lebesgue measurable and has positive measure. (Hint: the set

$$\psi(X - C) = \{\psi(y): y \varepsilon X - C\}$$

is countable and therefore has measure zero; consequently

$$\mu(f^{-1}(X - C)) = \tfrac{1}{2}.)$$

(3e) There exists a Lebesgue measurable set M, $M \subset \{y: 0 \leqq y \leqq 1\}$, such that $f^{-1}(M)$ is not Lebesgue measurable. (Hint: by 16.E, $f^{-1}(C)$ contains a non measurable set. Recall that every subset of a set of Lebesgue measure zero is Lebesgue measurable.)

(3f) If ϕ is the characteristic function of the set M mentioned in (3e) and if $\tilde{f}(x) = \phi(f(x))$, then ϕ is Lebesgue measurable but $\tilde{f}$ is not.

(4) The set M in (3e) is an example of a Lebesgue measurable set which is not a Borel set; (cf. 15.6).

§ 20. SEQUENCES OF MEASURABLE FUNCTIONS

Theorem A. *If $\{f_n\}$ is a sequence of extended real valued, measurable functions on a measurable space X, then each of the four functions h, g, f^*, and f_*, defined by*

$$h(x) = \sup\{f_n(x): n = 1, 2, \cdots\},$$

$$g(x) = \inf\{f_n(x): n = 1, 2, \cdots\},$$

$$f^*(x) = \limsup_n f_n(x),$$

$$f_*(x) = \liminf_n f_n(x),$$

is measurable.

Proof. It is easy to reduce the general case to the case of finite valued functions. The equation

$$\{x: g(x) < c\} = \bigcup_{n=1}^{\infty} \{x: f_n(x) < c\}$$

implies the measurability of g. The result for h follows from the relation

$$h(x) = -\inf\{-f_n(x): n = 1, 2, \cdots\}.$$

The measurability of f^* and f_* is a consequence of the relations

$$f^*(x) = \inf_{n \geq 1} \sup_{m \geq n} f_m(x), \quad f_*(x) = \sup_{n \geq 1} \inf_{m \geq n} f_m(x),$$

respectively. ∎

It follows from Theorem A that the set of points of convergence of a sequence $\{f_n\}$ of measurable functions, i.e. the set

$$\{x: \limsup_n f_n(x) = \liminf_n f_n(x)\},$$

has a measurable intersection with every measurable set, and, consequently, that the function f, defined by $f(x) = \lim_n f_n(x)$ at every x for which the limit exists, is a measurable function.

A very useful concept in the theory of measurable functions is that of a simple function. A function f, defined on a measurable space X, is called **simple** if there is a finite, disjoint class $\{E_1, \cdots,$

$E_n\}$ of measurable sets and a finite set $\{\alpha_1, \cdots, \alpha_n\}$ of real numbers such that

$$f(x) = \begin{cases} \alpha_i & \text{if} \quad x \in E_i, \quad i = 1, \cdots, n, \\ 0 & \text{if} \quad x \in' E_1 \cup \cdots \cup E_n. \end{cases}$$

(We emphasize the fact that the values of a simple function are to be *finite* real numbers: this will be essential in the sequel.) In other words a simple function takes on only a finite number of values different from zero, each on a measurable set.

The simplest example of a simple function is the characteristic function χ_E of a measurable set E. It is easy to verify that a simple function is always measurable; in fact we have, for the simple function f described above,

$$f(x) = \sum_{i=1}^{n} \alpha_i \chi_{E_i}(x).$$

The product of two simple functions, and any finite linear combination of simple functions, are again simple functions.

Theorem B. *Every extended real valued measurable function f is the limit of a sequence $\{f_n\}$ of simple functions; if f is non negative, then each f_n may be taken non negative and the sequence $\{f_n\}$ may be assumed increasing.*

Proof. Suppose first that $f \geqq 0$. For every $n = 1, 2, \cdots$, and for every x in X, we write

$$f_n(x) = \begin{cases} \dfrac{i-1}{2^n} & \text{if} \quad \dfrac{i-1}{2^n} \leqq f(x) < \dfrac{i}{2^n}, \quad i = 1, \cdots, 2^n n, \\ n & \text{if} \quad f(x) \geqq n. \end{cases}$$

Clearly f_n is a non negative simple function, and the sequence $\{f_n\}$ is increasing. If $f(x) < \infty$, then, for some n,

$$0 \leqq f(x) - f_n(x) \leqq \frac{1}{2^n};$$

if $f(x) = \infty$, then $f_n(x) = n$ for every n. This proves the second half of the theorem; the first half follows (recalling that the difference of two simple functions is a simple function) by applying the result just proved separately to f^+ and f^-. ▮

(1) All the concepts and results of this section and the preceding one (except, of course, the ones depending on such order properties of the real numbers as positiveness) can be extended to complex valued functions.

(2) If the function f in Theorem B is bounded, then the sequence $\{f_n\}$ may be made to converge to f uniformly.

(3) An **elementary function** is defined in the same way as a simple function, the only change being that the number of sets E_i, and therefore the number of corresponding values α_i, is allowed to be countably infinite. Every real valued measurable function f is the limit of a uniformly convergent sequence of elementary functions.

§ 21. POINTWISE CONVERGENCE

In the preceding three sections we have developed the theory of measurable functions about as far as it is convenient to do so without mentioning measure. From now on we shall suppose that the underlying space X is a measure space $(X,\mathbf{S},\mu)$.

If a certain proposition concerning the points of a measure space is true for every point, with the exception at most of a set of points which form a measurable set of measure zero, it is customary to say that the proposition is true for **almost every** point, or that it is true **almost everywhere.** The phrase "almost everywhere" is used so frequently that it is convenient to introduce the abbreviation **a.e.** Thus, for instance, we might say that a function is a constant a.e.—meaning that there exists a real number c such that $\{x: f(x) \neq c\}$ is a set of measure zero. A function f is called **essentially bounded** if it is bounded a.e., i.e. if there exists a positive, finite constant c such that $\{x: |f(x)| > c\}$ is a set of measure zero. The infimum of the values of c for which this statement is true is called the **essential supremum** of $|f|$, abbreviated to

$$\text{ess. sup. } |f|.$$

Let $\{f_n\}$ be a sequence of extended real valued functions which converges a.e. on the measure space X to a limit function f. This means, of course, that there exists a set E_0 of measure zero (which may be empty) such that, if $x \,\varepsilon'\, E_0$ and $\epsilon > 0$, then an integer $n_0 = n_0(x,\epsilon)$ can be found with the property that

$$f_n(x) < -\frac{1}{\epsilon}, \qquad \text{if } f(x) = -\infty,$$

$$|f_n(x) - f(x)| < \epsilon, \quad \text{if } -\infty < f(x) < \infty,$$

$$f_n(x) > \frac{1}{\epsilon}, \qquad \text{if } f(x) = \infty,$$

whenever $n \geqq n_0$. We shall say that a sequence $\{f_n\}$ of real valued functions is **fundamental a.e.** if there exists a set E_0 of measure zero such that, if $x \,\varepsilon'\, E_0$ and $\epsilon > 0$, then an integer $n_0 = n_0(x,\epsilon)$ can be found with the property that

$$|f_n(x) - f_m(x)| < \epsilon, \quad \text{whenever} \quad n \geqq n_0 \quad \text{and} \quad m \geqq n_0.$$

Similarly in the theory of real sequences one distinguishes between a sequence $\{a_n\}$ of extended real numbers which converges to an extended real number a, and a sequence $\{a_n\}$ of finite real numbers which is a fundamental sequence, i.e. which satisfies Cauchy's necessary and sufficient condition for convergence to a finite limit.

It is clear that if a sequence converges to a finite valued limit function a.e., then it is fundamental a.e., and, conversely, that corresponding to a sequence which is fundamental a.e. there always exists a finite valued limit function to which it converges a.e. If moreover the sequence converges a.e. to f and also converges a.e. to g, then $f(x) = g(x)$ a.e., i.e. the limit function is uniquely determined to within a set of measure zero.

We shall have occasion in the sequel to refer to several different kinds of convergence, and we shall consistently employ terminology similar to that of the preceding paragraphs. Thus, if we define a new kind of convergence of a sequence $\{f_n\}$ to a limit f, by specifying the sense in which f_n is to be near to f for large n, then we shall use without any further explanation the notion of a sequence which is fundamental in this new sense—meaning that, for large n and m, the differences $f_n - f_m$ are to be near to 0 in the specified sense of nearness.

An example of another kind of convergence for sequences of real valued functions is **uniform convergence a.e.** The sequence

$\{f_n\}$ converges to f uniformly a.e. if there exists a set E_0 of measure zero such that, for every $\epsilon > 0$, an integer $n_0 = n_0(\epsilon)$ can be found with the property that

$$|f_n(x) - f(x)| < \epsilon, \quad \text{if} \quad n \geqq n_0 \quad \text{and} \quad x \varepsilon' E_0,$$

in other words if the sequence of functions converges uniformly to f (in the ordinary sense of that phrase) on the set $X - E_0$. Once more it is true, and easily verified, that a sequence converges uniformly a.e. to some limit function if and only if it is uniformly fundamental a.e.

The following result (known as **Egoroff's theorem**) establishes an interesting and useful connection between convergence a.e. and uniform convergence.

Theorem A. *If E is a measurable set of finite measure, and if $\{f_n\}$ is a sequence of a.e. finite valued measurable functions which converges a.e. on E to a finite valued measurable function f, then, for every $\epsilon > 0$, there exists a measurable subset F of E such that $\mu(F) < \epsilon$ and such that the sequence $\{f_n\}$ converges to f uniformly on $E - F$.*

Proof. By omitting, if necessary, a set of measure zero from E, we may assume that the sequence $\{f_n\}$ converges to f everywhere on E. If

$$E_n^m = \bigcap_{i=n}^{\infty} \left\{x : |f_i(x) - f(x)| < \frac{1}{m}\right\},$$

then

$$E_1^m \subset E_2^m \subset \cdots,$$

and, since the sequence $\{f_n\}$ converges to f on E,

$$\lim_n E_n^m \supset E$$

for every $m = 1, 2, \cdots$. Hence $\lim_n \mu(E - E_n^m) = 0$, so that there exists a positive integer $n_0 = n_0(m)$ such that

$$\mu(E - E_{n_0(m)}^m) < \frac{\epsilon}{2^m}.$$

(To be sure n_0 depends also on ϵ, but ϵ remains fixed throughout the entire proof.) If

$$F = \bigcup_{m=1}^{\infty} (E - E_{n_0(m)}{}^m),$$

then F is a measurable set, $F \subset E$, and

$$\mu(F) = \mu(\bigcup_{m=1}^{\infty} (E - E_{n_0(m)}{}^m)) \leqq \sum_{m=1}^{\infty} \mu(E - E_{n_0(m)}{}^m) < \epsilon.$$

Since $E - F = E \cap \bigcap_{m=1}^{\infty} E_{n_0(m)}{}^m$, and since, therefore, for $n \geqq n_0(m)$ and for x in $E - F$, we have $x \varepsilon E_n{}^m$, it follows that $|f_n(x) - f(x)| < \frac{1}{m}$, which proves uniform convergence on $E - F$. ∎

Motivated by Egoroff's theorem we introduce the concept of **almost uniform convergence.** A sequence $\{f_n\}$ of a.e. finite valued measurable functions will be said to converge to the measurable function f almost uniformly if, for every $\epsilon > 0$, there exists a measurable set F such that $\mu(F) < \epsilon$ and such that the sequence $\{f_n\}$ converges to f uniformly on F'. In this language Egoroff's theorem asserts that on a set of finite measure convergence a.e. implies almost uniform convergence. The following result goes in the converse direction.

Theorem B. *If $\{f_n\}$ is a sequence of measurable functions which converges to f almost uniformly, then $\{f_n\}$ converges to f a.e.*

Proof. Let F_n be a measurable set such that $\mu(F_n) < \frac{1}{n}$ and such that the sequence $\{f_n\}$ converges to f uniformly on F_n', $n = 1, 2, \cdots$. If $F = \bigcap_{n=1}^{\infty} F_n$, then

$$\mu(F) \leqq \mu(F_n) < \frac{1}{n},$$

so that $\mu(F) = 0$, and it is clear that, for x in F', $\{f_n(x)\}$ converges to $f(x)$. ∎

We remark that the phrase "almost uniform convergence" is a somewhat confusing (but unfortunately standard) misnomer, which conflicts with the "almost everywhere" terminology. Some such phrase as "nearly uniform convergence" might come closer to suggesting the true state of affairs; as it stands, some care has to be exercised to distinguish between almost uniform convergence and almost everywhere uniform convergence.

(1) If f is any real valued, Lebesgue measurable function on the real line, then there exists a Borel measurable function g such that $f(x) = g(x)$ a.e. (Hint: write $E_r = \{x: f(x) < r\}$ for every rational number r, and use 13.B to express E_r in the form $F_r \Delta N_r$, where F_r is a Borel set and N_r has measure zero. Let N be a Borel set of measure zero containing $\bigcup_r N_r$ and define g by

$$g(x) = \begin{cases} 0 & \text{if } x \varepsilon N, \\ f(x) & \text{if } x \varepsilon' N. \end{cases}$$

Cf. 18.2.)

(2) If E is a measurable set of positive finite measure, and if $\{f_n\}$ is a sequence of a.e. finite valued measurable functions which is fundamental a.e., then there exists a positive finite constant c and a measurable subset F of E of positive measure such that, for every $n = 1, 2, \cdots$ and for every x in F, $|f_n(x)| \leq c$.

(3) If E is a measurable set of σ-finite measure, and if $\{f_n\}$ is a sequence of a.e. finite valued measurable functions which converges a.e. on E to a finite valued measurable function f, then there exists a sequence $\{E_i\}$ of measurable sets such that $\mu(E - \bigcup_{i=1}^{\infty} E_i) = 0$ and such that the sequence $\{f_n\}$ converges uniformly on each E_i, $i = 1, 2, \cdots$. (Hint: it is sufficient to prove the result if $\mu(E) < \infty$. In this case apply Egoroff's theorem to find E_i so that

$$\mu(E - \bigcup_{i=1}^{n} E_i) < \frac{1}{n}$$

and so that $\{f_n\}$ converges uniformly on E_i.)

(4) Let X be the set of positive integers, let $\mathbf{S}$ be the class of all subsets of X, and, for E in $\mathbf{S}$, let $\mu(E)$ be the number of points in E. If χ_n is the characteristic function of the set $\{1, \cdots, n\}$, then the sequence $\{\chi_n\}$ converges to 1 everywhere but it is not almost uniformly fundamental. In other words, Egoroff's theorem is not true if E is not of finite measure.

(5) For every essentially bounded function f, write $\|f\| = \text{ess. sup. } |f|$. If $\{f_n\}$ is a sequence of essentially bounded measurable functions, then the sequence $\{f_n\}$ converges to f uniformly a.e. if and only if $\lim_n \|f_n - f\| = 0$.

(6) Is the set $\mathfrak{M}$ of all essentially bounded measurable functions a Banach space with respect to the norm described in (5)?

§ 22. CONVERGENCE IN MEASURE

In this section, as in the preceding one, we shall work throughout with a fixed measure space $(X, \mathbf{S}, \mu)$.

Theorem A. *Suppose that f and f_n, $n = 1, 2, \cdots$, are real valued measurable functions on a set E of finite measure, and write, for every $\epsilon > 0$,*

$$E_n(\epsilon) = \{x: |f_n(x) - f(x)| \geqq \epsilon\}, \quad n = 1, 2, \cdots.$$

The sequence $\{f_n\}$ converges to f a.e. on E if and only if

$$\lim_n \mu(E \cap \bigcup_{m=n}^{\infty} E_m(\epsilon)) = 0$$

for every $\epsilon > 0$.

Proof. It follows from the definition of convergence that the sequence $\{f_n(x)\}$ of real numbers fails to converge to the real number $f(x)$ if and only if there is a positive number ϵ such that x belongs to $E_n(\epsilon)$ for an infinite number of values of n. In other words, if D is the set of those points x at which $\{f_n(x)\}$ does not converge to $f(x)$, then

$$D = \bigcup_{\epsilon>0} \limsup_n E_n(\epsilon) = \bigcup_{k=1}^{\infty} \limsup_n E_n\left(\frac{1}{k}\right).$$

Consequently a necessary and sufficient condition that $\mu(E \cap D) = 0$ (i.e. that the sequence $\{f_n\}$ converge to f a.e. on E) is that $\mu(E \cap \limsup_n E_n(\epsilon)) = 0$ for every $\epsilon > 0$. The desired conclusion follows from the relations

$$\mu(E \cap \limsup_n E_n(\epsilon)) = \mu(E \cap \bigcap_{n=1}^{\infty} \bigcup_{m=n}^{\infty} E_m(\epsilon)) =$$
$$= \lim_n \mu(E \cap \bigcup_{m=n}^{\infty} E_m(\epsilon)). \quad ∎$$

The desire to investigate the result of an obvious weakening of the condition of Theorem A motivates the definition of still another method of convergence which has frequent application. A sequence $\{f_n\}$ of a.e. finite valued, measurable functions **converges in measure** to the measurable function f if, for every $\epsilon > 0$, $\lim_n \mu(\{x: |f_n(x) - f(x)| \geqq \epsilon\}) = 0$. In accordance with our general comment on different kinds of convergence in the preceding section, we shall say that a sequence $\{f_n\}$ of a.e. finite valued measurable functions is **fundamental in measure** if, for every $\epsilon > 0$,

$$\mu(\{x: |f_n(x) - f_m(x)| \geqq \epsilon\}) \to 0 \quad \text{as} \quad n \text{ and } m \to \infty.$$

It follows trivially from Theorem A that if a sequence of finite valued measurable functions converges a.e. to a finite limit [or is fundamental a.e.] on a set E of finite measure, then it converges in measure [or is fundamental in measure] on E. The following theorem is a slight strengthening of this assertion in that it makes no assumptions of finiteness.

Theorem B. *Almost uniform convergence implies convergence in measure.*

Proof. If $\{f_n\}$ converges to f almost uniformly, then, for any two positive numbers ϵ and δ, there exists a measurable set F such that $\mu(F) < \delta$ and such that $|f_n(x) - f(x)| < \epsilon$, whenever x belongs to F' and n is sufficiently large. ∎

Theorem C. *If $\{f_n\}$ converges in measure to f, then $\{f_n\}$ is fundamental in measure. If also $\{f_n\}$ converges in measure to g, then $f = g$ a.e.*

Proof. The first assertion of the theorem follows from the relation

$$\{x: |f_n(x) - f_m(x)| \geqq \epsilon\} \subset$$
$$\subset \left\{x: |f_n(x) - f(x)| \geqq \frac{\epsilon}{2}\right\} \cup \left\{x: |f_m(x) - f(x)| \geqq \frac{\epsilon}{2}\right\}.$$

To prove the second assertion, we observe that, similarly,

$$\{x: |f(x) - g(x)| \geqq \epsilon\} \subset$$
$$\subset \left\{x: |f_n(x) - f(x)| \geqq \frac{\epsilon}{2}\right\} \cup \left\{x: |f_n(x) - g(x)| \geqq \frac{\epsilon}{2}\right\}.$$

Since, by proper choice of n, the measure of both sets on the right can be made arbitrarily small, we have

$$\mu(\{x: |f(x) - g(x)| \geqq \epsilon\}) = 0$$

for every $\epsilon > 0$; this implies, as asserted, that $f = g$ a.e. ∎

In addition to these comparatively elementary remarks, we shall present two slightly deeper properties of convergence in measure.

Theorem D. *If $\{f_n\}$ is a sequence of measurable functions which is fundamental in measure, then some subsequence $\{f_{n_k}\}$ is almost uniformly fundamental.*

Proof. For any positive integer k we may find an integer $\bar{n}(k)$ such that if $n \geqq \bar{n}(k)$ and $m \geqq \bar{n}(k)$, then

$$\mu\left(\left\{x: |f_n(x) - f_m(x)| \geqq \frac{1}{2^k}\right\}\right) < \frac{1}{2^k}.$$

We write

$$n_1 = \bar{n}(1), \quad n_2 = (n_1 + 1) \cup \bar{n}(2), \quad n_3 = (n_2 + 1) \cup \bar{n}(3), \quad \cdots;$$

then $n_1 < n_2 < n_3 < \cdots$, so that the sequence $\{f_{n_k}\}$ is indeed an infinite subsequence of $\{f_n\}$. If

$$E_k = \left\{x: |f_{n_k}(x) - f_{n_{k+1}}(x)| \geqq \frac{1}{2^k}\right\}$$

and $k \leqq i \leqq j$, then, for every x which does not belong to $E_k \cup E_{k+1} \cup E_{k+2} \cup \cdots$, we have

$$|f_{n_i}(x) - f_{n_j}(x)| \leqq \sum_{m=i}^{\infty} |f_{n_m}(x) - f_{n_{m+1}}(x)| < \frac{1}{2^{i-1}},$$

so that, in other words, the sequence $\{f_{n_i}\}$ is uniformly fundamental on $X - (E_k \cup E_{k+1} \cup \cdots)$. Since

$$\mu(E_k \cup E_{k+1} \cup \cdots) \leqq \sum_{m=k}^{\infty} \mu(E_m) < \frac{1}{2^{k-1}},$$

the proof of Theorem D is complete. ∎

Theorem E. *If $\{f_n\}$ is a sequence of measurable functions which is fundamental in measure, then there exists a measurable function f such that $\{f_n\}$ converges in measure to f.*

Proof. By Theorem D we can find a subsequence $\{f_{n_k}\}$ which is almost uniformly fundamental and therefore fundamental a.e.; we write $f(x) = \lim_k f_{n_k}(x)$ for every x for which the limit exists. We observe that, for every $\epsilon > 0$,

$$\{x: |f_n(x) - f(x)| \geqq \epsilon\} \subset$$

$$\subset \left\{x: |f_n(x) - f_{n_k}(x)| \geqq \frac{\epsilon}{2}\right\} \cup \left\{x: |f_{n_k}(x) - f(x)| \geqq \frac{\epsilon}{2}\right\}.$$

The measure of the first term on the right is by hypothesis arbitrarily small if n and n_k are sufficiently large, and the measure of the second term also approaches 0 (as $k \to \infty$), since almost uniform convergence implies convergence in measure. ▮

(1) Suppose that the measure space $(X,\mathbf{S},\mu)$ is totally finite, and let $\{f_n\}$ and $\{g_n\}$ be sequences of finite valued measurable functions converging in measure to f and g respectively.

(1a) If α and β are real constants, then $\{\alpha f_n + \beta g_n\}$ converges in measure to $\alpha f + \beta g$; $\{|f_n|\}$ converges in measure to $|f|$.

(1b) If $f = 0$ a.e., then $\{f_n^2\}$ converges in measure to f^2.

(1c) The sequence $\{f_n g\}$ converges in measure to fg. (Hint: given a positive number δ, find a constant c such that if $E = \{x: |g(x)| \leq c\}$, then $\mu(X - E) < \delta$, and consider the situation separately on E and $X - E$.)

(1d) The sequence $\{f_n^2\}$ converges in measure to f^2. (Hint: apply (1b) to $\{f_n - f\}$.)

(1e) The sequence $\{f_n g_n\}$ converges in measure to fg. (Hint: apply the identity which expresses a product in terms of sums and squares.)

(1f) Are the statements (1a)–(1e) valid for measure spaces which are not totally finite?

(2) Every subsequence of a sequence which is fundamental in measure is fundamental in measure.

(3) If $\{f_n\}$ is a sequence of measurable functions which is fundamental in measure, and if $\{f_{n_i}\}$ and $\{f_{m_j}\}$ are subsequences which converge a.e. to the limit functions f and g respectively, then $f = g$ a.e.

(4) If X is the set of positive integers, $\mathbf{S}$ is the class of all subsets of X, and, for every E in $\mathbf{S}$, $\mu(E)$ is the number of points in E, then, for the measure space $(X,\mathbf{S},\mu)$, convergence in measure is equivalent to uniform convergence everywhere.

(5) Is it necessarily true on a set of infinite measure that convergence a.e. implies convergence in measure? (Cf. 21.4 and (4).)

(6) Let the measure space X be the closed unit interval with Lebesgue measure. If, for $n = 1, 2, \cdots$,

$$E_n^i = \left[\frac{i-1}{n}, \frac{i}{n}\right], \quad i = 1, \cdots, n,$$

and if χ_n^i is the characteristic function of E_n^i, then the sequence $\{\chi_1^1, \chi_2^1, \chi_2^2, \chi_3^1, \chi_3^2, \chi_3^3, \cdots\}$ converges in measure to 0, but fails to converge at any point of X.

(7) Let $\{E_n\}$ be a sequence of measurable sets and let χ_n be the characteristic function of E_n, $n = 1, 2, \cdots$. The sequence $\{\chi_n\}$ is fundamental in measure if and only if $\rho(E_n,E_m) \to 0$ as n and $m \to \infty$. (For the definition of ρ see 9.4.)

Chapter V

INTEGRATION

§ 23. INTEGRABLE SIMPLE FUNCTIONS

A simple function $f = \sum_{i=1}^{n} \alpha_i \chi_{E_i}$ on a measure space $(X,\mathbf{S},\mu)$ is **integrable** if $\mu(E_i) < \infty$ for every index i for which $\alpha_i \neq 0$. The **integral** of f, in symbols

$$\int f(x)d\mu(x) \quad \text{or} \quad \int f d\mu$$

is defined by $\int f d\mu = \sum_{i=1}^{n} \alpha_i \mu(E_i)$. It follows easily from the additivity of μ that if f is also equal to $\sum_{j=1}^{m} \beta_j \chi_{F_j}$, then $\int f d\mu = \sum_{j=1}^{m} \beta_j \mu(F_j)$, i.e. that the value of the integral is independent of the representation of f and is therefore unambiguously defined. We observe that the absolute value of an integrable simple function, a finite, constant multiple of an integrable simple function, and the sum of two integrable simple functions are integrable simple functions.

If E is a measurable set and f is an integrable simple function, then it is easy to see that the function $\chi_E f$ is an integrable simple function also; we define the **integral** of f **over** E by

$$\int_E f d\mu = \int \chi_E f d\mu.$$

The simplest example of an integrable simple function is the characteristic function of a measurable set E of finite measure; we have $\int \chi_E d\mu = \int_E d\mu = \mu(E)$.

In the sequel we shall define the notions of integrability and integral on a wider domain than the class of integrable simple functions. Some useful definitions and the statements of several important results (but very few proofs) depend only on such elementary properties of integration as we have already explicitly mentioned. In order to avoid unnecessary duplication, we shall therefore proceed as follows. Throughout this section we shall use the word "function" as an abbreviation for "simple function." As a consequence of this policy all our definitions and theorems will make sense not only for simple functions but also for the wider class we shall subsequently consider. The proofs in this section will, however, apply to simple functions only; we shall complete the proofs, so that they will apply to the more general case also, a little later.

The proofs of Theorem A and B below are omitted; these results are immediate consequences of the definitions and, in the case of Theorem A, an obvious and simple computation.

Theorem A. *If f and g are integrable functions and α and β are real numbers, then*

$$\int (\alpha f + \beta g) d\mu = \alpha \int f d\mu + \beta \int g d\mu.$$

Theorem B. *If an integrable function f is non negative a.e., then* $\int f d\mu \geqq 0.$

Theorem C. *If f and g are integrable functions such that $f \geqq g$ a.e., then*

$$\int f d\mu \geqq \int g d\mu.$$

Proof. Apply Theorem B to $f - g$ in place of f. ▮

Theorem D. *If f and g are integrable functions, then*

$$\int |f + g| d\mu \leqq \int |f| d\mu + \int |g| d\mu.$$

Proof. Apply Theorem C to $|f| + |g|$ and $|f + g|$ in place of f and g, respectively. ▮

Theorem E. *If f is an integrable function, then*

$$\left|\int f d\mu\right| \leqq \int |f| d\mu.$$

Proof. Apply Theorem C first to $|f|$ and f and then to $|f|$ and $-f$. ▌

Theorem F. *If f is an integrable function, α and β are real numbers, and E is a measurable set such that, for x in E, $\alpha \leqq f(x) \leqq \beta$, then*

$$\alpha\mu(E) \leqq \int_E f d\mu \leqq \beta\mu(E).$$

Proof. Since the principal assumption is equivalent to the relation $\alpha\chi_E \leqq \chi_E f \leqq \beta\chi_E$, the desired result follows from Theorem C if $\mu(E) < \infty$; the case in which $\mu(E) = \infty$ is easily treated by direct application of the definition of integrability. ▌

The **indefinite integral** of an integrable function f is the set function ν, defined for every measurable set E by $\nu(E) = \int_E f d\mu$.

Theorem G. *If an integrable function f is non negative a.e., then its indefinite integral is monotone.*

Proof. If E and F are measurable sets such that $E \subset F$, then $\chi_E f \leqq \chi_F f$ a.e., and the desired result follows from Theorem C. ▌

A finite valued set function ν defined on the class of all measurable sets of a measure space $(X,\mathbf{S},\mu)$ is **absolutely continuous** if for every positive number ϵ there exists a positive number δ such that $|\nu(E)| < \epsilon$ for every measurable set E for which $\mu(E) < \delta$.

Theorem H. *The indefinite integral of an integrable function is absolutely continuous.*

Proof. If c is any positive number greater than all the values of $|f|$, then, for every measurable set E, we have

$$\left|\int_E f d\mu\right| \leqq c\mu(E). \quad ▌$$

Theorem I. *The indefinite integral of an integrable function is countably additive.*

Proof. If f is the characteristic function of a measurable set E of finite measure, then the assertion of countable additivity for the indefinite integral of f is just a restatement of the countable additivity of μ on measurable subsets of E. The assertion of the theorem for arbitrary integrable simple functions is a consequence of the fact that every such function is a finite linear combination of characteristic functions. ▮

If f and g are integrable functions, we define the **distance**, $\rho(f,g)$, between them by the equation

$$\rho(f,g) = \int |f - g| d\mu.$$

The function ρ deserves the name "distance" in every respect but one. It is true and trivial that

$$\rho(f,f) = 0, \quad \rho(f,g) = \rho(g,f), \quad \text{and} \quad \rho(f,g) \leq \rho(g,h) + \rho(h,f).$$

It is not true, however, that if $\rho(f,g) = 0$, then $f = g$. The distance between two integrable functions can, for instance, vanish if they are equal almost everywhere (but not necessarily everywhere). In a subsequent section we shall study this phenomenon in some detail.

(1) If one of two simple functions is integrable, then so is their product.

(2) If E and F are measurable sets of finite measure, then $\rho(\chi_E,\chi_F) = \mu(E \Delta F)$. Cf. 9.4 and 22.7.

(3) Let $(X,\mathbf{S},\mu)$ be the closed unit interval with Lebesgue measure, and, for some fixed point x_0 in X, write $\nu(E) = \chi_E(x_0)$. Is the set function ν absolutely continuous?

(4) If ν is an absolutely continuous set function on the class of all measurable sets of a measure space $(X,\mathbf{S},\mu)$, then $\nu(E) = 0$ for every measurable set E for which $\mu(E) = 0$.

(5) If a totally finite measure space X consists of a finite number of points, then every real valued measurable function on X is an integrable simple function, and the theory of integration specializes to the theory of finite sums.

§ 24. SEQUENCES OF INTEGRABLE SIMPLE FUNCTIONS

We shall continue in this section to work with a fixed measure space $(X,\mathbf{S},\mu)$, and to use the device of abbreviating "simple function" to "function." Since all the methods of this section (with only one minor exception, occurring at the end of the proof of Theorem D) are based on the general results of the preceding section, it will turn out that not only the statements but even the proofs of the following theorems will remain unaltered when we turn to the general case.

A sequence $\{f_n\}$ of integrable functions is **fundamental in the mean, or mean fundamental,** if

$$\rho(f_n, f_m) \to 0 \quad \text{as} \quad n \quad \text{and} \quad m \to \infty.$$

Theorem A. *A mean fundamental sequence $\{f_n\}$ of integrable functions is fundamental in measure.*

Proof. If, for any fixed positive number ϵ,

$$E_{nm} = \{x: |f_n(x) - f_m(x)| \geqq \epsilon\},$$

then

$$\rho(f_n, f_m) = \int |f_n - f_m| d\mu \geqq \int_{E_{nm}} |f_n - f_m| d\mu \geqq \epsilon\mu(E_{nm}),$$

so that $\mu(E_{nm}) \to 0$ as n and $m \to \infty$. ∎

Theorem B. *If $\{f_n\}$ is a mean fundamental sequence of integrable functions, and if the indefinite integral of f_n is ν_n, $n = 1, 2, \cdots$, then*

$$\nu(E) = \lim_n \nu_n(E)$$

exists for every measurable set E, and the set function ν is finite valued and countably additive.

Proof. Since $|\nu_n(E) - \nu_m(E)| \leqq \int |f_n - f_m| d\mu \to 0$ as n and $m \to \infty$, the existence, finiteness, and uniformity of the limit are clear, and it follows from the finite additivity of limits that ν is finitely additive. If $\{E_n\}$ is a disjoint sequence of measurable sets whose union is E, then we have, for every pair of positive integers n and k

$$|\nu(E) - \textstyle\sum_{i=1}^k \nu(E_i)| \leqq$$
$$\leqq |\nu(E) - \nu_n(E)| + |\nu_n(E) - \textstyle\sum_{i=1}^k \nu_n(E_i)| + |\nu_n(\bigcup_{i=1}^k E_i) - \nu(\bigcup_{i=1}^k E_i)|.$$

The first and third terms of the right side of this inequality may be made arbitrarily small by choosing n sufficiently large, and, for fixed n, the middle term may be made arbitrarily small by choosing k sufficiently large. This proves that

$$\nu(E) = \lim_k \sum_{i=1}^{k} \nu(E_i) = \sum_{i=1}^{\infty} \nu(E_i). \quad \blacksquare$$

If $\{\nu_n\}$ is a sequence of finite valued set functions defined for all measurable sets, we say that the terms of the sequence are **uniformly absolutely continuous** whenever for every positive number ϵ there exists a positive number δ such that $|\nu_n(E)| < \epsilon$ for every measurable set E for which $\mu(E) < \delta$, and for every positive integer n.

Theorem C. *If $\{f_n\}$ is a mean fundamental sequence of integrable functions, and if the indefinite integral of f_n is ν_n, $n = 1, 2, \cdots$, then the set functions ν_n are uniformly absolutely continuous.*

Proof. If $\epsilon > 0$, let n_0 be a positive integer such that, for $n \geqq n_0$ and $m \geqq n_0$,

$$\int |f_n - f_m| d\mu < \frac{\epsilon}{2},$$

and let δ be a positive number such that

$$\int_E |f_n| d\mu < \frac{\epsilon}{2}, \quad n = 1, \cdots, n_0$$

for every measurable set E for which $\mu(E) < \delta$; (cf. 23.H). If E is a measurable set for which $\mu(E) < \delta$ and if $n \leqq n_0$, then

$$|\nu_n(E)| \leqq \int_E |f_n| d\mu < \epsilon;$$

if, on the other hand, $n > n_0$, then

$$|\nu_n(E)| \leqq \int |f_n - f_{n_0}| d\mu + \int_E |f_{n_0}| d\mu < \epsilon. \quad \blacksquare$$

Since the following theorem is of no particular importance in the general case, we shall restrict its statement and proof to the case of simple functions only.

Theorem D. *If $\{f_n\}$ and $\{g_n\}$ are mean fundamental sequences of integrable simple functions which converge in measure to the same measurable function f, if the indefinite integrals of f_n and g_n are ν_n and λ_n respectively, and if, for every measurable set E,*

$$\nu(E) = \lim_n \nu_n(E) \quad \text{and} \quad \lambda(E) = \lim_n \lambda_n(E),$$

then the set functions ν and λ are identical.

Proof. Since, for every $\epsilon > 0$,

$$E_n = \{x: |f_n(x) - g_n(x)| \geqq \epsilon\} \subset$$
$$\subset \left\{x: |f_n(x) - f(x)| \geqq \frac{\epsilon}{2}\right\} \cup \left\{x: |f(x) - g_n(x)| \geqq \frac{\epsilon}{2}\right\},$$

it follows that $\lim_n \mu(E_n) = 0$. Hence, if E is a measurable set of finite measure, then in the relation

$$\int_E |f_n - g_n| d\mu \leqq \int_{E-E_n} |f_n - g_n| d\mu + \int_{E\cap E_n} |f_n| d\mu + \int_{E\cap E_n} |g_n| d\mu$$

the first term on the right is dominated by $\epsilon\mu(E)$, and the last two terms can be made arbitrarily small by choosing n sufficiently large, because of the uniform absolute continuity proved in Theorem C. It follows that

$$\lim_n |\nu_n(E) - \lambda_n(E)| = 0,$$

and hence that $\nu(E) = \lambda(E)$. Since ν and λ are both countably additive, it follows that $\nu(E) = \lambda(E)$ for every measurable set E of σ-finite measure.

Since the f_n and g_n are simple functions, each of them is defined in terms of a finite class of measurable sets of finite measure. If E_0 is the union of all sets in all these finite classes, then E_0 is a measurable set of σ-finite measure, and we have, for every measurable set E,

$$\nu_n(E - E_0) = \lambda_n(E - E_0) = 0$$

and therefore $\nu(E - E_0) = \lambda(E - E_0) = 0$. Since this implies that $\nu(E) = \nu(E \cap E_0)$ and $\lambda(E) = \lambda(E \cap E_0)$, the proof of Theorem D is complete. ∎

(1) Is the set of all integrable simple functions a complete metric space with respect to the distance ρ?

(2) In the notation of Theorem B, if $\{E_n\}$ is a disjoint sequence of measurable sets, then the series $\sum_{n=1}^{\infty} \nu(E_n)$ converges absolutely. (Hint: the series converges unconditionally.)

§ 25. INTEGRABLE FUNCTIONS

An a.e. finite valued, measurable function f on a measure space $(X,\mathbf{S},\mu)$ is **integrable** if there exists a mean fundamental sequence $\{f_n\}$ of integrable simple functions which converges in measure to f. The **integral** of f, in symbols

$$\int f(x)d\mu(x) \quad \text{or} \quad \int f d\mu$$

is defined by $\int f d\mu = \lim_n \int f_n d\mu$. It follows from 24.D (with $E = \bigcup_n N(f_n)$) that the value of the integral of f is uniquely determined by any particular such sequence. We emphasize the fact that the value of the integral is always finite. We observe that it follows from the known and obvious properties of mean convergence and convergence in measure that the absolute value of an integrable function, a finite constant multiple of an integrable function, and the sum of two integrable functions are integrable functions. The relations

$$f^+ = \tfrac{1}{2}(|f| + f) \quad \text{and} \quad f^- = \tfrac{1}{2}(|f| - f)$$

show also that if f is integrable, then f^+ and f^- are integrable.

If E is a measurable set and if $\{f_n\}$ is a mean fundamental sequence of integrable simple functions converging in measure to the integrable function f, then it is easy to see that the sequence $\{\chi_E f_n\}$ is mean fundamental and converges in measure to $\chi_E f$. We define the **integral of f over E** by

$$\int_E f d\mu = \int \chi_E f d\mu.$$

We recall that the theorems of §§ 23 and 24 were stated for general integrable functions but were proved for integrable simple functions only. We are now in a position to complete their proofs.

The results 23.A and 23.B follow immediately from elementary properties of limits; 23.C–23.G follow from 23.B verbatim as before.

To prove the absolute continuity of an indefinite integral, 23.H, let $\{f_n\}$ be a mean fundamental sequence of integrable simple functions which converges in measure to the integrable function f. We have

$$\left|\int_E f d\mu\right| \leqq \left|\int_E f_n d\mu\right| + \left|\int_E f_n d\mu - \int_E f d\mu\right|,$$

for every measurable set E. Since the f_n are simple functions, the theorem 24.C on uniform absolute continuity may be applied to prove that the first term on the right becomes arbitrarily small if the measure of E is taken sufficiently small. The second term on the right approaches 0 as $n \to \infty$, by the definition of $\int_E f d\mu$; this completes the proof of 23.H.

The proof of the countable additivity of an indefinite integral is even simpler. Indeed, using the notation of the preceding paragraph, the fact that the f_n are simple functions justifies the application of 24.B, which then yields exactly the assertion of 23.I.

The proofs of 24.A–24.C were based on the statements, and not on the proofs, of the results of § 23, and are therefore valid in the general case. This remark completes the proofs of all the theorems of the preceding two sections.

We shall say that a sequence $\{f_n\}$ of integrable functions **converges in the mean,** or **mean converges,** to an integrable function f if

$$\rho(f_n, f) = \int |f_n - f| d\mu \to 0 \quad \text{as} \quad n \to \infty.$$

Our first result concerning this concept is extremely similar, in statement and in proof, to 24.A.

Theorem A. *If $\{f_n\}$ is a sequence of integrable functions which converges in the mean to f, then $\{f_n\}$ converges to f in measure.*

Proof. If, for any fixed positive number ϵ,

$$E_n = \{x: |f_n(x) - f(x)| \geqq \epsilon\},$$

then

$$\int |f_n - f| d\mu \geqq \int_{E_n} |f_n - f| d\mu \geqq \epsilon\mu(E_n),$$

so that $\mu(E_n) \to 0$ as $n \to \infty$. ∎

Theorem B. *If f is an a.e. non negative integrable function, then a necessary and sufficient condition that $\int f d\mu = 0$ is that $f = 0$ a.e.*

Proof. If $f = 0$ a.e., then the sequence each of whose terms is identically zero is a mean fundamental sequence of integrable simple functions which converges in measure to f, and it follows that $\int f d\mu = 0$. To prove the converse, we observe that if $\{f_n\}$ is a mean fundamental sequence of integrable simple functions which converges in measure to f, then we may assume that $f_n \geqq 0$, since we may replace each f_n by its absolute value. The assumption $\int f d\mu = 0$ implies that $\lim_n \int f_n d\mu = 0$, i.e. that $\{f_n\}$ mean converges to 0. It follows from Theorem A that $\{f_n\}$ converges to 0 in measure and hence the desired result is implied by 22.C. ∎

Theorem C. *If f is an integrable function and E is a set of measure zero, then*

$$\int_E f d\mu = 0.$$

Proof. Since $\int_E f d\mu = \int \chi_E f d\mu$, and since the characteristic function of a set of measure zero vanishes a.e., the desired result follows from Theorem B. ∎

Theorem D. *If f is an integrable function which is positive a.e. on a measurable set E, and if $\int_E f d\mu = 0$, then $\mu(E) = 0$.*

Proof. We write $F_0 = \{x: f(x) > 0\}$ and $F_n = \left\{x: f(x) \geqq \frac{1}{n}\right\}$, $n = 1, 2, \cdots$; since the assumption of positiveness implies that $E - F_0$ is a set of measure zero, we have merely to prove that $E \cap F_0$ is one also. Since

$$0 = \int_{E \cap F_n} f d\mu \geqq \frac{1}{n} \mu(E \cap F_n) \geqq 0,$$

and since $F_0 = \bigcup_{n=1}^{\infty} F_n$, the desired result follows from the relation $\mu(E \cap F_0) \leqq \sum_{n=1}^{\infty} \mu(E \cap F_n)$. ∎

Theorem E. *If f is an integrable function such that $\int_F f d\mu = 0$ for every measurable set F, then $f = 0$ a.e.*

Proof. If $E = \{x: f(x) > 0\}$, then, by hypothesis, $\int_E f d\mu = 0$, and therefore, by Theorem D, E is a set of measure zero. Applying the same reasoning to $-f$ shows that $\{x: f(x) < 0\}$ is a set of measure zero. ∎

Theorem F. *If f is an integrable function, then the set $N(f) = \{x: f(x) \neq 0\}$ has σ-finite measure.*

Proof. Let $\{f_n\}$ be a mean fundamental sequence of integrable simple functions which converges in measure to f. For every $n = 1, 2, \cdots$, $N(f_n)$ is a measurable set of finite measure. If $E = N(f) - \bigcup_{n=1}^{\infty} N(f_n)$, and if F is any measurable subset of E, then it follows from the relation

$$\int_F f d\mu = \lim_n \int_F f_n d\mu = 0$$

and Theorem E that $f = 0$ a.e. on E. In view of the definition of $N(f)$ this implies that $\mu(E) = 0$; we have

$$N(f) \subset \bigcup_{n=1}^{\infty} N(f_n) \cup E. \quad ∎$$

It is frequently useful to define the symbol $\int f d\mu$ for certain non integrable functions f. If, for instance, f is an extended real valued, measurable function such that $f \geqq 0$ a.e. and if f is *not* integrable, then we write

$$\int f d\mu = \infty.$$

The most general class of functions f for which it is convenient to define $\int f d\mu$ is the class of all those extended real valued measurable functions f for which at least one of the two functions f^+ and f^- is integrable; in that case we write

$$\int f d\mu = \int f^+ d\mu - \int f^- d\mu.$$

Since at most one of the two numbers, $\int f^+ d\mu$ and $\int f^- d\mu$, is infinite, the value of $\int f d\mu$ is always $+\infty$, $-\infty$, or a finite real number—it is never the indeterminate form $\infty - \infty$. We shall make free use of this extended notion of integration, but we shall continue to apply the adjective "integrable" to such functions only as are integrable in the sense of our former definitions.

(1) If X is the space of positive integers (described, for instance, in 22.4), then a function f is integrable if and only if the series $\sum_{n=1}^{\infty} |f(n)|$ is convergent, and, if this condition is satisfied, then $\int f d\mu = \sum_{n=1}^{\infty} f(n)$.

(2) If f is a non negative integrable function, then its indefinite integral is a finite measure on the class of all measurable sets.

(3) If f is integrable, then, for every positive number ϵ,

$$\mu(\{x: |f(x)| \geqq \epsilon\}) < \infty.$$

(4) If g is a finite, increasing, and continuous function of a real variable, and $\bar{\mu}_g$ is the Lebesgue–Stieltjes measure induced by g (cf. 15.9), and if f is a function which is integrable with respect to this measure, then the integral $\int f(x) d\bar{\mu}_g(x)$ is called the **Lebesgue–Stieltjes integral** of f with respect to g and is denoted by $\int_{-\infty}^{+\infty} f(x) dg(x)$. If, in particular, $g(x) = x$, then we obtain the **Lebesgue integral,** denoted by $\int_{-\infty}^{+\infty} f(x) dx$. If f is a continuous function such that $N(f)$ is a bounded set, then f is Lebesgue integrable.

§ 26. SEQUENCES OF INTEGRABLE FUNCTIONS

Theorem A. *If $\{f_n\}$ is a mean fundamental sequence of integrable simple functions which converges in measure to the integrable function f, then*

$$\rho(f,f_n) = \int |f - f_n| d\mu \to 0 \quad as \quad n \to \infty;$$

hence, to every integrable function f and to every positive number ϵ, there corresponds an integrable simple function g such that $\rho(f,g) < \epsilon$.

Proof. For any fixed positive integer m, $\{|f_n - f_m|\}$ is a mean fundamental sequence of integrable simple functions which converges in measure to $|f - f_m|$, and, therefore,

$$\int |f - f_m| d\mu = \lim_n \int |f_n - f_m| d\mu.$$

The fact that the sequence $\{f_n\}$ is mean fundamental implies the desired result. ∎

Theorem B. *If $\{f_n\}$ is a mean fundamental sequence of integrable functions, then there exists an integrable function f such that $\rho(f_n,f) \to 0$ (and consequently $\int f_n d\mu \to \int f d\mu$) as $n \to \infty$.*

Proof. By Theorem A, for each positive integer n there is an integrable simple function g_n such that $\rho(f_n,g_n) < \frac{1}{n}$. It follows that $\{g_n\}$ is a mean fundamental sequence of integrable simple functions; let f be a measurable (and therefore integrable) function such that $\{g_n\}$ converges in measure to f. Since

$$0 \leqq \left| \int f_n d\mu - \int f d\mu \right| \leqq \int |f_n - f| d\mu = \rho(f_n,f) \leqq$$

$$\leqq \rho(f_n,g_n) + \rho(g_n,f),$$

the desired result follows from Theorem A. ∎

In order to phrase our next result in a concise and intuitive fashion, we recall first the definition of a certain kind of continuity for set functions. A finite valued set function ν on a class $\mathbf{E}$ of sets is continuous from above at 0 (cf. § 9) if, for every decreasing sequence $\{E_n\}$ of sets in $\mathbf{E}$ for which $\lim_n E_n = 0$, we have $\lim_n \nu(E_n) = 0$. If $\{\nu_n\}$ is a sequence of such finite valued set functions on $\mathbf{E}$, we shall say that the terms of the sequence are **equicontinuous** from above at 0 if, for every decreasing sequence $\{E_n\}$ of sets in $\mathbf{E}$ for which $\lim_n E_n = 0$, and for every positive number ϵ, there exists a positive integer m_0 such that if $m \geqq m_0$, then $|\nu_n(E_m)| < \epsilon$, $n = 1, 2, \cdots$.

Theorem C. *A sequence $\{f_n\}$ of integrable functions converges in the mean to the integrable function f if and only if $\{f_n\}$ converges in measure to f and the indefinite integrals of $|f_n|$, $n = 1, 2, \cdots$, are uniformly absolutely continuous and equicontinuous from above at* 0.

Proof. We prove first the necessity of the conditions. Since convergence in measure and uniform absolute continuity follow from 25.A and 24.C respectively, we have only to prove the asserted equicontinuity.

The mean convergence of $\{f_n\}$ to f implies that to every positive number ϵ there corresponds a positive integer n_0 such that if $n \geqq n_0$, then $\int |f_n - f| d\mu < \frac{\epsilon}{2}$. Since the indefinite integral of a non negative integrable function is a finite measure (23.I), it follows from 9.E that such an indefinite integral is continuous from above at 0. Consequently, if $\{E_m\}$ is a decreasing sequence of measurable sets with an empty intersection, then there exists a positive integer m_0 such that, for $m \geqq m_0$,

$$\int_{E_m} |f| d\mu < \frac{\epsilon}{2} \quad \text{and} \quad \int_{E_m} |f_n - f| d\mu < \frac{\epsilon}{2}, \quad n = 1, \cdots, n_0.$$

Hence, if $m \geqq m_0$, then we have

$$\int_{E_m} |f_n| d\mu \leqq \int_{E_m} |f_n - f| d\mu + \int_{E_m} |f| d\mu < \epsilon$$

for every positive integer n, and this is exactly the desired equicontinuity result.

We turn to the proof of sufficiency. Since a countable union of measurable sets of σ–finite measure is a measurable set of σ–finite measure, it follows from 25.F that

$$E_0 = \bigcup_{n=1}^{\infty} \{x : f_n(x) \neq 0\}$$

is such a set. If $\{E_n\}$ is an increasing sequence of measurable sets of finite measure such that $\lim_n E_n = E_0$, and if $F_n = E_0 - E_n$, $n = 1, 2, \cdots$, then $\{F_n\}$ is a decreasing sequence and $\lim_n F_n = 0$. The assumed equicontinuity implies that, for every positive number δ, there exists a positive integer k such that $\int_{F_k} |f_n| d\mu < \frac{\delta}{2}$, and consequently

$$\int_{F_k} |f_m - f_n| d\mu \leqq \int_{F_k} |f_m| d\mu + \int_{F_k} |f_n| d\mu < \delta.$$

If for any fixed $\epsilon > 0$ we write

$$G_{mn} = \{x : |f_m(x) - f_n(x)| \geqq \epsilon\},$$

then it follows that

$$\int_{E_k} |f_m - f_n| d\mu \leqq \int_{E_k - G_{mn}} |f_m - f_n| d\mu + \int_{E_k \cap G_{mn}} |f_m - f_n| d\mu \leqq$$

$$\leqq \epsilon\mu(E_k) + \int_{E_k \cap G_{mn}} |f_m - f_n| d\mu.$$

By convergence in measure and uniform absolute continuity, the second term on the dominant side of this chain of inequalities may be made arbitrarily small by choosing m and n sufficiently large, so that

$$\limsup_{m,n} \int_{E_k} |f_m - f_n| d\mu \leqq \epsilon\mu(E_k).$$

Since ϵ is arbitrary, it follows that

$$\lim_{m,n} \int_{E_k} |f_m - f_n| d\mu = 0.$$

Since

$$\int |f_m - f_n| d\mu = \int_{E_0} |f_m - f_n| d\mu =$$

$$= \int_{E_k} |f_m - f_n| d\mu + \int_{F_k} |f_m - f_n| d\mu,$$

we have

$$\limsup_{m,n} \int |f_m - f_n| d\mu < \delta,$$

and therefore, since δ is arbitrary,

$$\lim_{m,n} \int |f_m - f_n| d\mu = 0.$$

We have proved, in other words, that the sequence $\{f_n\}$ is fundamental in the mean; it follows from Theorem B that there exists an integrable function g such that $\{f_n\}$ mean converges to g. Since mean convergence implies convergence in measure, we must have $f = g$ a.e. ∎

The following result is known as Lebesgue's **bounded convergence theorem.**

Theorem D. *If $\{f_n\}$ is a sequence of integrable functions which converges in measure to f [or else converges to f a.e.], and if g is an integrable function such that $|f_n(x)| \leqq |g(x)|$ a.e., $n = 1, 2, \cdots$, then f is integrable and the sequence $\{f_n\}$ converges to f in the mean.*

Proof. In the case of convergence in measure, the theorem is an immediate corollary of Theorem C—the uniformities required in that theorem are all consequences of the inequality

$$\int_E |f_n| d\mu \leqq \int_E |g| d\mu.$$

The case of convergence a.e. may be reduced to that of convergence in measure (even though the integrals are not necessarily over a set of finite measure, cf. 22.4 and 22.5) by making use of the existence of g. If we assume, as we may without any loss of generality, that $|f_n(x)| \leqq |g(x)|$ and $|f(x)| \leqq |g(x)|$ for every x in X, then we have, for every fixed positive number ϵ,

$$E_n = \bigcup_{i=n}^{\infty} \{x: |f_i(x) - f(x)| \geqq \epsilon\} \subset \left\{x: |g(x)| \geqq \frac{\epsilon}{2}\right\},$$

and therefore $\mu(E_n) < \infty$, $n = 1, 2, \cdots$. Since the assumption of convergence a.e. implies that $\mu(\bigcap_{n=1}^{\infty} E_n) = 0$, it follows from 9.E that

$$\limsup_n \mu(\{x: |f_n(x) - f(x)| \geqq \epsilon\}) \leqq \lim_n \mu(E_n) =$$

$$= \mu(\lim_n E_n) = 0.$$

In other words, convergence a.e., together with being bounded by an integrable function, implies convergence in measure, and the proof of the theorem is complete. ▮

(1) Is the set of all integrable functions a Banach space with respect to the norm defined by $\|f\| = \int |f| d\mu$?

(2) If $\{f_n\}$ is a uniformly fundamental sequence of functions, integrable over a measurable set E of finite measure, then the function f, defined by $f(x) = \lim_n f_n(x)$, is integrable over E and $\int_E |f_n - f| d\mu \to 0$ as $n \to \infty$.

(3) If the measure space $(X, \mathbf{S}, \mu)$ is finite, then Theorem C remains true even if the equicontinuity condition is omitted.

(4) Let $(X, \mathbf{S}, \mu)$ be the space of positive integers (cf. 22.4).

(4a) Write

$$f_n(k) = \begin{cases} \dfrac{1}{n} & \text{if } 1 \leqq k \leqq n, \\ 0 & \text{if } k > n. \end{cases}$$

The sequence $\{f_n\}$ may be used to show that the equicontinuity condition may not, in general, be omitted from Theorem C.

(4b) The sequence described in (4a) may be used to show also that if $\{f_n\}$ is a uniformly convergent sequence of integrable functions whose limit function f is also integrable, then we do not necessarily have $\lim_n \int f_n d\mu = \int f d\mu$; (cf. (2) above).

(4c) Write

$$f_n(k) = \begin{cases} \dfrac{1}{k} & \text{if } 1 \leqq k \leqq n, \\ 0 & \text{if } k > n. \end{cases}$$

The sequence $\{f_n\}$ may be used to show that the limit of a uniformly convergent sequence of integrable functions need not be integrable.

(5) Let X be the closed unit interval with Lebesgue measure and let $\{E_n\}$ be a decreasing sequence of open intervals such that $\mu(E_n) = \frac{1}{n}$, $n = 1, 2, \cdots$. The sequence $\{n\chi_{E_n}\}$ may be used to show that the boundedness condition cannot be omitted from Theorem D.

(6) If $\{f_n\}$ is a sequence of integrable functions which converges in the mean to the integrable function f, and if g is an essentially bounded measurable function, then $\{f_n g\}$ mean converges to fg.

(7) If $\{f_n\}$ is a sequence of non negative integrable functions which converges a.e. to an integrable function f, and if $\int f_n d\mu = \int f d\mu$, $n = 1, 2, \cdots$, then $\{f_n\}$ converges to f in the mean. (Hint: write $g_n = f_n - f$ and observe that the trivial inequality $|f_n - f| \leqq f_n + f$ implies that $0 \leqq g_n^- \leqq f$. The bounded convergence theorem may therefore be applied to the sequence $\{g_n^-\}$; the desired result follows from the fact that $\int g_n^+ d\mu - \int g_n^- d\mu = 0$, $n = 1, 2, \cdots$.)

§ 27. PROPERTIES OF INTEGRALS

Theorem A. *If f is measurable, g is integrable, and $|f| \leqq |g|$ a.e., then f is integrable.*

Proof. Consideration of the positive and negative parts of f shows that it is sufficient to prove the theorem for non negative functions f. If f is a simple function, the result is clear. In the general case there is an increasing sequence $\{f_n\}$ of non negative simple functions such that $\lim_n f_n(x) = f(x)$ for all x in X. Since $0 \leqq f_n \leqq |g|$, each f_n is integrable and the desired result follows from the bounded convergence theorem. ∎

Theorem B. *If $\{f_n\}$ is an increasing sequence of extended real valued non negative measurable functions and if* $\lim_n f_n(x) = f(x)$ *a.e., then* $\lim_n \int f_n d\mu = \int f d\mu$.

Proof. If f is integrable, then the result follows from the bounded convergence theorem and Theorem A. The only novel feature of the present theorem is its application to the not necessarily integrable case; we have to prove that if $\int f d\mu = \infty$, then $\lim_n \int f_n d\mu = \infty$, or, in other words, that if $\lim_n \int f_n d\mu < \infty$, then

f is integrable. From the finiteness of the limit we may conclude that

$$\lim_{m,n} \left| \int f_m d\mu - \int f_n d\mu \right| = 0.$$

Since $f_m - f_n$ is of constant sign for each fixed m and n, we have

$$\left| \int f_m d\mu - \int f_n d\mu \right| = \int | f_m - f_n | d\mu,$$

so that the sequence $\{f_n\}$ is mean convergent and therefore (26.B) mean converges to an integrable function g. Since mean convergence implies convergence in measure, and therefore a.e. convergence for some subsequence, we have $f = g$ a.e. ∎

Theorem C. *A measurable function is integrable if and only if its absolute value is integrable.*

Proof. The new part of this theorem is the assertion that the integrability of $|f|$ implies that of f, and this follows from Theorem A with $|f|$ in place of g. ∎

Theorem D. *If f is integrable and g is an essentially bounded measurable function, then fg is integrable.*

Proof. If $|g| \leqq c$ a.e., then $|fg| \leqq c|f|$ a.e. and therefore the result follows from Theorem C. ∎

Theorem E. *If f is an essentially bounded measurable function and E is a measurable set of finite measure, then f is integrable over E.*

Proof. Since the characteristic function of a measurable set of finite measure is an integrable function, the result follows from Theorem D with χ_E and f in place of f and g. ∎

Our next and final result is known as **Fatou's lemma.**

Theorem F. *If $\{f_n\}$ is a sequence of non negative integrable functions for which*

$$\liminf_n \int f_n d\mu < \infty,$$

then the function f, defined by

$$f(x) = \lim\inf_n f_n(x),$$

is integrable and

$$\int f d\mu \leqq \lim\inf_n \int f_n d\mu.$$

Proof. If $g_n(x) = \inf\{f_i(x): n \leqq i < \infty\}$, then $g_n \leqq f_n$ and the sequence $\{g_n\}$ is increasing. Since $\int g_n d\mu \leqq \int f_n d\mu$, it follows that

$$\lim_n \int g_n d\mu \leqq \lim\inf_n \int f_n d\mu < \infty.$$

Since $\lim_n g_n(x) = \lim\inf_n f_n(x) = f(x)$, it follows from Theorem B that f is integrable and

$$\int f d\mu = \lim_n \int g_n d\mu \leqq \lim\inf_n \int f_n d\mu. \quad \blacksquare$$

(1) If f is a measurable function, g is an integrable function, and α and β are real numbers such that $\alpha \leqq f(x) \leqq \beta$ a.e., then there exists a real number γ, $\alpha \leqq \gamma \leqq \beta$, such that $\int f |g| d\mu = \gamma \cdot \int |g| d\mu$. (Hint:

$$\alpha \cdot \int |g| d\mu \leqq \int f |g| d\mu \leqq \beta \cdot \int |g| d\mu.)$$

This result is known as the **mean value theorem** for integrals.

(2) If $\{f_n\}$ is a sequence of integrable functions such that

$$\sum_{n=1}^{\infty} \int |f_n| d\mu < \infty,$$

then the series $\sum_{n=1}^{\infty} f_n(x)$ converges a.e. to an integrable function f and

$$\int f d\mu = \sum_{n=1}^{\infty} \int f_n d\mu.$$

(Hint: apply Theorem B to the sequence of partial sums of the series $\sum_{n=1}^{\infty} |f_n(x)|$ and recall that absolute convergence implies convergence.)

(3) If f and f_n are integrable functions, $n = 1, 2, \cdots$, such that $|f_n(x)| \leqq |f(x)|$ a.e., then the functions f^* and f_*, defined by

$$f^*(x) = \lim\sup_n f_n(x) \quad \text{and} \quad f_*(x) = \lim\inf_n f_n(x),$$

are integrable and

$$\int f^* d\mu \geqq \lim\sup_n \int f_n d\mu \geqq \lim\inf_n \int f_n d\mu \geqq \int f_* d\mu.$$

(Hint: by considering separately the positive and negative parts, reduce the general case to the case of non negative f_n, and then apply Fatou's lemma to $\{f + f_n\}$ and $\{f - f_n\}$.)

(4) A measurable function f is integrable over a measurable set E of finite measure if and only if the series

$$\sum_{n=1}^{\infty} \mu(E \cap \{x: |f(x)| \geq n\})$$

converges. (Hint: use Abel's method of partial summation.) What can be said if $\mu(E) = \infty$, or if the summation is extended from $n = 0$?

(5) Suppose that $\{E_n\}$ is a sequence of measurable sets and m is any fixed positive integer, and let G be the set of all those points which belong to E_n for at least m values of n. Then G is measurable and

$$m\mu(G) \leq \sum_{n=1}^{\infty} \mu(E_n).$$

(Hint: consider $\sum_{n=1}^{\infty} \int_G \chi_{E_n}(x) d\mu(x)$.)

(6) Suppose that f is a finite valued, measurable function on a totally finite measure space $(X,\mathbf{S},\mu)$, and write

$$s_n = \sum_{k=-\infty}^{+\infty} \frac{k}{2^n} \mu\left(\left\{x: \frac{k}{2^n} < f(x) \leq \frac{k+1}{2^n}\right\}\right), \quad n = 1, 2, \cdots.$$

Then

$$\int f d\mu = \lim_n s_n,$$

in the sense that if f is integrable, then each series s_n is absolutely convergent, the limit exists, and is equal to the integral, and, conversely, if any one of the series s_n converges absolutely, then all others do, the limit exists, f is integrable, and the equality holds. (Hint: it is sufficient to prove the result for non negative functions. Write

$$f_n(x) = \begin{cases} \frac{k}{2^n} & \text{if } \frac{k}{2^n} < f(x) \leq \frac{k+1}{2^n}, \quad k = 0, 1, 2, \cdots, \\ 0 & \text{if } f(x) = 0, \end{cases}$$

and apply Theorem B. For the converse direction observe that

$$f(x) \leq 2f_n(x) + \mu(X),$$

so that f is integrable and therefore the preceding reasoning applies.)

(7) The following considerations are at the basis of an alternative popular approach to integration. Let f be a non negative integrable function on a measure space $(X,\mathbf{S},\mu)$. For every measurable set E we write

$$a(E) = \inf\{f(x): x \varepsilon E\},$$

and for every finite, disjoint class $\mathbf{C} = \{E_1, \cdots, E_n\}$ of measurable sets we write

$$s(\mathbf{C}) = \sum_{i=1}^{n} a(E_i)\mu(E_i).$$

We assert that the supremum of all numbers of the form $s(\mathbf{C})$ is equal to $\int f d\mu$.

If f is a simple function, the result is clear. If g is a non negative simple function such that $g \leqq f$, say $g = \sum_{i=1}^{n} \alpha_i \chi_{E_i}$, we write $\mathbf{C} = \{E_1, \cdots, E_n\}$. Then

$$\int g d\mu = \sum_{i=1}^{n} \alpha_i \mu(E_i) \leqq \sum_{i=1}^{n} a(E_i)\mu(E_i) = s(\mathbf{C}).$$

It follows that if $\{g_n\}$ is an increasing sequence of non negative simple functions converging to f, then

$$\lim_n \int g_n d\mu \leqq \sup s(\mathbf{C}),$$

and therefore $\int f d\mu \leqq \sup s(\mathbf{C})$. On the other hand, for every $\mathbf{C}$, $s(\mathbf{C}) \leqq \int f d\mu$, since $s(\mathbf{C})$ is, in fact, the integral of a function such as g.

(7a) Does the result of the preceding paragraph extend to non integrable, non negative functions?

(7b) If f is an integrable function on a totally finite measure space $(X,\mathbf{S},\mu)$, and if its distribution function g is continuous, (cf. 18.11), then

$$\int f d\mu = \int_{-\infty}^{+\infty} x dg(x)$$

(cf. 25.4). (Hint: assume $f \geqq 0$, and make use of (7) above by considering the "approximating sums" $s(\mathbf{C})$ of both integrals.)

Chapter VI

GENERAL SET FUNCTIONS

§ 28. SIGNED MEASURES

In this chapter we shall discuss a not too difficult but rather useful generalization of the notion of measure; the principal difference between measures and the set functions we now propose to treat is that the latter are not required to be non negative.

Suppose that μ_1 and μ_2 are two measures on a σ-ring $\mathbf{S}$ of subsets of a set X. If we define, for every set E in $\mathbf{S}$, $\mu(E) = \mu_1(E) + \mu_2(E)$, then it is clear that μ is a measure, and this result, on the possibility of adding two measures, extends immediately to any finite sum. Another way of manufacturing new measures is to multiply a given measure by an arbitrary non negative constant. Combining these two methods, we see that if $\{\mu_1, \cdots, \mu_n\}$ is a finite set of measures and $\{\alpha_1, \cdots, \alpha_n\}$ is a finite set of non negative real numbers, then the set function μ, defined for every set E in $\mathbf{S}$ by

$$\mu(E) = \sum_{i=1}^{n} \alpha_i \mu_i(E),$$

is a measure.

The situation is different if we allow negative coefficients. If, for instance, μ_1 and μ_2 are two measures on $\mathbf{S}$, and if we define μ by $\mu(E) = \mu_1(E) - \mu_2(E)$, then we face two new possibilities. The first of these, namely that μ may be negative on some sets, is not only not a serious objection but in fact an interesting phenomenon worth investigating. The second possibility presents, however, a difficulty that has to be overcome before the investigation can begin. It can, namely, happen that $\mu_1(E) = \mu_2(E) = \infty$; what sense, in this case, can we make of the expression for $\mu(E)$?

To avoid the difficulty of indeterminate forms, we shall agree to subtract two measures only if at least one of them is finite. This convention is analogous to the one we adopted in presenting the most general definition of the symbol $\int f d\mu$. (We recall that $\int f d\mu$ is defined for a measurable function f if and only if at least one of the two functions f^+ and f^- is integrable, i.e. if and only if at least one of the two set functions ν^+ and ν^-, defined by

$$\nu^+(E) = \int_E f^+ d\mu \quad \text{and} \quad \nu^-(E) = \int_E f^- d\mu,$$

is a finite measure.) The analogy can be carried further: if f is a measurable function such that $\int f d\mu$ is defined, then the set function ν, defined by $\nu(E) = \int_E f d\mu$, is the difference of two measures.

The definition that we want to make is sufficiently motivated by the preceding paragraphs. We define a **signed measure** as an extended real valued, countably additive set function μ on the class of all measurable sets of a measurable space $(X, \mathbf{S})$, such that $\mu(0) = 0$, and such that μ assumes at most one of the values $+\infty$ and $-\infty$.

We observe that implicit in the requirement of countable additivity is the requirement that if $\{E_n\}$ is a disjoint sequence of measurable sets, then the series $\sum_{n=1}^{\infty} \mu(E_n)$ is either convergent or definitely divergent (to $+\infty$ or $-\infty$)—in any case that the symbol $\sum_{n=1}^{\infty} \mu(E_n)$ makes sense.

The words "[totally] finite" and "[totally] σ-finite" will be used for signed measures just as for measures, except that $\mu(E)$ has to be replaced by $|\mu(E)|$, or, equivalently, $\mu(E) < \infty$ has to be replaced by $-\infty < \mu(E) < \infty$. For instance, a signed measure μ is totally finite if X is measurable and $|\mu(X)| < \infty$.

One of our objectives in the following study is to prove that every signed measure is the difference of two measures. If this result is granted, it follows that we could have defined the concept of signed measure on a ring and then attempted to copy the extension procedure for measures; and it follows equally that it would have been a waste of time to do so, since we may, instead, reduce the discussion of signed measures to that of measures.

It follows from the definition of signed measures, just as for measures, that a signed measure is finitely additive and, therefore, subtractive.

Theorem A. *If E and F are measurable sets and μ is a signed measure such that*

$$E \subset F \quad \textit{and} \quad |\mu(F)| < \infty,$$

then $|\mu(E)| < \infty$.

Proof. We have $\mu(F) = \mu(F - E) + \mu(E)$. If exactly one of the summands is infinite, then so is $\mu(F)$; if they are both infinite, then (since μ assumes at most one of the values $+\infty$ and $-\infty$) they are equal and again $\mu(F)$ is infinite. Only one possibility remains, namely that both summands are finite, and this proves that every measurable subset of a set of finite signed measure has finite signed measure. ∎

Theorem B. *If μ is a signed measure and $\{E_n\}$ is a disjoint sequence of measurable sets such that $|\mu(\bigcup_{n=1}^{\infty} E_n)| < \infty$, then the series $\sum_{n=1}^{\infty} \mu(E_n)$ is absolutely convergent.*

Proof. Write

$$E_n^+ = \begin{cases} E_n & \text{if} \quad \mu(E_n) \geqq 0, \\ 0 & \text{if} \quad \mu(E_n) < 0, \end{cases}$$

and

$$E_n^- = \begin{cases} E_n & \text{if} \quad \mu(E_n) \leqq 0, \\ 0 & \text{if} \quad \mu(E_n) > 0. \end{cases}$$

Then

$$\mu(\textstyle\bigcup_{n=1}^{\infty} E_n^+) = \sum_{n=1}^{\infty} \mu(E_n^+)$$

and

$$\mu(\textstyle\bigcup_{n=1}^{\infty} E_n^-) = \sum_{n=1}^{\infty} \mu(E_n^-).$$

Since the terms of both the last written series are of constant sign, and since μ takes on at most one of the values $+\infty$ and $-\infty$, it follows that at least one of these series is convergent. Since the sum of the two series is the convergent series $\sum_{n=1}^{\infty} \mu(E_n)$, it follows that they both converge, and, since the convergence of the series of positive terms and the series of negative terms is equivalent to absolute convergence, the proof of the theorem is complete. ∎

Theorem C. *If μ is a signed measure, if $\{E_n\}$ is a monotone sequence of measurable sets, and if, in case $\{E_n\}$ is a decreasing sequence, $|\mu(E_n)| < \infty$ for at least one value of n, then*

$$\mu(\lim_n E_n) = \lim_n \mu(E_n).$$

Proof. The proof of the assertion concerning increasing sequences is the same as for measures (replacing $\{E_n\}$ by the disjoint sequence $\{E_i - E_{i-1}\}$ of differences, cf. 9.D); the same is true for decreasing sequences (reduction to the preceding case by complementation, cf. 9.E), except that Theorem A has to be used to ensure the finiteness of the subtrahends that occur. ∎

(1) The sum of two [totally] σ-finite measures is a [totally] σ-finite measure. Is this assertion valid for infinite sums?

(2) A **complex measure** on the class of all measurable sets of a measurable space is a set function μ such that, for every measurable set E, $\mu(E) = \mu_1(E) + i\mu_2(E)$, where $i = \sqrt{-1}$, and where μ_1 and μ_2 are signed measures in the sense of this section. Are Theorems A, B, and C true for complex measures?

(3) If a signed measure μ is the difference of two measures in two ways, $\mu = \mu_1 - \mu_2$ and $\mu = \nu_1 - \nu_2$, then is it true that $\mu_1 = \nu_1$ and $\mu_2 = \nu_2$?

(4) The fact that a signed measure assumes at most one of the values $+\infty$ and $-\infty$ follows from the requirement of additivity. (Hint: if $\mu(E) = +\infty$ and $\mu(F) = -\infty$, then the right side of at least one of the relations

$$\mu(E) = \mu(E - F) + \mu(E \cap F),$$

$$\mu(F) = \mu(F - E) + \mu(E \cap F),$$

and

$$\mu(E \Delta F) = \mu(E - F) + \mu(F - E)$$

is indeterminate.)

§ 29. HAHN AND JORDAN DECOMPOSITIONS

If μ is a signed measure on the class of all measurable sets of a measurable space $(X, \mathbf{S})$, we shall call a set E **positive** (with respect to μ) if, for every measurable set F, $E \cap F$ is measurable and $\mu(E \cap F) \geqq 0$; similarly we shall call E **negative** if, for every measurable set F, $E \cap F$ is measurable and $\mu(E \cap F) \leqq 0$. The empty set is both positive and negative in this sense; we do not assert that any other, non trivial, positive sets or negative sets necessarily exist.

Theorem A. *If μ is a signed measure, then there exist two disjoint sets A and B whose union is X, such that A is positive and B is negative with respect to μ.*

The sets A and B are said to form a **Hahn decomposition** of X with respect to μ.

Proof. Since μ assumes at most one of the values $+\infty$ and $-\infty$, we may assume that, say

$$-\infty < \mu(E) \leqq \infty$$

for every measurable set E. Since the difference of two negative sets, and a disjoint, countable union of negative sets are obviously negative, it follows that every countable union of negative sets is negative. We write $\beta = \inf \mu(B)$ for all measurable negative sets B. Let $\{B_i\}$ be a sequence of measurable negative sets such that $\lim_i \mu(B_i) = \beta$; if $B = \bigcup_{i=1}^{\infty} B_i$, then B is a measurable negative set for which $\mu(B)$ is minimal.

We shall prove that the set $A = X - B$ is a positive set. Suppose that, on the contrary, E_0 is a measurable subset of A for which $\mu(E_0) < 0$. The set E_0 cannot be a negative set, for then $B \cup E_0$ would be a negative set with a smaller value of μ than $\mu(B)$, which is impossible. Let k_1 be the smallest positive integer with the property that E_0 contains a measurable set E_1 for which $\mu(E_1) \geqq \frac{1}{k_1}$. (Observe that, since $\mu(E_0) < 0$, $\mu(E_0)$ and $\mu(E_1)$ are both finite.) Since

$$\mu(E_0 - E_1) = \mu(E_0) - \mu(E_1) \leqq \mu(E_0) - \frac{1}{k_1} < 0,$$

the argument just applied to E_0 is applicable to $E_0 - E_1$ also. Let k_2 be the smallest positive integer with the property that $E_0 - E_1$ contains a measurable subset E_2 with $\mu(E_2) \geqq \frac{1}{k_2}$, and proceed so on ad infinitum. Since μ is finite valued for measurable subsets of E_0 (28.A), we must have $\lim_n \frac{1}{k_n} = 0$. It follows that, for every measurable subset F of

$$F_0 = E_0 - \bigcup_{j=1}^{\infty} E_j,$$

we have $\mu(F) \leqq 0$, i.e. that F_0 is a measurable negative set. Since F_0 is disjoint from B, and since

$$\mu(F_0) = \mu(E_0) - \sum_{j=1}^{\infty} \mu(E_j) \leqq \mu(E_0) < 0,$$

this contradicts the minimality of B, and we conclude that the hypothesis $\mu(E_0) < 0$ is untenable. ∎

It is not difficult to construct examples to show that a Hahn decomposition is *not* unique. If, however,

$$X = A_1 \cup B_1 \quad \text{and} \quad X = A_2 \cup B_2$$

are two Hahn decompositions of X, then we can prove that, for every measurable set E,

$$\mu(E \cap A_1) = \mu(E \cap A_2) \quad \text{and} \quad \mu(E \cap B_1) = \mu(E \cap B_2).$$

To see this, we observe that

$$E \cap (A_1 - A_2) \subset E \cap A_1,$$

so that $\mu(E \cap (A_1 - A_2)) \geqq 0$, and

$$E \cap (A_1 - A_2) \subset E \cap B_2,$$

so that $\mu(E \cap (A_1 - A_2)) \leqq 0$. Hence $\mu(E \cap (A_1 - A_2)) = 0$ and, by symmetry, $\mu(E \cap (A_2 - A_1)) = 0$; it follows that

$$\mu(E \cap A_1) = \mu(E \cap (A_1 \cup A_2)) = \mu(E \cap A_2).$$

It follows from this result that the equations

$$\mu^+(E) = \mu(E \cap A) \quad \text{and} \quad \mu^-(E) = -\mu(E \cap B)$$

unambiguously define two set functions μ^+ and μ^- on the class of all measurable sets, called, respectively, the **upper variation** and the **lower variation** of μ. The set function $|\mu|$, defined for every measurable set E by $|\mu|(E) = \mu^+(E) + \mu^-(E)$, is the **total variation** of μ. (Observe the important notational distinction between $|\mu|(E)$ and $|\mu(E)|$.)

Theorem B. *The upper, lower, and total variations of a signed measure μ are measures and $\mu(E) = \mu^+(E) - \mu^-(E)$ for every measurable set E. If μ is [totally] finite or σ-finite, then so also are μ^+ and μ^-; at least one of the measures μ^+ and μ^- is always finite.*

Proof. The variations of μ are clearly non negative; if every measurable set is a countable union of measurable sets for which μ is finite, it follows from 28.A that the same is true for μ^+ and μ^-. The equation $\mu = \mu^+ - \mu^-$ follows from the definitions of μ^+ and μ^-; the fact that μ takes on at most one of the values $+\infty$ and $-\infty$ implies that at least one of the set functions μ^+ and μ^- is always finite. Since the countable additivity of μ^+ and μ^- is evident, the proof is complete. ∎

It follows from Theorem B that every signed measure is the difference of two measures (of which at least one is finite); the representation of μ as the difference of its upper and lower variations is called the **Jordan decomposition** of μ.

(1) If μ is a finite signed measure and if $\{E_n\}$ is a sequence of measurable sets such that $\lim_n E_n$ exists, (i.e. such that $\limsup_n E_n = \liminf_n E_n$), then

$$\mu(\lim_n E_n) = \lim_n \mu(E_n).$$

(2) A finite signed measure, together with its variations, is bounded. For this reason finite signed measures are often said to be of **bounded variation.**

(3) If μ is a signed measure and if E is a measurable set, then

$$\mu^+(E) = \sup\{\mu(F): E \supset F \varepsilon \mathbf{S}\} \quad \text{and} \quad \mu^-(E) = -\inf\{\mu(F): E \supset F \varepsilon \mathbf{S}\}.$$

An alternative and frequently used proof of the validity of the Jordan decomposition may be given by treating these equations as the definitions of μ^+ and μ^-.

(4) Does the set of all totally finite signed measures on a σ-algebra form a Banach space with respect to the norm defined by $\|\mu\| = |\mu|(X)$?

(5) If $(X,\mathbf{S},\mu)$ is a measure space and f is an integrable function on X, then the set function ν, defined by $\nu(E) = \int_E f(x)d\mu(x)$, is a finite signed measure, and

$$\nu^+(E) = \int_E f^+ d\mu, \quad \nu^-(E) = \int_E f^- d\mu.$$

What is $|\nu|(E)$ in terms of f?

(6) If μ and ν are totally finite measures on a σ-algebra $\mathbf{S}$ and if E is a set in $\mathbf{S}$, then, corresponding to every real number t, there exists a set A_t in $\mathbf{S}$ such that $A_t \subset E$ and such that, for every set F in $\mathbf{S}$ for which $F \subset A_t$ [or for which $F \subset E - A_t$] we have $\nu(F) \leqq t\mu(F)$, [or $\nu(F) \geqq t\mu(F)$].

(7) If μ is a signed measure and f is a measurable function such that f is integrable with respect to $|\mu|$, then we may write, by definition,

$$\int f d\mu = \int f d\mu^+ - \int f d\mu^-.$$

This integral has many of the essential properties of the "positive" integrals discussed in Chapter V. If μ is a finite signed measure, then, for every measurable set E,

$$|\mu|(E) = \sup \left| \int_E f d\mu \right|,$$

where the supremum is extended over all measurable functions f such that $|f| \leqq 1$.

(8) By the separate consideration of real and imaginary parts, integrals such as $\int f d\mu$ may be defined for complex valued functions f and complex measures μ; (cf. 28.2). Motivated by (7) above, we define the total variation of a finite complex measure μ by $|\mu|(E) = \sup \left| \int_E f d\mu \right|$, where the supremum is extended over all (possibly complex valued) measurable functions f such that $|f| \leqq 1$. What is the relation between $|\mu|$ and the total variations of the real and imaginary parts of μ?

§ 30. ABSOLUTE CONTINUITY

Motivated by the properties of indefinite integrals, we introduced the abstract concept of signed measure, and we showed that the abstraction had several of the important properties of the concrete concept which it generalized. Indefinite integrals have, however, certain additional properties (or, rather, certain relations to the measures in terms of which they are defined) that are not shared by general signed measures. In a special case we have already discussed one such property of very great significance (absolute continuity, § 23); we propose now to examine a more general framework in which the discussion of absolute continuity still makes sense.

If $(X,\mathbf{S})$ is a measurable space and μ and ν are signed measures on $\mathbf{S}$, we say that ν is **absolutely continuous** with respect to μ, in symbols $\nu \ll \mu$, if $\nu(E) = 0$ for every measurable set E for which $|\mu|(E) = 0$. In a suggestively imprecise phrase, $\nu \ll \mu$ means that ν is small whenever μ is small. We call attention, however, to the lack of symmetry in the precise form of the definition; the smallness of μ is expressed by a condition on its total variation. Our first result concerning absolute continuity asserts that this asymmetry is only apparent.

Theorem A. *If μ and ν are signed measures, then the conditions*

(a) $$\nu \ll \mu,$$

(b) $$\nu^+ \ll \mu \quad \textit{and} \quad \nu^- \ll \mu,$$

(c) $$|\nu| \ll |\mu|,$$

are mutually equivalent.

Proof. If (a) is valid, then $\nu(E) = 0$ whenever $|\mu|(E) = 0$. If $X = A \cup B$ is a Hahn decomposition with respect to ν, then we have, whenever $|\mu|(E) = 0$,

$$0 \leqq |\mu|(E \cap A) \leqq |\mu|(E) = 0$$

and

$$0 \leqq |\mu|(E \cap B) \leqq |\mu|(E) = 0,$$

and therefore

$$\nu^+(E) = \nu(E \cap A) = 0 \quad \text{and} \quad \nu^-(E) = \nu(E \cap B) = 0;$$

this proves the validity of (b).

The facts that (b) implies (c) and (c) implies (a) follow from the relations

$$|\nu|(E) = \nu^+(E) + \nu^-(E) \quad \text{and} \quad 0 \leqq |\nu(E)| \leqq |\nu|(E)$$

respectively. ▮

The following theorem establishes the relation between our present form of the definition of absolute continuity and the one we used (for finite valued set functions) in § 23. The theorem asserts essentially that another precise interpretation of "ν is small whenever μ is small," which is apparently quite different from the definition of absolute continuity, is in the presence of a finiteness condition equivalent to it.

Theorem B. *If ν is a finite signed measure and if μ is a signed measure such that $\nu \ll \mu$, then, corresponding to every positive number ϵ, there is a positive number δ such that $|\nu|(E) < \epsilon$ for every measurable set E for which $|\mu|(E) < \delta$.*

Proof. Suppose that it is possible, for some $\epsilon > 0$, to find a sequence $\{E_n\}$ of measurable sets such that $|\mu|(E_n) < \frac{1}{2^n}$ and $|\nu|(E_n) \geqq \epsilon$, $n = 1, 2, \cdots$. If $E = \limsup_n E_n$, then

$$|\mu|(E) \leqq \sum_{i=n}^{\infty} |\mu|(E_i) < \frac{1}{2^{n-1}}, \quad n = 1, 2, \cdots,$$

and therefore $|\mu|(E) = 0$. On the other hand (since ν is finite)

$$|\nu|(E) = \lim_n |\nu|(E_n \cup E_{n+1} \cup \cdots) \geqq \limsup_n |\nu|(E_n) \geqq \epsilon.$$

Since this contradicts the relation $\nu \ll \mu$, the proof of the theorem is complete. ▮

It is easy to verify that the relation "$\ll$" is reflexive (i.e. $\mu \ll \mu$) and transitive (i.e. $\mu_1 \ll \mu_2$ and $\mu_2 \ll \mu_3$ imply that $\mu_1 \ll \mu_3$). Two signed measures μ and ν for which both $\nu \ll \mu$ and $\mu \ll \nu$ are called **equivalent,** in symbols $\mu \equiv \nu$.

The antithesis of the relation of absolute continuity is the relation of singularity. If $(X,\mathbf{S})$ is a measurable space and μ and ν are signed measures on $\mathbf{S}$, we say that μ and ν are **mutually singular,** or more simply that μ and ν are **singular,** in symbols $\mu \perp \nu$, if there exist two disjoint sets A and B whose union is X such that, for every measurable set E, $A \cap E$ and $B \cap E$ are measurable and $|\mu|(A \cap E) = |\nu|(B \cap E) = 0$. Despite the symmetry of the relation, it is occasionally more natural to use an unsymmetric expression such as "ν is singular with respect to μ" instead of "μ and ν are singular."

It is clear that singularity is indeed an extreme form of non absolute continuity. If ν is singular with respect to μ, then not only is it false that the vanishing of $|\mu|$ implies that of $|\nu|$, but in fact essentially the only sets for which $|\nu|$ does not necessarily vanish are the ones for which $|\mu|$ does.

We conclude this section with the introduction of a new notation. We have already used the traditional and suggestive "almost everywhere" terminology on measure spaces; this is perfectly satisfactory as long as we restrict our attention to one measure at a time. Since, however, in the discussion of absolute continuity and singularity we have necessarily to deal with several measures simultaneously, and since it is clumsy to say "almost everywhere with respect to μ" very often, we shall adopt the following convention. If, for each point x of a measurable space $(X,\mathbf{S})$, $\pi(x)$ is a proposition concerning x, and if μ is a signed measure on $\mathbf{S}$, then the symbol

$$\pi(x)\ [\mu] \quad \text{or} \quad \pi\ [\mu]$$

shall mean that $\pi(x)$ is true for almost every x with respect to the measure $|\mu|$. Thus, for instance, if f and g are two functions on X, we shall write $f = g\ [\mu]$ for the statement that $\{x: f(x) \neq g(x)\}$ is a measurable set of measure zero with respect to $|\mu|$. The symbol $[\mu]$ may be read as "modulo μ."

(1) If μ is a signed measure and f is a function integrable with respect to $|\mu|$, and if ν is defined for every measurable set E by $\nu(E) = \int_E f d\mu$ (cf. 29.7), then $\nu \ll \mu$.

(2) Let the measure space $(X,\mathbf{S},\mu)$ be the unit interval with Lebesgue measure. Write $F = \{x: 0 \leqq x \leqq \frac{1}{2}\}$, and let f_1 and f_2 be the functions defined by $f_1(x) = 2\chi_F(x) - 1$ and $f_2(x) = x$. If the set functions μ_i are defined by $\mu_i(E) = \int_E f_i d\mu$, $i = 1, 2$, then $\mu_2 \ll \mu_1$. It is not, however, true that $\mu_2(E) = 0$ whenever $\mu_1(E) = 0$. If μ_2 were defined by $\mu_2(E) = \int_E (f_2 - \frac{1}{2})d\mu$, then even this stronger condition would be satisfied.

(3) For every signed measure μ, the variations μ^+ and μ^- are mutually singular, and they are each absolutely continuous with respect to μ.

(4) For every signed measure μ, $\mu \equiv |\mu|$.

(5) If μ is a signed measure and E is a measurable set, then $|\mu|(E) = 0$ if and only if $\mu(F) = 0$ for every measurable subset F of E.

(6) If μ and ν are any two measures on a σ-ring $\mathbf{S}$, then $\nu \ll \mu + \nu$.

(7) Let f_1 and f_2 be integrable functions on a totally finite measure space $(X,\mathbf{S},\mu)$ and let μ_i be the indefinite integral of f_i, $i = 1, 2$. If $\mu(\{x: f_1(x) = 0\} \Delta \{x: f_2(x) = 0\}) = 0$, then $\mu_1 \equiv \mu_2$.

(8) Let ψ be the Cantor function (cf. 19.3), and let μ_0 be the Lebesgue–Stieltjes measure, on the Borel subsets of the unit interval, induced by ψ; (cf. 15.9). If μ is Lebesgue measure, then μ_0 and μ are mutually singular.

(9) If μ and ν are signed measures such that ν is both absolutely continuous and singular with respect to μ, then $\nu = 0$.

(10) If ν_1, ν_2, and μ are finite signed measures such that both ν_1 and ν_2 are singular with respect to μ, then $\nu = \nu_1 + \nu_2$ is also singular with respect to μ. (Hint: if $X = A_1 \cup B_1$ and $X = A_2 \cup B_2$ are decompositions such that $|\mu|$ is identically zero for measurable subsets of A_i and $|\nu_i|$ is identically zero for measurable subsets of B_i, $i = 1, 2$, then

$$X = [(A_1 \cap A_2) \cup (A_1 \cap B_2) \cup (A_2 \cap B_1)] \cup (B_1 \cap B_2)$$

is such a decomposition for μ and ν.)

(11) If μ and ν are measures on a σ-algebra $\mathbf{S}$ such that μ is finite and $\nu \ll \mu$, then there exists a measurable set E such that $X - E$ is of σ-finite measure with respect to ν, and such that, for every measurable subset F of E, $\nu(F)$ is either 0 or ∞. (Hint: use the method of exhaustion (cf. 17.3), to find a measurable set E with the property that, for every measurable subset F of E, $\nu(F)$ is either 0 or ∞, and such that $\mu(E)$ is maximal; another application of the method of exhaustion shows that $X - E$ is of σ-finite measure with respect to ν.)

(12) Theorem B is not necessarily true if ν is not finite. (Hint: let X be the set of all positive integers, and, for every subset E of X, write

$$\mu(E) = \textstyle\sum_{n \varepsilon E} 2^{-n}, \quad \nu(E) = \sum_{n \varepsilon E} 2^n.)$$

§ 31. THE RADON–NIKODYM THEOREM

Theorem A. *If μ and ν are totally finite measures such that $\nu \ll \mu$ and ν is not identically zero, then there exists a positive number ϵ and a measurable set A such that $\mu(A) > 0$ and such that A is a positive set for the signed measure $\nu - \epsilon\mu$.*

Proof. Let $X = A_n \cup B_n$ be a Hahn decomposition with respect to the signed measure $\nu - \frac{1}{n}\mu$, $n = 1, 2, \cdots$, and write

$$A_0 = \textstyle\bigcup_{n=1}^{\infty} A_n, \quad B_0 = \bigcap_{n=1}^{\infty} B_n.$$

Since $B_0 \subset B_n$, we have

$$0 \leqq \nu(B_0) \leqq \frac{1}{n}\mu(B_0), \quad n = 1, 2, \cdots,$$

and consequently $\nu(B_0) = 0$. It follows that $\nu(A_0) > 0$ and therefore, by absolute continuity, that $\mu(A_0) > 0$. Hence we must have $\mu(A_n) > 0$ for at least one value of n; if, for such a value of n, we write $A = A_n$ and $\epsilon = \frac{1}{n}$, the requirements of the theorem are all satisfied. ∎

We proceed now to establish the fundamental result (known as the **Radon–Nikodym theorem**) concerning absolute continuity.

Theorem B. *If $(X,\mathbf{S},\mu)$ is a totally σ-finite measure space and if a σ-finite signed measure ν on $\mathbf{S}$ is absolutely continuous* *with respect to μ, then there exists a finite valued measurable function f on X such that*

$$\nu(E) = \int_E f d\mu$$

for every measurable set E. The function f is unique in the sense that if also $\nu(E) = \int_E g d\mu$, $E \in \mathbf{S}$, then $f = g$ $[\mu]$.

We emphasize the fact that f is not asserted to be integrable; it is, in fact, clear that a necessary and sufficient condition that f be integrable is that ν be finite. The use of the symbol $\int f d\mu$ implicitly asserts, however (cf. § 25), that either the positive or the negative part of f is integrable, corresponding to the fact that either the upper or the lower variation of ν is finite.

Proof. Since X is a countable, disjoint union of measurable sets on which both μ and ν are finite, there is no loss of generality (for both the existence and the uniqueness proofs) in assuming finiteness in the first place. Since if ν is finite, f is integrable, uniqueness follows from 25.E. Since, finally, the assumption $\nu \ll \mu$ is equivalent to the simultaneous validity of the conditions

$$\nu^+ \ll \mu \quad \text{and} \quad \nu^- \ll \mu,$$

it remains only to prove the existence of f in the case in which both μ and ν are finite measures.

Let $\mathcal{K}$ be the class of all non negative functions f, integrable with respect to μ, such that $\int_E f d\mu \leq \nu(E)$ for every measurable set E, and write

$$\alpha = \sup \left\{ \int f d\mu : f \in \mathcal{K} \right\}.$$

Let $\{f_n\}$ be a sequence of functions in $\mathcal{K}$ such that

$$\lim_n \int f_n d\mu = \alpha.$$

If E is any fixed measurable set, n is any fixed positive integer, and $g_n = f_1 \cup \cdots \cup f_n$, then E may be written as a finite, disjoint union of measurable sets, $E = E_1 \cup \cdots \cup E_n$, so that $g_n(x) = f_j(x)$ for x in E_j, $j = 1, \cdots, n$. Consequently we have

$$\int_E g_n d\mu = \sum_{j=1}^n \int_{E_j} f_j d\mu \leqq \sum_{j=1}^n \nu(E_j) = \nu(E).$$

If we write $f_0(x) = \sup \{f_n(x): n = 1, 2, \cdots\}$, then $f_0(x) = \lim_n g_n(x)$ and it follows from 27.B that $f_0 \varepsilon \mathcal{K}$ and $\int f_0 d\mu = \alpha$. Since f_0 is integrable, there exists a finite valued function f such that $f_0 = f$ $[\mu]$; we shall prove that if $\nu_0(E) = \nu(E) - \int_E f d\mu$, then the measure ν_0 is identically zero.

If ν_0 is not identically zero, then, by Theorem A, there exists a positive number ϵ and a measurable set A such that $\mu(A) > 0$ and such that

$$\epsilon\mu(E \cap A) \leqq \nu_0(E \cap A) = \nu(E \cap A) - \int_{E \cap A} f d\mu$$

for every measurable set E. If $g = f + \epsilon\chi_A$, then

$$\int_E g d\mu = \int_E f d\mu + \epsilon\mu(E \cap A) \leqq \int_{E-A} f d\mu + \nu(E \cap A) \leqq \nu(E)$$

for every measurable set E, so that $g \varepsilon \mathcal{K}$. Since, however,

$$\int g d\mu = \int f d\mu + \epsilon\mu(A) > \alpha,$$

this contradicts the maximality of $\int f d\mu$, and the proof of the theorem is complete. ∎

(1) If $(X,\mathbf{S},\mu)$ is a measure space and if $\nu(E) = \int_E f d\mu$ for every measurable set E, then

$$X = \{x: f(x) > 0\} \cup \{x: f(x) \leqq 0\}$$

is a Hahn decomposition with respect to ν.

(2a) Suppose that $(X,\mathbf{S})$ is a measurable space and μ and ν are totally finite measures on $\mathbf{S}$ such that $\nu \ll \mu$. If $\bar{\mu} = \mu + \nu$ and if $\nu(E) = \int_E f d\bar{\mu}$ for every measurable set E, then $0 \leqq f(x) < 1$ $[\mu]$.

(2b) If $\int g d\nu = \int f g d\bar{\mu}$ for every non negative measurable function g, then

$\nu(E) = \int_E \frac{f}{1-f} d\mu$ for every measurable set E. (Hint: rewrite the hypothesis in the form $\int g(1-f)d\nu = \int fg d\mu$ and, given E, write $g = \frac{\chi_E}{1-f}$.)

(3) Let $(X,\mathbf{S},\mu)$ be the unit interval with Lebesgue measure and let M be a non measurable set. Let (α_1,β_1) and (α_2,β_2) be two pairs of positive real numbers such that $\alpha_1 + \beta_1 = \alpha_2 + \beta_2 = 1$, and let $\tilde{\mu}_i$ be the extension of μ, determined by (α_i,β_i), to the σ-ring $\tilde{\mathbf{S}}$ generated by $\mathbf{S}$ and M, $i = 1, 2$ (cf. 16.2). There exist measurable functions f_1 and f_2 such that

$$\tilde{\mu}_1(E) = \int_E f_1 d\tilde{\mu}_2 \quad \text{and} \quad \tilde{\mu}_2(E) = \int_E f_2 d\tilde{\mu}_1$$

for every measurable set E. What are the functions f_1 and f_2?

(4) The Radon-Nikodym theorem remains true even if μ is only a signed measure. (Hint: let $X = A \cup B$ be a Hahn decomposition with respect to μ and apply the Radon-Nikodym theorem separately to ν and μ^+ in A and to ν and μ^- in B.)

(5) Let μ be a totally σ-finite signed measure. Since both μ^+ and μ^- are absolutely continuous with respect to both μ and $|\mu|$, we have

$$\mu^+(E) = \int_E f_+ d\mu = \int_E g_+ d|\mu| \quad \text{and} \quad \mu^-(E) = \int_E f_- d\mu = \int_E g_- d|\mu|.$$

The functions f_+, g_+, f_-, and g_- satisfy the relations $f_+ = g_+ [\mu]$ and $f_- = -g_- [\mu]$. What are these functions?

(6) If μ is a signed measure and if $\nu(E) = \int_E f d\mu$ and $|\nu|(E) = \int_E g d|\mu|$ for every measurable set E, then $g = |f| [\mu]$.

(7) The Radon-Nikodym theorem remains true even if ν is not σ-finite, but, in this case, the integrand f is not necessarily finite valued. (Hint: it is sufficient to consider the case in which ν is a measure and μ is finite; in this case apply 30.11.)

(8) The Radon-Nikodym theorem is not necessarily true if μ is not totally σ-finite, even if ν remains finite. (Hint: let X be an uncountable set and let $\mathbf{S}$ be the class of all those sets which are either countable or have countable complements. For every E in $\mathbf{S}$, let $\mu(E)$ be the number of points in E and let $\nu(E)$ be 0 or 1 according as E is countable or not.)

(9) If $(X,\mathbf{S})$ is a measurable space and μ and ν are σ-finite measures on $\mathbf{S}$ such that $\nu \ll \mu$, then the Radon-Nikodym theorem may be applied to each measurable set separately. The question might be raised whether or not a function f may be defined once for all on the whole space X so as to serve as a suitable integrand simultaneously for every measurable set. The answer is no, as the following pathological example shows.

Let A be any uncountable set (with, say, cardinal number α), and let B be a set of cardinal number $\beta > \alpha$. Let X be the set of all ordered pairs (a,b) with $a \varepsilon A$ and $b \varepsilon B$. It is convenient to call a set of the form $\{(a,b_0): a \varepsilon A\}$ a **horizontal line,** and a set of the form $\{(a_0,b): b \varepsilon B\}$ a **vertical line.** We shall call a set E **full** on a horizontal or vertical line L if $L - E$ is countable; (cf. 12.1). Let $\mathbf{S}$ be the class of all sets which may be covered by countably many horizontal and vertical lines, and which are such that on every horizontal and vertical line they are either countable or full. For every E in $\mathbf{S}$ let $\mu(E)$ be the number of horizontal and vertical lines on which E is full, and let $\nu(E)$ be the number of vertical lines on which E is full. Clearly μ and ν are σ-finite measures and $\nu \ll \mu$. Suppose now that there exists a function f on X such that $\nu(E) = \int_E f d\mu$ for every E in $\mathbf{S}$. It is easy to see that the set $\{x: f(x) = 0\}$ has to be countable on every vertical line and full on every horizontal line. The first requirement implies that the cardinal number of this set is at most $\alpha\aleph_0 = \alpha$, and the second requirement implies that the cardinal number of this set is at least $\beta(\alpha - \aleph_0) \geqq \beta$.

(10) There is a condition on measure spaces, which is more general than total σ-finiteness and more restrictive than σ-finiteness, in the presence of which the Radon–Nikodym theorem is still true. The condition is that the space be the union of a disjoint class $\mathbf{D}$ of measurable sets of finite measure with the property that every measurable set may be covered by countably many sets of $\mathbf{D}$ and a set of measure zero. The following is an example of a non totally σ-finite measure space satisfying this condition.

Let X be the Euclidean plane, and let $\mathbf{S}$ be the class of all those sets which may be covered by countably many horizontal lines and which are Lebesgue measurable on each such line. If E is a Lebesgue measurable subset of a horizontal line, define $\mu(E)$ to be the Lebesgue measure of E; for the general E in $\mathbf{S}$, μ is thereby uniquely determined by the requirement of countable additivity.

(11) If, in (9) above, $B = A$ and the cardinal number of this set is $\aleph_1$ (= the smallest uncountable cardinal), then the proof breaks down, i.e. there exists in that case a subset E of X which is countable on every vertical and full on every horizontal line. (Hint: well order A, i.e. assign to every a in A an ordinal number $\xi(a) < \Omega$(= the smallest uncountable ordinal) so that the correspondence is one to one between all points of A and all ordinals less than Ω, and write $E = \{(a,b): \xi(a) > \xi(b)\}$.

(12) If μ is a totally finite measure and $\nu(E) = \int_E f d\mu$ for every measurable set E, then the set

$$B(t) = \{x: f(x) \leqq t\}$$

is a negative set for the signed measure $\nu - t\mu$; (cf. (1) above). A proof of the Radon–Nikodym theorem may be based on an attempt to reconstruct f from the sets $B(t)$; (cf. 18.10). The main complication of this approach is the non uniqueness of negative sets. A tool for partially dealing with this complication is to select $B(t)$, for each t, so as to maximize the value of $\mu(B(t))$.

§ 32. DERIVATIVES OF SIGNED MEASURES

There is a special notation for the functions which occur as integrands in the Radon–Nikodym theorem, which is frequently very suggestive. If μ is a totally σ-finite measure and if $\nu(E) = \int_E f d\mu$ for every measurable set E, we shall write

$$f = \frac{d\nu}{d\mu} \quad \text{or} \quad d\nu = f d\mu.$$

All the properties of Radon–Nikodym integrands (which we may also call Radon–Nikodym **derivatives**), which are suggested by the well known differential formalism, correspond to true theorems. Some of these are trivial $\left(\text{e.g. } \frac{d(\nu_1 + \nu_2)}{d\mu} = \frac{d\nu_1}{d\mu} + \frac{d\nu_2}{d\mu}\right)$, while others are more or less deep properties of integration. Examples of the latter kind of result are the chain rule for differentiation and, as an easy corollary, the substitution rule for the differentials occurring under an integral sign; both these results are precisely stated and proved below. It is, of course, important to remember that a Radon–Nikodym derivative $\frac{d\nu}{d\mu}$ is unique only a.e. with respect to μ, and that, therefore, in the detailed verbal interpretation of a differential formula, frequent use has to be made of the qualifying "almost everywhere."

Theorem A. *If λ and μ are totally σ-finite measures such that $\mu \ll \lambda$ and if ν is a totally σ-finite signed measure such that $\nu \ll \mu$, then*

$$\frac{d\nu}{d\lambda} = \frac{d\nu}{d\mu}\frac{d\mu}{d\lambda} \quad [\lambda].$$

Proof. Since the validity of the desired equation for the upper and lower variations of ν implies its validity for ν itself, we may and do assume that ν is a measure; for simplicity of notation we write $\frac{d\nu}{d\mu} = f$ and $\frac{d\mu}{d\lambda} = g$. Since ν is non negative, it follows from 25.D that $f \geqq 0$ $[\mu]$ and therefore that there is no loss of generality in assuming that f is everywhere non negative.

Let $\{f_n\}$ be an increasing sequence of non negative simple functions converging at every point to f, (20.B); then, by 27.B, we have

$$\lim_n \int_E f_n d\mu = \int_E f d\mu \quad \text{and} \quad \lim_n \int_E f_n g d\lambda = \int_E f g d\lambda$$

for every measurable set E. Since, for every measurable set F,

$$\int_E \chi_F d\mu = \mu(E \cap F) = \int_{E \cap F} g d\lambda = \int_E \chi_F g d\lambda,$$

it follows that $\int_E f_n d\mu = \int_E f_n g d\lambda$, $n = 1, 2, \cdots$, and therefore that $\nu(E) = \int_E f d\mu = \int_E f g d\lambda$. ∎

Theorem B. *If λ and μ are totally σ-finite measures such that $\mu \ll \lambda$, and if f is a finite valued measurable function for which $\int f d\mu$ is defined, then $\int f d\mu = \int f \frac{d\mu}{d\lambda} d\lambda$.*

Proof. We write $\nu(E) = \int_E f d\mu$ for every measurable set E, and apply Theorem A. It follows that $\nu(E) = \int_E f \frac{d\mu}{d\lambda} d\lambda$ for every measurable set E; the desired result follows by putting $E = X$. ∎

Our next and final result concerning the relations among signed measures treats the **Lebesgue decomposition** of a totally σ-finite signed measure into an absolutely continuous part and a singular part with respect to another totally σ-finite signed measure.

Theorem C. *If $(X, \mathbf{S})$ is a measurable space and μ and ν are totally σ-finite signed measures on $\mathbf{S}$, then there exist two uniquely determined totally σ-finite signed measures ν_0 and ν_1 whose sum is ν, such that $\nu_0 \perp \mu$ and $\nu_1 \ll \mu$.*

Proof. As usual we may assume that μ and ν are finite. Since $\nu_i (i = 0, 1)$ will be absolutely continuous or singular with respect to μ according as it is absolutely continuous or singular with respect to $|\mu|$, we may assume that μ is a measure. Since, finally, we may treat ν^+ and ν^- separately, we may also assume that ν is a measure.

The proof of the theorem for totally finite measures is a useful trick, based on the elementary observation that ν is absolutely continuous with respect to $\mu + \nu$. There exists, accordingly, a measurable function f such that

$$\nu(E) = \int_E f d\mu + \int_E f d\nu$$

for every measurable set E. Since $0 \leqq \nu(E) \leqq \mu(E) + \nu(E)$, we have $0 \leqq f \leqq 1 \, [\mu + \nu]$ and therefore $0 \leqq f \leqq 1 \, [\nu]$. If we write $A = \{x : f(x) = 1\}$ and $B = \{x : 0 \leqq f(x) < 1\}$, then

$$\nu(A) = \int_A d\mu + \int_A d\nu = \mu(A) + \nu(A)$$

and therefore (since ν is finite) $\mu(A) = 0$. If

$$\nu_0(E) = \nu(E \cap A) \quad \text{and} \quad \nu_1(E) = \nu(E \cap B)$$

for every measurable set E, then it is clear that $\nu_0 \perp \mu$; it remains to prove that $\nu_1 \ll \mu$.

If $\mu(E) = 0$, then

$$\int_{E \cap B} d\nu = \nu(E \cap B) = \int_{E \cap B} f d\nu$$

and therefore $\int_{E \cap B} (1 - f) d\nu = 0$. Since $1 - f \geqq 0 \, [\nu]$, it follows that $\nu_1(E) = \nu(E \cap B) = 0$; this completes the proof of the existence of ν_0 and ν_1.

If $\nu = \nu_0 + \nu_1$ and $\nu = \bar{\nu}_0 + \bar{\nu}_1$ are two Lebesgue decompositions of ν, then $\nu_0 - \bar{\nu}_0 = \bar{\nu}_1 - \nu_1$. Since $\nu_0 - \bar{\nu}_0$ is singular (cf. 30.10) and $\bar{\nu}_1 - \nu_1$ is absolutely continuous with respect to μ, it follows that $\nu_0 = \bar{\nu}_0$ and $\nu_1 = \bar{\nu}_1$; (cf. 30.9). ∎

(1) Using the concept of integration with respect to a signed measure, the definition of Radon–Nikodym derivatives may be extended to the case in which μ is a signed measure, and Theorem A remains true if λ and μ are signed measures. (Hint: consider a Hahn decomposition with respect to each of the three signed measures λ, μ, and ν, and construct the decomposition of X into the eight sets obtained by taking one set from each decomposition and forming the intersection of these three sets. On measurable subsets of each of the eight sets each of the functions, λ, μ, and ν, is of constant sign and therefore, after an obvious, slight modification, Theorem A applies.)

(2) If μ and ν are totally σ-finite signed measures such that $\mu \equiv \nu$, then $\frac{d\mu}{d\nu} = 1/\frac{d\nu}{d\mu}$.

(3) If μ and ν are totally σ-finite signed measures such that $\nu \ll \mu$, then $\nu\left(\left\{x: \frac{d\nu}{d\mu}(x) = 0\right\}\right) = 0$.

(4) If μ_0, μ_1, and μ_2 are totally finite measures, and if $d\mu_0 = f_1 d(\mu_0 + \mu_1) = f_2 d(\mu_0 + \mu_2) = f d(\mu_0 + \mu_1 + \mu_2)$, then we have, almost everywhere with respect to $\mu_0 + \mu_1 + \mu_2$,

$$f(x) = \begin{cases} \dfrac{f_1(x)f_2(x)}{f_1(x) + f_2(x) - f_1(x)f_2(x)} & \text{if } f_1(x)f_2(x) \neq 0, \\ 0 & \text{if } f_1(x) = f_2(x) = 0. \end{cases}$$

(5) Given two sequences $\{\mu_n\}$ and $\{\nu_n\}$ of totally finite measures, write

$$\bar{\mu}_n = \sum_{i=1}^n \mu_i, \quad \bar{\nu}_n = \sum_{i=1}^n \nu_i, \quad \mu = \sum_{i=1}^\infty \mu_i, \quad \nu = \sum_{i=1}^\infty \nu_i,$$

and assume that μ and ν are finite measures. If $\bar{\nu}_n \ll \bar{\mu}_n$, $n = 1, 2, \cdots$, then $\nu \ll \mu$ and

$$\lim_n \frac{d\bar{\nu}_n}{d\bar{\mu}_n} = \frac{d\nu}{d\mu} \; [\mu].$$

The proof of this assertion may be based on the following lemmas.

(5a) If $\{E_n\}$ is a sequence of measurable sets such that $\bar{\mu}_n(E_n) = 0$, $n = 1, 2, \cdots$, then $\mu(\limsup_n E_n) = 0$. (Hint: $\bar{\mu}_n(\bigcup_{k=n}^\infty E_k) \leq \sum_{k=n}^\infty \bar{\mu}_k(E_k)$.)

(5b) If $\{\phi_n\}$ and $\{\psi_n\}$ are sequences of functions such that $\phi_n = \psi_n \; [\mu_n]$, $n = 1, 2, \cdots$, then, for a.e. x $[\mu]$,

$$\limsup_n \phi_n(x) = \limsup_n \psi_n(x) \quad \text{and} \quad \liminf_n \phi_n(x) = \liminf_n \psi_n(x).$$

(Hint: write $E_n = \{x: \phi_n(x) \neq \psi_n(x)\}$ and apply (5a).)

In view of the result (5b) it is sufficient to prove (5) for any fixed determination of the derivatives $\frac{d\bar{\nu}_n}{d\bar{\mu}_n}$. If

$$\frac{d\nu_n}{d\mu} = f_n \quad \text{and} \quad \frac{d\mu_n}{d\mu} = g_n, \quad n = 1, 2, \cdots,$$

then it follows from Theorem A that one such determination is

$$\frac{d\bar{\nu}_n}{d\bar{\mu}_n} = \frac{f_1 + \cdots + f_n}{g_1 + \cdots + g_n} \; [\mu_n], \quad n = 1, 2, \cdots.$$

(5c) $\sum_{n=1}^\infty f_n = \frac{d\nu}{d\mu}$ and $\sum_{n=1}^\infty g_n = 1 \; [\mu]$. (Hint: since

$$\sum_{i=1}^n \mu_i(E) = \int_E (g_1 + \cdots + g_n) d\mu$$

and

$$\sum_{i=1}^n \nu_i(E) = \int_E (f_1 + \cdots + f_n) d\mu, \quad n = 1, 2, \cdots,$$

the desired result follows from 27.B and 25.E.)

Chapter VII

PRODUCT SPACES

§ 33. CARTESIAN PRODUCTS

If X and Y are any two sets (not necessarily subsets of the same space), the **Cartesian product** $X \times Y$ is the set of all ordered pairs (x,y), where $x \,\varepsilon\, X$ and $y \,\varepsilon\, Y$. The best known example of a Cartesian product is the Euclidean plane, which is most often viewed as the product of two coordinate axes. Most of the development in the sequel uses the words and concepts suggested by this example. Thus, for instance, if $A \subset X$ and $B \subset Y$, we shall call the set $E = A \times B$ (a subset of $X \times Y$) a **rectangle** and we shall refer to the component sets A and B as its **sides.** (Observe that our usage here differs from the classical terminology which speaks of rectangles only if the sides are intervals.)

Theorem A. *A rectangle is empty if and only if one of its sides is empty.*

Proof. If $A \times B \neq 0$, say $(x,y) \,\varepsilon\, A \times B$, then $x \,\varepsilon\, A$ and $y \,\varepsilon\, B$, so that $A \neq 0$ and $B \neq 0$. If, on the other hand, neither A nor B is empty, then there is a point (x,y) such that $(x,y) \,\varepsilon$ $A \times B$, so that $A \times B \neq 0$. ▮

Theorem B. *If* $E_1 = A_1 \times B_1$ *and* $E_2 = A_2 \times B_2$ *are non empty rectangles, then* $E_1 \subset E_2$ *if and only if*

$$A_1 \subset A_2 \quad and \quad B_1 \subset B_2.$$

Proof. The "if" is obvious. To prove the converse, let (x,y) be a point in $A_1 \times B_1$ and suppose that there exists a point x_1

in A_1 such that $x_1 \varepsilon' A_2$. Then

$$(x_1,y) \varepsilon A_1 \times B_1 \quad \text{and} \quad (x_1,y) \varepsilon' A_2 \times B_2;$$

it follows that no such point x_1 can exist and therefore $A_1 \subset A_2$. The same proof with only notational changes shows that $B_1 \subset B_2$. ∎

Theorem C. *If $A_1 \times B_1 = A_2 \times B_2$ is a non empty rectangle, then $A_1 = A_2$ and $B_1 = B_2$.*

Proof. It follows from Theorem B that

$$A_1 \subset A_2 \subset A_1 \quad \text{and} \quad B_1 \subset B_2 \subset B_1. \blacksquare$$

Theorem D. *If $E = A \times B$, $E_1 = A_1 \times B_1$, and $E_2 = A_2 \times B_2$ are non empty rectangles, then a necessary and sufficient condition that E be the disjoint union of E_1 and E_2, is that either A is the disjoint union of A_1 and A_2, and $B = B_1 = B_2$, or else B is the disjoint union of B_1 and B_2, and $A = A_1 = A_2$.*

Proof. We prove first that the condition is necessary. Since $E_1 \subset E$ and $E_2 \subset E$, it follows from Theorem B that $A_1 \subset A$ and $A_2 \subset A$, and therefore that $A_1 \cup A_2 \subset A$; similarly $B_1 \cup B_2 \subset B$. Since

$$E_1 \cup E_2 \subset (A_1 \cup A_2) \times (B_1 \cup B_2),$$

it follows that $A \subset A_1 \cup A_2$ and $B \subset B_1 \cup B_2$, and therefore $A = A_1 \cup A_2$ and $B = B_1 \cup B_2$. Since, finally, a similar argument shows that

$$0 = E_1 \cap E_2 \supset (A_1 \cap A_2) \times (B_1 \cap B_2),$$

it follows from Theorem A that at least one of the two sets $A_1 \cap A_2$ and $B_1 \cap B_2$ is empty.

Suppose, for instance, that $A_1 \cap A_2 = 0$; we are to show that in this case $B = B_1 = B_2$. (The case $B_1 \cap B_2 = 0$ is treated similarly.) Suppose on the contrary that there exists a point y in $B - B_1$. Then, if x is any point in A_1, we have $(x,y) \varepsilon E$, but (since $y \varepsilon' B_1$), $(x,y) \varepsilon' E_1$, and (since $x \varepsilon' A_2$), $(x,y) \varepsilon' E_2$. Since this contradicts the assumption $E = E_1 \cup E_2$, it follows that $B - B_1 = 0$ and also, by a similar argument, $B - B_2 = 0$.

The sufficiency of the condition is easier. If, for instance, A is the disjoint union of A_1 and A_2 and $B = B_1 = B_2$, then $A \supset A_1$, $A \supset A_2$, $B \supset B_1$, $B \supset B_2$, so that $E \supset E_1 \cup E_2$. Also if $(x,y) \varepsilon E$, then

$$(x,y) \varepsilon E_1 \quad \text{or} \quad (x,y) \varepsilon E_2$$

according as $x \varepsilon A_1$ or $x \varepsilon A_2$, so that E is indeed the disjoint union of E_1 and E_2. ▮

Theorem E. *If* **S** *and* **T** *are rings of subsets of* X *and* Y *respectively, then the class* **R** *of all finite, disjoint unions of rectangles of the form* $A \times B$, *where* $A \varepsilon$ **S** *and* $B \varepsilon$ **T**, *is a ring.*

Proof. We observe first that the intersection of two sets of the form $A \times B$ is another set of that form. If either of the two given sets, or their intersection, is empty, this result is trivial. If

$$E_1 = A_1 \times B_1, \quad E_2 = A_2 \times B_2, \quad \text{and} \quad (x,y) \varepsilon E_1 \cap E_2,$$

then $x \varepsilon A_1 \cap A_2$ and $y \varepsilon B_1 \cap B_2$, so that

$$E_1 \cap E_2 \subset (A_1 \cap A_2) \times (B_1 \cap B_2).$$

On the other hand, by Theorem B, $(A_1 \cap A_2) \times (B_1 \cap B_2)$ is contained in E_1 and E_2 and therefore in $E_1 \cap E_2$, so that

$$E_1 \cap E_2 = (A_1 \cap A_2) \times (B_1 \cap B_2).$$

Since **S** and **T** are rings, $A_1 \cap A_2 \varepsilon$ **S** and $B_1 \cap B_2 \varepsilon$ **T**. It follows immediately that the class **R** is closed under the formation of finite intersections.

Since

$$(A_1 \times B_1) - (A_2 \times B_2) =$$
$$= [(A_1 \cap A_2) \times (B_1 - B_2)] \cup [(A_1 - A_2) \times B_1],$$

we see that the difference of two sets of the given form is a disjoint union of two other sets of that form; since

$$\bigcup_{i=1}^{n} E_i - \bigcup_{j=1}^{m} F_j = \bigcup_{i=1}^{n} \bigcap_{j=1}^{m} (E_i - F_j),$$

it follows, using the result of the preceding paragraph, that the class **R** is closed under the formation of differences. Since **R** is obviously closed under the formation of finite, disjoint unions, the proof of the theorem is complete. ▌

Suppose now that in addition to the two sets X and Y we are also given two σ-rings $\mathbf{S}$ and $\mathbf{T}$ of subsets of X and Y respectively. We shall denote by $\mathbf{S} \times \mathbf{T}$ the σ-ring of subsets of $X \times Y$ generated by the class of all sets of the form $A \times B$, where $A \in \mathbf{S}$ and $B \in \mathbf{T}$.

Theorem F. *If* $(X,\mathbf{S})$ *and* $(Y,\mathbf{T})$ *are measurable spaces, then* $(X \times Y, \mathbf{S} \times \mathbf{T})$ *is a measurable space.*

The measurable space $(X \times Y, \mathbf{S} \times \mathbf{T})$ is the **Cartesian product** of the two given measurable spaces.

Proof. If $(x,y) \in X \times Y$, then there exist sets A and B such that $x \in A \in \mathbf{S}$ and $y \in B \in \mathbf{T}$; it follows that $(x,y) \in A \times B \in \mathbf{S} \times \mathbf{T}$. ▌

We observe that this is the first time we ever referred to the fact that a measurable space is the union of its measurable sets; in the present chapter we shall make essential use of this property of measurable spaces.

We shall frequently use the concept of **measurable rectangle.** Two equally obvious and natural definitions of this phrase suggest themselves. According to one, a rectangle in the Cartesian product of two measurable spaces $(X,\mathbf{S})$ and $(Y,\mathbf{T})$ is measurable if it belongs to $\mathbf{S} \times \mathbf{T}$, and, according to the other, $A \times B$ is measurable if $A \in \mathbf{S}$ and $B \in \mathbf{T}$. It is an easy consequence of the results we shall obtain that for non empty rectangles the two concepts coincide; for the time being we adopt the second of our proposed definitions. We may say, accordingly, that the class of measurable sets in the Cartesian product of two measurable spaces is the σ-ring generated by the class of all measurable rectangles.

(1) The intersection of any countable class of [measurable] rectangles is a [measurable] rectangle. Does this statement remain true if the word "countable" is omitted?

(2) The "only if" part of Theorem B, Theorem C, and the necessity of the condition in Theorem D are all false for empty rectangles.

(3) Under the hypotheses of Theorem E, the class $\mathbf{P}$ of all sets of the form $A \times B$, where $A \in \mathbf{S}$ and $B \in \mathbf{T}$, is a semiring. Is this statement true if $\mathbf{S}$ and $\mathbf{T}$ are not necessarily rings, but merely semirings?

(4) If the rings **S** and **T** in (3) each contain at least two different non empty sets, then **P** is not a ring.

(5) A necessary and sufficient condition that $\mathbf{S} \times \mathbf{T}$ be a σ-algebra is that both **S** and **T** be σ-algebras.

(6) If $(X,\mathbf{S})$ and $(Y,\mathbf{T})$ are measurable spaces, then every measurable set in $X \times Y$ is contained in a measurable rectangle. (Hint: the class of all those sets which may be covered by a measurable rectangle is a σ-ring.)

§ 34. SECTIONS

Let $(X,\mathbf{S})$ and $(Y,\mathbf{T})$ be measurable spaces and let $(X \times Y, \mathbf{S} \times \mathbf{T})$ be their Cartesian product. If E is any subset of $X \times Y$ and x is any point of X, we shall call the set $E_x = \{y: (x,y) \in E\}$ a **section** of E, or, more precisely, the section **determined** by x. At times when it is important to call attention not so much to the particular point which determines the section as merely to the fact that the section is determined by some point of the space X (and is therefore a subset of Y), we shall use the phrase X-**section.** The main point is to distinguish such a section from a Y-**section** determined by a point y in Y; the latter is defined, of course, as the set $E^y = \{x: (x,y) \in E\}$. We emphasize that a section of a set in a product space is *not* a set in that product space but a subset of one of the component spaces.

If f is any function defined on a subset E of the product space $X \times Y$ and x is any point of X, we shall call the function f_x, defined on the section E_x by

$$f_x(y) = f(x,y),$$

a **section** of f, or, more precisely an X-**section** of f, or, still more precisely, the section **determined** by x. The concept of a Y-**section** of f, determined by a point y in Y is defined similarly by $f^y(x) = f(x,y)$.

Theorem A. *Every section of a measurable set is a measurable set.*

Proof. Let **E** be the class of all those subsets of $X \times Y$ which have the property that each of their sections is measurable. If $E = A \times B$ is a measurable rectangle, then every section of E is either empty or else equal to one of the sides, (A or B according as the section is a Y–section or an X–section), and therefore $E \varepsilon \mathbf{E}$. Since it is easy to verify that $\mathbf{E}$ is a σ–ring, it follows that $\mathbf{S} \times \mathbf{T} \subset \mathbf{E}$. ▌

Theorem B. *Every section of a measurable function is a measurable function.*

Proof. If f is a measurable function on $X \times Y$, if x is a point of X, and if M is any Borel set on the real line, then the measurability of $N(f_x) \cap f_x^{-1}(M)$ follows from Theorem A and the relations

$$f_x^{-1}(M) = \{y : f_x(y) \varepsilon M\} = \{y : f(x,y) \varepsilon M\} =$$
$$= \{y : (x,y) \varepsilon f^{-1}(M)\} = (f^{-1}(M))_x.$$

(Observe that $N(f_x) = (N(f))_x$.) The proof of the measurability of an arbitrary Y–section of f is similar. ▌

(1) If χ is the characteristic function of a subset E of $X \times Y$, then χ_x and χ^y are the characteristic functions of E_x and E^y respectively. If, in particular, χ is the characteristic function of a rectangle $A \times B$, then

$$\chi(x,y) = \chi_A(x)\chi_B(y).$$

Every section of a simple function is a simple function.

(2) Let $X = Y =$ any uncountable set, and $\mathbf{S} = \mathbf{T} =$ the class of all countable subsets. If $D = \{(x,y): x = y\}$ is the "diagonal" in $X \times Y$, then every section of D is measurable but D is not; in other words the converse of Theorem A is not true.

(3) If an extended real valued function f defined on the Cartesian product of two measurable spaces X and Y has the property that, for every Borel set M on the real line, $f^{-1}(M)$ intersects every measurable set in a measurable set, then every section of f also has that property. Does this assertion remain valid if the definition of measurable space is altered by omitting from it the requirement that the space be the union of its measurable sets? What are the implication relations between this property and measurability?

(4) A non empty rectangle is a measurable set if and only if it is a measurable rectangle. (Hint: if $A \times B$ is measurable, then every section of $A \times B$ is measurable.)

(5) Let $(X,\mathbf{S})$ be a measurable space such that $X \varepsilon \mathbf{S}$ (i.e. such that $\mathbf{S}$ is a σ–algebra); let Y be the real line, and let $\mathbf{T}$ be the class of all Borel sets. If f is a real valued, non negative function on X, then the **upper ordinate set** of f is defined to be the subset

$$V^*(f) = \{(x,y): x \varepsilon X, \quad 0 \leq y \leq f(x)\}$$

of $X \times Y$, and the **lower ordinate set** of f is

$$V_*(f) = \{(x,y): x \varepsilon X, \quad 0 \leq y < f(x)\}.$$

(Observe that, for instance, the lower ordinate set of the function identically equal to zero is empty.) The following considerations are at the basis of an alternative treatment of measurable functions.

(5a) If f is a non negative simple function, then $V^*(f)$ and $V_*(f)$ are measurable. (Hint: each of the sets is the union of a finite number of measurable rectangles.)

(5b) If f and g are non negative functions such that $f(x) \leqq g(x)$ for all x, then $V^*(f) \subset V^*(g)$ and $V_*(f) \subset V_*(g)$.

(5c) If $\{f_n\}$ is an increasing sequence of non negative functions converging at every point to f, then $\{V_*(f_n)\}$ is an increasing sequence of sets whose union is $V_*(f)$; similarly if $\{f_n\}$ is decreasing to f, then $\{V^*(f_n)\}$ is a decreasing sequence of sets whose intersection is $V^*(f)$.

(5d) If f is a non negative measurable function, then $V^*(f)$ and $V_*(f)$ are measurable sets. (Hint: if f is bounded, then there exist sequences $\{g_n\}$ and $\{h_n\}$ of simple functions such that

$$0 \leqq g_n \leqq g_{n+1} \leqq f \leqq h_{n+1} \leqq h_n, \quad n = 1, 2, \cdots$$

and such that $\lim_n g_n = \lim_n h_n = f$.)

(5e) If E is any measurable set in $X \times Y$, and if α and β are real numbers such that $\alpha > 0$, then the set $\{(x,y): (x, \alpha y + \beta) \,\varepsilon\, E\}$ is a measurable subset of $X \times Y$. (Hint: the conclusion is true if E is a measurable rectangle, and the class of all sets for which the conclusion is true is a σ-ring.)

(5f) If f is a non negative function such that $V^*(f)$ [or $V_*(f)$] is measurable, then f is measurable. (Hint: it is sufficient, for the proof of the unparenthetical statement, to show that $\{x: f(x) > c\}$ is measurable for every positive real number c. If $E = V^*(f)$, then

$$\bigcup_{n=1}^{\infty} \left\{(x,y): \left(x, \frac{1}{n}y + c\right) \varepsilon\, E, y > 0\right\} = \{(x,y): f(x) > c, y > 0\};$$

the desired result follows from the fact that the sides of a measurable rectangle are measurable.)

(5g) If the **graph** of a (not necessarily non negative) function f is defined as the set $\{(x,y): f(x) = y\}$, then the graph of a measurable function is a measurable set.

§ 35. PRODUCT MEASURES

Continuing our study of Cartesian products, we turn now to the case where the component spaces are not merely measurable spaces but measure spaces.

Theorem A. *If $(X, \mathbf{S}, \mu)$ and $(Y, \mathbf{T}, \nu)$ are σ-finite measure spaces, and if E is any measurable subset of $X \times Y$, then the functions f and g, defined on X and Y respectively by $f(x) = \nu(E_x)$ and $g(y) = \mu(E^y)$, are non negative measurable functions such that $\int f d\mu = \int g d\nu$.*

Proof. If $\mathbf{M}$ is the class of all those sets E for which the conclusion of the theorem is true, then it is easy to see that $\mathbf{M}$ is closed under the formation of countable, disjoint unions. We observe that the σ-finiteness of μ and ν implies that every set in $\mathbf{S} \times \mathbf{T}$ may be covered by a countable disjoint union of measurable rectangles, both sides of each of which have finite measure. If, therefore, we could prove that every measurable subset of every measurable rectangle with sides of finite measure belongs to $\mathbf{M}$, it would follow (as stated) that every measurable set belongs to $\mathbf{M}$. In other words, we have reduced the proof to the case of finite measures; we shall complete the proof (for finite measures) by showing that every measurable rectangle (and therefore every finite, disjoint union of measurable rectangles) belongs to $\mathbf{M}$, and that $\mathbf{M}$ is a monotone class.

If $E = A \times B$ is a non empty measurable rectangle, then $f = \nu(B)\chi_A$ and $g = \mu(A)\chi_B$. It follows that f and g are measurable and that $\int f d\mu = \mu(A) \cdot \nu(B) = \int g d\nu$.

The fact that $\mathbf{M}$ is a monotone class is a consequence of the standard theorems on the integration of sequences of functions, specifically 26.D and 27.B. (The finiteness of the measures μ and ν is used in justifying the application of these results.)

Since the class of all finite, disjoint unions of measurable rectangles is a ring (33.E), and since, by definition, the class of measurable sets is the σ-ring generated by this ring, it follows (6.B) that every measurable set is in $\mathbf{M}$, and the proof of the theorem is complete. ∎

Theorem B. *If* $(X,\mathbf{S},\mu)$ *and* $(Y,\mathbf{T},\nu)$ *are σ-finite measure spaces, then the set function* λ, *defined for every set E in* $\mathbf{S} \times \mathbf{T}$ *by*

$$\lambda(E) = \int \nu(E_x) d\mu(x) = \int \mu(E^y) d\nu(y),$$

is a σ-finite measure with the property that, for every measurable rectangle $A \times B$,

$$\lambda(A \times B) = \mu(A) \cdot \nu(B).$$

The latter condition determines λ *uniquely.*

The measure λ is called the **product** of the given measures μ and ν, in symbols $\lambda = \mu \times \nu$; the measure space $(X \times Y, \mathbf{S} \times \mathbf{T}, \mu \times \nu)$ is the **Cartesian product** of the given measure spaces.

Proof. The fact that λ is a measure is a consequence of the theorem on the integration of monotone sequences (27.B; cf. also 27.2). The σ-finiteness of λ follows from the fact that every measurable subset of $X \times Y$ may be covered by countably many measurable rectangles of finite measure; uniqueness is implied by 13.A. ▮

(1) Let $X = Y$ be the unit interval, and let $\mathbf{S} = \mathbf{T}$ be the class of Borel sets; let $\mu(E)$ be the Lebesgue measure of E, and let $\nu(E)$ be the number of points in E. If $D = \{(x,y): x = y\}$, then D is a measurable subset of $X \times Y$ such that $\int \nu(D_x)d\mu(x) = 1$ and $\int \mu(D^y)d\nu(y) = 0$. In other words, Theorem A is not true if the condition of σ-finiteness is omitted.

(2) The Cartesian product of two σ-finite and complete measure spaces need not be complete. (Hint: let $X = Y$ be the unit interval, let M be a non measurable subset of X, let y be any point of Y, and consider the set $M \times \{y\}$; cf. 34.4.)

(3) Suppose that $(X,\mathbf{S},\mu)$ is a totally σ-finite measure space and that $(Y,\mathbf{T},\nu)$ is the real line with $\mathbf{T}$ = the class of all Borel sets and ν = Lebesgue measure; let λ be the product measure $\mu \times \nu$. We have already seen (34.5) that for any non negative, measurable function, and *a fortiori* for any non negative, integrable function f on X, the ordinate sets $V^*(f)$ and $V_*(f)$ are measurable subsets of $X \times Y$; we now assert that $\lambda(V_*(f)) = \lambda(V^*(f)) = \int f d\mu$. (Hint: in view of the known results on approximation of functions by simple functions and integration of sequences of functions, it is sufficient to establish the equation for simple functions f.) This equation is sometimes used, in an alternative approach to integration theory, as the definition of $\int f d\mu$; it is a precise formulation of the statement that "the integral is the area under the curve."

(4) Under the hypotheses of (3), the graph of a measurable function has measure zero. (Hint: it is sufficient to consider non negative, bounded, measurable functions on totally finite measure spaces, and to these the result of (3) applies.)

(5) If $(X,\mathbf{S},\mu)$ and $(Y,\mathbf{T},\nu)$ are σ-finite measure spaces, and if $\lambda = \mu \times \nu$, then, for every set E in $\mathbf{H}(\mathbf{S} \times \mathbf{T})$, $\lambda^*(E)$ is the infimum of sums of the type $\sum_{n=1}^{\infty} \lambda(E_n)$, where $\{E_n\}$ is a sequence of measurable rectangles covering E. (Hint: cf. 33.3, 10.A, and 8.5.)

§ 36. FUBINI'S THEOREM

In this section we shall study the relations between integrals on a product space and integrals on the component spaces.

Throughout this section we shall assume that

$(X,\mathbf{S},\mu)$ and $(Y,\mathbf{T},\nu)$ are σ–finite measure spaces and λ is the product measure $\mu \times \nu$ on $\mathbf{S} \times \mathbf{T}$.

If a function h on $X \times Y$ is such that its integral is defined (i.e. if, for instance, h is an integrable function or a non negative measurable function), then the integral is denoted by

$$\int h(x,y)d\lambda(x,y) \quad \text{or} \quad \int h(x,y)d(\mu \times \nu)(x,y)$$

and is called the **double integral** of h. If h_x is such that

$$\int h_x(y)d\nu(y) = f(x)$$

is defined, and if it happens that $\int fd\mu$ is also defined, it is customary to write

$$\int fd\mu = \iint h(x,y)d\nu(y)d\mu(x) = \int d\mu(x)\int h(x,y)d\nu(y).$$

The symbols $\iint h(x,y)d\mu(x)d\nu(y)$ and $\int d\nu(y)\int h(x,y)d\mu(x)$ are defined similarly, as the integral (if it exists) of the function g on Y, defined by $g(y) = \int h^y(x)d\mu(x)$. The integrals $\iint hd\mu d\nu$ and $\iint hd\nu d\mu$ are called the **iterated integrals** of h. To indicate the double and iterated integrals of h over a measurable subset E of $X \times Y$, i.e. the integrals of $\chi_E h$, we shall use the symbols

$$\int_E hd\lambda, \quad \iint_E hd\mu d\nu, \quad \text{and} \quad \iint_E hd\nu d\mu.$$

Since X–sections (of sets or functions) are determined by points in X, it makes sense to assert that a proposition is true for almost every X–section, meaning, of course, that the set of those points x for which the proposition is not true is a set of measure zero in X. The phrase "almost every Y–section" is defined similarly; if a proposition is true simultaneously for a.e. X–section and a.e.

Y–section, we shall simply say that it is true for almost every section.

We begin with an elementary but important result.

Theorem A. *A necessary and sufficient condition that a measurable subset E of $X \times Y$ have measure zero is that almost every X–section [or almost every Y–section] have measure zero.*

Proof. By the definition of product measure we have

$$\lambda(E) = \begin{cases} \int \nu(E_x)d\mu(x), \\ \int \mu(E^y)d\nu(y). \end{cases}$$

If $\lambda(E) = 0$, then the integrals on the right are in particular finite and hence (by 25.B) their non negative integrands must vanish a.e. If, on the other hand, either of the integrands vanishes a.e., then $\lambda(E) = 0$. ▮

Theorem B. *If h is a non negative, measurable function on $X \times Y$, then*

$$\int hd(\mu \times \nu) = \iint hd\mu d\nu = \iint hd\nu d\mu.$$

Proof. If h is the characteristic function of a measurable set E, then

$$\int h(x,y)d\nu(y) = \nu(E_x) \quad \text{and} \quad \int h(x,y)d\mu(x) = \mu(E^y),$$

and the desired result follows from 35.B. In the general case we may find an increasing sequence $\{h_n\}$ of non negative simple functions converging to h everywhere, (20.B). Since a simple function is a finite linear combination of characteristic functions, the conclusion of the theorem is valid for every h_n in place of h.

By 27.B, $\lim_n \int h_n d\lambda = \int hd\lambda$. If $f_n(x) = \int h_n(x,y)d\nu(y)$, then it follows from the properties of the sequence $\{h_n\}$ that $\{f_n\}$ is an increasing sequence of non negative measurable functions converging for every x to $f(x) = \int h(x,y)d\nu(y)$; (cf. 27.B). Hence f is measurable (and obviously non negative); one more application of 27.B yields the conclusion that

$$\lim_n \int f_n d\mu = \int f d\mu.$$

This proves the equality of the double integral and one of the iterated integrals; the truth of the other equality follows similarly. ∎

Both Theorems A and B are sometimes referred to as parts of **Fubini's theorem**; the following result is, however, the one most commonly known by that name.

Theorem C. *If h is an integrable function on $X \times Y$, then almost every section of h is integrable. If the functions f and g are defined by $f(x) = \int h(x,y)d\nu(y)$ and $g(y) = \int h(x,y)d\mu(x)$, then f and g are integrable and*

$$\int h d(\mu \times \nu) = \int f d\mu = \int g d\nu.$$

Proof. Since a real valued function is integrable if and only if its positive and negative parts are integrable, it is sufficient to consider only non negative functions h. The asserted identity follows in this case from Theorem B. Since, therefore, the non negative, measurable functions f and g have finite integrals, it follows that they are integrable. Since, finally, this implies that f and g are finite valued almost everywhere, the sections of h have the desired integrability properties, and the proof is complete. ∎

(1) Let X be a set of cardinal number $\aleph_1$, let **S** be the class of all countable sets and their complements, and, for A in **S**, let $\mu(A)$ be 0 or 1 according as A is countable or not. If $(Y,\mathbf{T},\nu) = (X,\mathbf{S},\mu)$, if E is a set in $X \times Y$ which is countable on every vertical line and full on every horizontal line (cf. 31.11), and if h is the characteristic function of E, then h is a non negative function such that

$$\int h(x,y)d\mu(x) = 1 \quad \text{and} \quad \int h(x,y)d\nu(y) = 0.$$

Why is this not a counter example to Theorem B?

(2) If $(X,\mathbf{S},\mu)$ and $(Y,\mathbf{T},\nu)$ are the unit interval with Lebesgue measure, and if E is a subset of $X \times Y$ such that E_x and $X - E^y$ are countable for every x and y (cf. (1)), then E is not measurable.

(3) The following considerations indicate an interesting extension of the results of this section. Let $(X,\mathbf{S},\mu)$ be a totally finite measure space and let $(Y,\mathbf{T})$ be a measurable space such that $Y \varepsilon \mathbf{T}$. Suppose that to almost every x in X there corresponds a finite measure ν_x on $\mathbf{T}$ so that if $\phi(x) = \nu_x(B)$, then, for each measurable subset B of Y, ϕ is a measurable function on X. If $\nu(B) = \int \nu_x(B)d\mu(x)$, if g is a non negative measurable function on Y, and if $f(x) = \int g(y)d\nu_x(y)$, then f is a non negative measurable function on X and $\int f d\mu = \int g d\nu$.

(4) The proof of Fubini's theorem sometimes appears to be slightly more complicated than the one we gave—the complication is caused by completing the measure λ. In other words, the theorems of this section remain true if λ is replaced by $\bar{\lambda}$. (Hint: every function which is measurable $(\overline{\mathbf{S} \times \mathbf{T}})$ is equal a.e. $[\bar{\lambda}]$ to a function which is measurable $(\mathbf{S} \times \mathbf{T})$; (cf. 21.1)).

In (5)–(9) below, we shall assume that the measure spaces $(X,\mathbf{S},\mu)$ and $(Y,\mathbf{T},\nu)$ are totally finite. It is easy to verify that the results obtained may be extended to totally σ-finite measure spaces, and, therefore, to each measurable set in the product of two σ-finite measure spaces.

(5) If E and F are measurable subsets of $X \times Y$ such that $\nu(E_x) = \nu(F_x)$ for [almost] every x in X, then $\lambda(E) = \lambda(F)$. (Certain usually not rigorously stated special cases of this assertion are known as **Cavalieri's principle.**)

(6) If f and g are integrable functions on X and Y respectively, then the function h, defined by $h(x,y) = f(x)g(y)$, is an integrable function on $X \times Y$ and

$$\int h d(\mu \times \nu) = \int f d\mu \cdot \int g d\nu.$$

(7) Suppose that $\mu(X) = \nu(Y) = 1$ and that A_0 and B_0 are measurable subsets of X and Y respectively such that $\mu(A_0) = \nu(B_0) = \frac{1}{2}$. Let χ be the characteristic function of $(A_0 \times Y) \Delta (X \times B_0)$ and write $f(x,y) = 2\chi(x,y)$. If, for every measurable set E in $X \times Y$,

$$\bar{\lambda}(E) = \int_E f(x,y)d\lambda(x,y),$$

then $\bar{\lambda}$ is a finite measure on $\mathbf{S} \times \mathbf{T}$ with the property that $\bar{\lambda}(A \times Y) = \mu(A)$ and $\bar{\lambda}(X \times B) = \nu(B)$, whenever $A \varepsilon \mathbf{S}$ and $B \varepsilon \mathbf{T}$. In other words, the product measure λ is not uniquely determined by its values on such special rectangles.

(8) The existence of the product measure is often proved by the following direct but combinatorially somewhat complicated method. The class of all finite, disjoint unions of measurable rectangles is a ring $\mathbf{R}$, (33.E); if

$$\bigcup_{i=1}^{n} (A_i \times B_i) \quad \text{and} \quad \bigcup_{j=1}^{m} (C_j \times D_j)$$

are two representations of the same set in $\mathbf{R}$, then since

$$\bigcup_{i=1}^{n} \bigcup_{j=1}^{m} [(A_i \cap C_j) \times (B_i \cap D_j)]$$

is another representation of the same set, we have

$$\sum_{i=1}^{n} \mu(A_i)\cdot\nu(B_i) = \sum_{j=1}^{m} \mu(C_j)\cdot\nu(D_j).$$

In other words, a set function λ is unambiguously defined on $\mathbf{R}$ by

$$\lambda(\bigcup_{i=1}^{n} (A_i \times B_i)) = \sum_{i=1}^{n} \mu(A_i)\cdot\nu(B_i).$$

It can be shown (essentially by proving a weakened form of Fubini's theorem for the sets of $\mathbf{R}$) that λ is a measure to which the extension theorem (13.A) may be applied.

(9) If A and B are arbitrary (not necessarily measurable) subsets of X and Y respectively, then

$$\lambda^*(A \times B) = \mu^*(A)\cdot\nu^*(B).$$

(Hint: if A^* and B^* are measurable covers of A and B respectively, then the relation $A \times B \subset A^* \times B^*$ implies that $\lambda^*(A \times B) \leq \mu^*(A)\cdot\nu^*(B)$. The reverse inequality may be proved by considering a measurable cover E^* of $A \times B$. Since $E^* \cap (A^* \times B^*)$ is also a measurable cover of $A \times B$, it is permissible to assume that $E^* \subset A^* \times B^*$. It follows from Fubini's theorem that

$$\lambda(E^*) \geq \int_{A^*} \nu(E_x^*)d\mu(x) \geq \mu(A^*)\cdot\nu^*(B).)$$

§ 37. FINITE DIMENSIONAL PRODUCT SPACES

In the preceding sections we have developed the theory of product spaces for two factors; our next task is to investigate how this theory may be extended to any finite number of factors. We suppose that n (>1) is a positive integer, and that $X_1, \cdots, X_n$ are sets; we define the **Cartesian product** of these sets to be the set of all ordered n-tuples of the form $(x_1, \cdots, x_n)$, where $x_i \in X_i$, $i = 1, \cdots, n$. We shall denote this Cartesian product by

$$X_1 \times \cdots \times X_n \quad \text{or} \quad \times_{i=1}^{n} X_i \quad \text{or} \quad \times \{X_i \colon i = 1, \cdots, n\}.$$

If A_i is any subset of X_i, $i = 1, \cdots, n$, the set $\times_{i=1}^{n} A_i$ is a **rectangle.**

It is worth while to ask about Cartesian product, as about every algebraic operation, whether or not it is associative. If, for instance, X_1, X_2, and X_3 are three sets, then, without changing the order in which they are presented, we may form the three new sets $(X_1 \times X_2) \times X_3$, $X_1 \times (X_2 \times X_3)$, and $X_1 \times X_2 \times X_3$. In what sense may we consider these three Cartesian products to be equal? Clearly they do not consist of the same elements; it is incorrect to confuse the ordered pair $((x_1,x_2),x_3)$, whose first element is itself an ordered pair, with the ordered triple (x_1,x_2,x_3). Just as clearly, however, there is a "natural" one to one correspondence between any two of the three Cartesian products under discussion, namely the one which makes the points

$$((x_1,x_2),x_3), \quad (x_1,(x_2,x_3)), \quad \text{and} \quad (x_1,x_2,x_3)$$

correspond to each other. Since it will turn out that this correspondence preserves all those structural properties of product spaces which are of interest to us, we shall wilfully fall into the trap we just pointed out and we shall consistently treat the three products described above as identical. We shall carry this identification procedure to its logical conclusion, and in the case, for instance, of seven factors, we shall consider the element

$$(((x_1,x_2),x_3), ((x_4,x_5), (x_6,x_7)))$$

of the set $((X_1 \times X_2) \times X_3) \times ((X_4 \times X_5) \times (X_6 \times X_7))$ to be the same as the element

$$(x_1,x_2,x_3,x_4,x_5,x_6,x_7)$$

of the set $X_1 \times X_2 \times X_3 \times X_4 \times X_5 \times X_6 \times X_7$.

The identification just described simplifies the language of many proofs. Since, for instance, we may view $X_1 \times \cdots \times X_n$ as a repeated product

$$(\cdots((X_1 \times X_2) \times X_3) \times \cdots) \times X_n$$

of two factors at a time, we may prove the analogs of the theorems of § 33 by mathematical induction on n. Some slight care has to be exercised in the formulation of the results. The correct version of the generalization of 33.D, for example, is the assertion that if

$$E = \times_{i=1}^{n} A_i, \quad F = \times_{i=1}^{n} B_i, \quad \text{and} \quad G = \times_{i=1}^{n} C_i$$

are non empty rectangles, then E is the disjoint union of F and G if and only if there is a j, $1 \leq j \leq n$, such that A_j is the disjoint union of B_j and C_j and such that

$$A_i = B_i = C_i, \quad \text{for} \quad i \neq j.$$

The concept of a section (of a set or a function) also requires a minor modification; an X_j–section, determined by a point x_j in X_j, of a set in $\times_{i=1}^n X_i$ is a subset of

$$\times\{X_i: 1 \leq i \leq n, \quad i \neq j\}.$$

If $(X_i, \mathbf{S}_i)$, $i = 1, \cdots, n$, are measurable spaces, we shall denote by

$$\mathbf{S}_1 \times \cdots \times \mathbf{S}_n, \quad \text{or} \quad \times_{i=1}^n \mathbf{S}_i, \quad \text{or} \quad \times\{\mathbf{S}_i: i = 1, \cdots, n\}$$

the σ–ring generated by the class of all those rectangles $\times_{i=1}^n A_i$ for which $A_i \in \mathbf{S}_i$, $i = 1, \cdots, n$, and we define the Cartesian product of the given measurable spaces as the measurable space $(X_1 \times \cdots \times X_n, \mathbf{S}_1 \times \cdots \times \mathbf{S}_n)$. It follows that every section of a measurable set [or a measurable function] is a measurable set [or a measurable function]. Proceeding by mathematical induction, it is now trivial to define the Cartesian product of σ–finite measure spaces $(X_i, \mathbf{S}_i, \mu_i)$, $i = 1, \cdots, n$; there is one and only one measure μ (denoted by $\mu_1 \times \cdots \times \mu_n$) on $\mathbf{S}_1 \times \cdots \times \mathbf{S}_n$ such that

$$\mu(A_1 \times \cdots \times A_n) = \prod_{i=1}^n \mu_i(A_i)$$

for every measurable rectangle $A_1 \times \cdots \times A_n$. The extension of Fubini's theorem is also immediate, so that the integral of any integrable function in a product space may be evaluated by forming the iterated integral in any order.

It is customary to refer to a product space $X = \times_{i=1}^n X_i$ as **n–dimensional.** This terminology is not meant to define dimension, nor to assert that n–dimensionality is an intrinsic structural property of a space; it serves merely to remind us of the way in which X was built from the components X_i. A measure space might appear as three dimensional in one context and two dimensional in another; if, for instance, $n = 3$, then we may view X as $X_1 \times X_2 \times X_3$ or as $X_0 \times X_3$, (where $X_0 = X_1 \times X_2$).

In (1)–(5) below we shall assume that X_i is the real line, $\mathbf{S}_i$ is the class of all Borel sets, and μ_i is Lebesgue measure, $i = 1, \cdots, n$; we write

$$(X, \mathbf{S}, \mu) = \times_{i=1}^n (X_i, \mathbf{S}_i, \mu_i).$$

(1) Sets of the σ-ring $\mathbf{S}$ are called the **Borel sets** of n-dimensional Euclidean space. The class of all Borel sets coincides with the σ-ring generated by the class of all open sets.

(2) If ϕ is a Borel measurable function on X, and if $f_1, \cdots, f_n$ are real valued, measurable functions on a measurable space $(Y,\mathbf{T})$ for which $Y \varepsilon \mathbf{T}$, then the function $\tilde{f}$, defined by $\tilde{f}(y) = \phi(f_1(y), \cdots, f_n(y))$, is a measurable function on Y; (cf. 19.B).

(3) The completed measure $\bar{\mu}$ is called **n-dimensional Lebesgue measure**; most of the results of §§ 15 and 16 are valid for $\bar{\mu}$. If, in particular, $\mathbf{U}$ and $\mathbf{C}$ are the class of all open sets and the class of all closed sets respectively, then, for every set E in X,

$$\mu^*(E) = \inf \{\mu(U): E \subset U \varepsilon \mathbf{U}\} \quad \text{and} \quad \mu_*(E) = \sup \{\mu(C): E \supset C \varepsilon \mathbf{C}\}.$$

(4) If T is any linear transformation defined by

$$T(x_1, \cdots, x_n) = (y_1, \cdots, y_n), \quad y_i = \textstyle\sum_{j=1}^n a_{ij}x_j + b_i, \quad i = 1, \cdots, n,$$

then, for every set E in X,

$$\mu^*(T(E)) = |\Delta| \cdot \mu^*(E) \quad \text{and} \quad \mu_*(T(E)) = |\Delta| \cdot \mu_*(E),$$

where Δ is the determinant of the matrix (a_{ij}). (Hint: it is sufficient to prove the assertion for measurable rectangles E whose sides are intervals. Treat first the following special cases.

(4a) $y_i = x_i + b_i$, $i = 1, \cdots, n$.

(4b) $y_i = x_i$ if $i \neq j$ and $i \neq k$; $y_j = x_k$ and $y_k = x_j$.

(4c) $y_i = x_i$ if $i \neq j$; $y_j = x_j \pm x_k$, where $k \neq j$.

(4d) $y_i = x_i$ if $i \neq j$; $y_j = cx_j$.

The general case follows from the fact that T may be written as the product of transformations of the types (4a)–(4d).)

(5) The function ϕ_j on X, defined by

$$\phi_j(x_1, \cdots, x_n) = x_j, \quad j = 1, \cdots, n,$$

is measurable.

(6) There is a way of defining n-dimensional Lebesgue measure which does not make use of the general theory of product spaces. To indicate this method, we shall consider the space $X_1 \times \cdots \times X_n$, where $X_i = X =$ the unit interval. For every x in X, let $x = .\alpha_1\alpha_2\alpha_3\cdots$ be a binary expansion of x and write

$$x_i = .\alpha_i\alpha_{n+i}\alpha_{2n+i}\cdots, \quad i = 1, \cdots, n.$$

(For each x which has two binary expansions, select a definite one of them, say for instance the terminating one.) The transformation T from X to $X_1 \times \cdots \times X_n$, defined by $T(x) = (x_1, \cdots, x_n)$, has the property that if a set E in $X_1 \times \cdots \times X_n$ is measurable, then $T^{-1}(E) = \{x: T(x) \varepsilon E\}$ is a measurable subset of X. (For the proof, consider the case in which E is a rectangle whose sides are intervals with binary rational end points.) The equation $(\mu_1 \times \cdots \times \mu_n)(E) = \mu(T^{-1}(E))$ (where μ is Lebesgue measure in X) may be used as the definition of the product measure $\mu_1 \times \cdots \times \mu_n$; this definition is consistent with our earlier one.

(7) By the familiar zig-zag diagonal process, i.e. by writing

$$x_1 = .\alpha_1\alpha_2\alpha_4\alpha_7\cdots,$$

$$x_2 = .\alpha_3\alpha_5\alpha_8\alpha_{12}\cdots,$$

$$x_3 = .\alpha_6\alpha_9\alpha_{13}\alpha_{18}\cdots,$$

$$x_4 = .\alpha_{10}\alpha_{14}\alpha_{19}\alpha_{25}\cdots,$$

$$\cdots$$

the procedure of (6) may be extended to yield a definition of product measure in an "infinite dimensional" analog of Euclidean space.

§ 38. INFINITE DIMENSIONAL PRODUCT SPACES

The first step of an extension of product space theory to infinitely many dimensions suggests itself naturally. If $\{X_i\}$ is a sequence of sets, the Cartesian product

$$X = \times_{i=1}^{\infty} X_i$$

is defined as the set of all sequences of the form $(x_1, x_2, \cdots)$ where $x_i \varepsilon X_i$, $i = 1, 2, \cdots$. If each X_i is a measure space, with a σ-ring $\mathbf{S}_i$ of measurable sets and a measure μ_i, it is not quite clear, however, how the concepts of measurability and measure should be defined in X. In this section we shall show how this may be done, under the assumption that the spaces X_i are totally finite measure spaces such that $\mu_i(X_i) = 1$, $i = 1, 2, \cdots$. We observe that the measure on every totally finite measure space $(X,\mathbf{S},\mu)$, for which $\mu(X) \neq 0$, may be trivially altered (by dividing the measure of every measurable set by $\mu(X)$) so that the measure of the entire space is 1. We shall see, however, that since the number 1 plays a distinguished role in the formation of products (particularly of infinite products), the condition $\mu_i(X_i) = 1$ is not merely an inessential normalization.

Suppose then that, for each $i = 1, 2, \cdots$, X_i is a set, $\mathbf{S}_i$ is a σ-algebra of subsets of X_i, and μ_i is a measure on $\mathbf{S}_i$ such that $\mu_i(X_i) = 1$. In this case we define a **rectangle** as a set of the form $\times_{i=1}^{\infty} A_i$, where $A_i \subset X_i$ for all i and $A_i = X_i$ for all but a finite number of values of i. We define a **measurable rectangle** as a rectangle $\times_{i=1}^{\infty} A_i$ for which each A_i is a measurable subset of X_i; in view of the preceding definition, this condition is a restriction on only a finite number of the A's. A subset of $\times_{i=1}^{\infty} X_i$ will be called **measurable** if it belongs to the σ-ring $\mathbf{S}$ (which is in fact a σ-algebra) generated by the class of all measurable rectangles; we shall write $\mathbf{S} = \times_{i=1}^{\infty} \mathbf{S}_i$.

Suppose that J is any subset of the set I of all positive integers; we shall say that two points

$$x = (x_1, x_2, \cdots) \quad \text{and} \quad y = (y_1, y_2, \cdots)$$

agree on J, in symbols $x \equiv y \ (J)$, if $x_j = y_j$ for every j in J. A set E in X is called a **J-cylinder** if $x \equiv y \ (J)$ implies that x and y belong or do not belong to E simultaneously. In other words, E is a J-cylinder if altering those coordinates of any point whose index is *not* in J cannot remove the point from E, nor insert it into E if it was not already there (cf. 6.5d). If, for instance, $J = \{1, \cdots, n\}$ and A_j is an arbitrary subset of $X_j, j = 1, \cdots, n$, then the rectangle $A_1 \times \cdots \times A_n \times X_{n+1} \times X_{n+2} \times \cdots$ is a J-cylinder.

We shall write

$$X^{(n)} = \times_{i=n+1}^{\infty} X_i, \quad n = 0, 1, 2, \cdots;$$

in view of our identification convention for product spaces, we may write $X = \times_{i=1}^{\infty} X_i = (X_1 \times \cdots \times X_n) \times X^{(n)}$. Since each space $X^{(n)}$ is an infinite dimensional product space such as $X(= X^{(0)})$, the considerations applied (above and in the sequel) to X, may also be applied to $X^{(n)}$. For every point $(x_1, \cdots, x_n)$ in $X_1 \times \cdots \times X_n$ and every set E in X, we shall denote by $E(x_1, \cdots, x_n)$ the section of E (in $X^{(n)}$) determined by $(x_1, \cdots, x_n)$. We observe that every such section of a [measurable] rectangle in X is a [measurable] rectangle in $X^{(n)}$.

Theorem A. *If $J = \{1, \cdots, n\}$ and if a subset E of X is a [measurable] J-cylinder, then $E = A \times X^{(n)}$, where A is a [measurable] subset of $X_1 \times \cdots \times X_n$.*

Proof. Let $(\bar{x}_{n+1}, \bar{x}_{n+2}, \cdots)$ be an arbitrary point of $X^{(n)}$ and let A be the $X^{(n)}$-section (in $X_1 \times \cdots \times X_n$) of E, determined by this point. Since both the sets E and $A \times X^{(n)}$ are J-cylinders, it follows that if a point $(x_1, x_2, \cdots)$ of X belongs to either of them, then so does the point $(x_1, \cdots, x_n, \bar{x}_{n+1}, \bar{x}_{n+2}, \cdots)$. It is clear, however, that if a point of this latter form belongs to either one of the sets E and $A \times X^{(n)}$, then it belongs also to the other. Using once more the fact that both these sets are J-cylinders, and hence that if $(x_1, \cdots, x_n, \bar{x}_{n+1}, \bar{x}_{n+2}, \cdots)$ belongs to either of them, then so does $(x_1, \cdots, x_n, x_{n+1}, x_{n+2}, \cdots)$, we conclude that E and $A \times X^{(n)}$ consist of the same points. The fact that the measurability of E implies that of A follows from 34.A. ▮

If m and n are positive integers, $m < n$, then it may happen that a non empty subset E of X is simultaneously a $\{1, \cdots, m\}$-cylinder and a $\{1, \cdots, n\}$-cylinder. By Theorem A we conclude that

$$E = A \times X^{(m)} \quad \text{and} \quad E = B \times X^{(n)},$$

where $A \subset X_1 \times \cdots \times X_m$ and $B \subset X_1 \times \cdots \times X_n$. Since we may rewrite the first of these relations in the form

$$E = (A \times X_{m+1} \times \cdots \times X_n) \times X^{(n)},$$

it follows from 33.C that $B = A \times X_{m+1} \times \cdots \times X_n$. Consequently if E is measurable, so that both A and B are measurable, then

$$(\mu_1 \times \cdots \times \mu_m)(A) = (\mu_1 \times \cdots \times \mu_n)(B).$$

It follows that a set function μ is unambiguously defined for every measurable $\{1, \cdots, n\}$-cylinder $A \times X^{(n)}$ by the equation

$$\mu(A \times X^{(n)}) = (\mu_1 \times \cdots \times \mu_n)(A).$$

We shall denote the domain of definition of μ, i.e. the class of all measurable sets which are $\{1, \cdots, n\}$-cylinders for some value of n, by $\mathbf{F}$; the sets of $\mathbf{F}$ may be referred to as the finite dimensional subsets of X. It is easy to verify that $\mathbf{F}$ is an algebra, that $\mathbf{S}(\mathbf{F}) = \mathbf{S}$, and that the set function μ on $\mathbf{F}$ is finite, non negative, and finitely additive.

We shall denote the analogs of $\mathbf{F}$ and μ in the space $X^{(n)}$, $n = 1, 2, \cdots$, by $\mathbf{F}^{(n)}$ and $\mu^{(n)}$ respectively. It follows from our results for finite dimensional product spaces that if E belongs to $\mathbf{F}$, then every section of the form $E(x_1, \cdots, x_n)$ belongs to $\mathbf{F}^{(n)}$, and

$$\mu(E) = \int \cdots \int \mu^{(n)}(E(x_1, \cdots, x_n)) d\mu_1(x_1) \cdots d\mu_n(x_n).$$

Theorem B. *If $\{(X_i, \mathbf{S}_i, \mu_i)\}$ is a sequence of totally finite measure spaces with $\mu_i(X_i) = 1$, then there exists a unique measure μ on the σ-algebra $\mathbf{S} = \times_{i=1}^{\infty} \mathbf{S}_i$ with the property that, for every measurable set E of the form $A \times X^{(n)}$,*

$$\mu(E) = (\mu_1 \times \cdots \times \mu_n)(A).$$

The measure μ is called the **product** of the given measures μ_i, $\mu = \times_{i=1}^{\infty} \mu_i$; the measure space

$$(\times_{i=1}^{\infty} X_i, \quad \times_{i=1}^{\infty} \mathbf{S}_i, \quad \times_{i=1}^{\infty} \mu_i)$$

is the **Cartesian product** of the given measure spaces.

Proof. In view of 9.F and 13.A, all we have to prove is that the set function μ on the algebra $\mathbf{F}$ of all finite dimensional measurable sets is continuous from above at 0, i.e. that if $\{E_n\}$ is a decreasing sequence of sets in $\mathbf{F}$ such that $0 < \epsilon \leqq \mu(E_j)$, $j = 1, 2, \cdots$, then $\bigcap_{j=1}^{\infty} E_j \neq 0$.

If $F_j = \left\{x_1 : \mu^{(1)}(E_j(x_1)) > \frac{\epsilon}{2}\right\}$, then it follows from the relation

$$\mu(E_j) = \int \mu^{(1)}(E_j(x_1)) d\mu_1(x_1) =$$

$$= \int_{F_j} \mu^{(1)}(E_j(x_1)) d\mu_1(x_1) + \int_{F_j'} \mu^{(1)}(E_j(x_1)) d\mu_1(x_1)$$

that $\mu(E_j) \leqq \mu_1(F_j) + \frac{\epsilon}{2}$, and therefore that

$$\mu_1(F_j) \geqq \frac{\epsilon}{2}.$$

Since $\{F_j\}$ is a decreasing sequence of measurable subsets of X_1, and since μ_1 (being countably additive) is continuous from above at 0, it follows that there exists at least one point $\bar{x}_1$ in X_1 such that $\mu^{(1)}(E_j(\bar{x}_1)) \geqq \frac{\epsilon}{2}$, $j = 1, 2, \cdots$. Since $\{E_j(\bar{x}_1)\}$ is a decreasing sequence of measurable subsets of $X^{(1)}$, the argument just applied to X, $\{E_j\}$, and ϵ may be repeated for $X^{(1)}$, $\{E_j(\bar{x}_1)\}$, and $\frac{\epsilon}{2}$. We obtain a point $\bar{x}_2$ in X_2 such that $\mu^{(2)}(E_j(\bar{x}_1, \bar{x}_2)) \geqq \frac{\epsilon}{4}$,

$j = 1, 2, \cdots$. Continuing in this manner, we obtain a sequence $\{\bar{x}_1, \bar{x}_2, \cdots\}$ such that $\bar{x}_n \in X_n$, $n = 1, 2, \cdots$, and

$$\mu^{(n)}(E_j(\bar{x}_1, \cdots, \bar{x}_n)) \geqq \frac{\epsilon}{2^n}, \quad j = 1, 2, \cdots.$$

The point $(\bar{x}_1, \bar{x}_2, \cdots)$ belongs to $\bigcap_{j=1}^{\infty} E_j$. To prove this assertion, we consider any particular E_j and we select the positive integer n so that E_j is a $\{1, \cdots, n\}$–cylinder. The fact that $\mu^{(n)}(E_j(\bar{x}_1, \cdots, \bar{x}_n)) > 0$ implies that E_j contains at least one point $(x_1, x_2, \cdots)$ such that $x_i = \bar{x}_i$ for $i = 1, \cdots, n$. The fact that E_j is a $\{1, \cdots, n\}$–cylinder implies then that $(\bar{x}_1, \bar{x}_2, \cdots)$ itself belongs to E_j. ∎

(1) It is not essential for the results of this section that the index set I is the set of positive integers; any countably infinite set may be used for I. (The space $X = \times \{X_i: i \in I\}$ consists, by definition, of all functions x defined on I and such that their value $x(i)$ at each index i is a point of X_i.) The proof of this assertion may be carried out by an enumeration of I, i.e. by establishing an arbitrary but fixed one to one correspondence between the given set I and the set of positive integers. The case, for instance, in which I is the set of *all* integers has many applications.

(2) The generalization of product space theory to uncountably many factors is surprisingly easy. If I is an arbitrary index set, and if, for each i in I, $(X_i, \mathbf{S}_i, \mu_i)$ is a totally finite measure space with $\mu_i(X_i) = 1$, then we may define $X = \times \{X_i: i \in I\}$ as in (1), and the concepts of rectangle, measurable rectangle, and measurable set verbatim as in the countable case. Since the class of all those sets which are J–cylinders for a *countable* subset J of I is a σ–algebra containing all measurable rectangles, it follows that each measurable set E is a J–cylinder for a suitable J. If $\mu(E)$ is defined to be $(\times_{j \in J} \mu_j)(E)$, then μ is a measure on the class of all measurable sets and μ has the product property which justifies its being denoted by $\times_{i \in I} \mu_i$.

(3) It is trivial to combine the theories of finite and infinite dimensional product spaces and thus to produce a theory of product spaces in which a finite number of the factors is not required to be a totally finite measure space but allowed to be σ–finite.

(4) If $X = \times_{i=1}^{\infty} X_i$ is a product space such as the one described in Theorem B, and if, for each i, E_i is a measurable set in X_i, then $E = \times_{i=1}^{\infty} E_i$ is a measurable set in X and

$$\mu(E) = \prod_{i=1}^{\infty} \mu_i(E_i) = \lim_n \prod_{i=1}^{n} \mu_i(E_i).$$

(Hint: if $F_n = E_1 \times \cdots \times E_n \times X^{(n)}$, then $\{F_n\}$ is a decreasing sequence of measurable sets in X such that

$$\bigcap_{n=1}^{\infty} F_n = \times_{i=1}^{\infty} E_i \quad \text{and} \quad \mu(F_n) = \prod_{i=1}^{n} \mu_i(E_i).)$$

(5) It is possible to use the theory of product spaces to give a completely non topological construction of Lebesgue measure on the real line (cf. the proof of 8.C), and hence on n-dimensional Euclidean space (cf. 37.6). To obtain such a construction let $(X_0,\mathbf{S}_0,\mu_0)$ be the measure space whose points are the two real numbers 0 and 1, with $\mathbf{S}_0$ = the class of all subsets of X_0, and $\mu_0(\{0\}) = \mu_0(\{1\}) = \frac{1}{2}$. For each $i = 1, 2, \cdots$, write $(X_i,\mathbf{S}_i,\mu_i) = (X_0,\mathbf{S}_0,\mu_0)$, and form the product space

$$(X,\mathbf{S},\mu) = (\times_{i=1}^{\infty} X_i, \quad \times_{i=1}^{\infty} \mathbf{S}_i, \quad \times_{i=1}^{\infty} \mu_i).$$

(5a) For each point $x = (x_1, x_2, \cdots)$ in X, the set $\{x\}$ is measurable and $\mu(\{x\}) = 0$. (Hint: cf. (4).)

(5b) The set $\bar{E}$ of all points $x = (x_1, x_2, \cdots)$ in X for which $x_i = 1$ for all but a finite number of values of i is measurable and has measure zero. (Hint: $\bar{E}$ is countable.) We shall write $\bar{X} = X - \bar{E}$, and, in what follows, we shall consider the measure space $(\bar{X},\bar{\mathbf{S}},\bar{\mu})$, where $\bar{\mathbf{S}} = \mathbf{S} \cap \bar{X}$ and $\bar{\mu}(E \cap \bar{X}) = \mu(E)$, $E \in \mathbf{S}$.

(5c) If for each $x = (x_1, x_2, \cdots)$ in $\bar{X}$ we write $z(x) = \sum_{i=1}^{\infty} x_i/2^i$, then the function z establishes a one to one correspondence between $\bar{X}$ and the interval $Z = \{z: 0 \leq z < 1\}$. (Hint: consider the binary expansion of each z in Z with the agreement that, if the expansion is not unique, the terminating expansion is selected in preference to the infinite one.)

(5d) If $A = \{z: 0 \leq a \leq z < b \leq 1\}$, and $E = \{x: z(x) \in A\}$, then E is measurable and $\bar{\mu}(E) = b - a$. (Hint: it is sufficient to consider the case in which a and b are binary rational numbers.)

(5e) If A is any Borel set in Z and $E = \{x: z(x) \in A\}$, then E is measurable and $\bar{\mu}(E)$ is equal to the Lebesgue measure of A. (Hint: the set function ν, defined by $\nu(A) = \bar{\mu}(E)$, is a measure which coincides with Lebesgue measure on intervals.)

The considerations of (5a)–(5e) serve to construct Lebesgue measure on the interval Z. Lebesgue measure on the entire real line may be obtained by considering the line as a countable, disjoint union of such intervals. Alternatively we may consider the space I of all integers (with the class of all subsets of I playing the role of the class of measurable sets and the measure of a set defined to be the number of its points), and observe the existence of an obvious one to one correspondence between the real line and the product space $I \times Z$.

(6) A construction similar to the one in (5) may be obtained by considering the space $(X_0,\mathbf{S}_0,\mu_0)$, where X_0 is the set of all positive integers, $\mathbf{S}_0$ is the class of all subsets of X_0, and $\mu_0(E) = \sum_{i \in E} 2^{-i}$. We form as before the product space $X = \times_{i=1}^{\infty} X_i$, whose points this time are sequences of positive integers. For each $x = (x_1, x_2, \cdots)$ in X we write

$$z(x) = \sum_{n=1}^{\infty} 2^{-(x_1+\cdots+x_n)}.$$

By the consideration of binary expansions it may be proved that the conclusions of (5c), (5d), and (5e) are valid for this z.

(7) Suppose that $X_0 = \{x: 0 \leq x_0 < 1\}$ is the semiclosed unit interval; let $\mathbf{S}_0$ be the class of all Borel sets in X_0 and let μ_0 be Lebesgue measure on $\mathbf{S}_0$. We write

$$(X_i,\mathbf{S}_i,\mu_i) = (X_0,\mathbf{S}_0,\mu_0), \quad i = 1, 2, \cdots,$$

and we form the product space $X = \times_{i=1}^{\infty} X_i$. There exists a one to one correspondence between X and X_0 such that every Borel set in X_0 corresponds to a measurable set (i.e. to a set belonging to $\times_{i=1}^{\infty} \mathbf{S}_i$) in X, and such that corresponding sets have equal measures. (Hint: if Y_0 is the two-point space described in (5) and denoted there by X_0, and if $Y_{ij} = Y_0$ for $i = 0, 1, 2, \cdots$ and $j = 1, 2, \cdots$, then $X_i = \times_{j=1}^{\infty} Y_{ij}$, $i = 0, 1, 2, \cdots$. The correspondence is based on the usual correspondence between doubly infinite sequences, i.e. elements of $X = \times_{i=1}^{\infty} X_i = \times_{i=1}^{\infty} \times_{j=1}^{\infty} Y_{ij}$, and simple sequences, i.e. elements of $X_0 = \times_{j=1}^{\infty} Y_{0j}$.)

Chapter VIII

TRANSFORMATIONS AND FUNCTIONS

§ 39. MEASURABLE TRANSFORMATIONS

In every mathematical system it is of interest to investigate the transformations that leave some or all structural properties of the system invariant. While it is not our intention to study in great detail the transformations that occur in measure theory, we shall in this section discuss some of their fundamental properties.

A **transformation** is a function T defined for every point of a set X and taking values in a set Y. The set X is called the **domain** of T; the set of those points of Y which are of the form $T(x)$ for some x in X is the **range** of T. A transformation whose domain is X and whose range is in Y is often described as a transformation from X **into** Y; if the range of T is Y, T is called a transformation from X **onto** Y. For every subset E of X, the **image** of E under T, in symbols $T(E)$, is the range of the transformation T from E into Y; for every subset F of Y, the **inverse image** of F under T, in symbols $T^{-1}(F)$, is defined to be the set of all those points of X whose image is in F; i.e.

$$T^{-1}(F) = \{x: T(x) \varepsilon F\}.$$

A transformation T is **one to one** if $T(x_1) = T(x_2)$ occurs when and only when $x_1 = x_2$. The inverse of a one to one transformation T, denoted by T^{-1}, is the transformation which is defined for every $y = T(x)$ in the range of T by $T^{-1}(y) = x$.

If T is a transformation from X into Y and S is a transformation from Y into Z, the **product** of S and T, in symbols ST, is the transformation from X into Z defined by $(ST)(x) = S(T(x))$.

A transformation T from X into Y assigns in an obvious way a function f on X to every function g on Y; f is defined by $f(x) = g(T(x))$. It is convenient and natural to write $f = gT$.

Theorem A. *If T is a transformation from X into Y, if g is a function on Y, and if M is any subset of the space in which the values of g lie, then*

$$\{x: (gT)(x) \in M\} = T^{-1}(\{y: g(y) \in M\}).$$

Proof. The following statements are mutually equivalent: (a) $x_0 \in \{x: (gT)(x) \in M\}$, (b) $g(T(x_0)) \in M$, (c) if $y_0 = T(x_0)$, then $g(y_0) \in M$, and (d) $T(x_0) \in \{y: g(y) \in M\}$. The equivalence of the first and last ones of these statements is exactly the assertion of the theorem. ▮

If $(X,\mathbf{S})$ and $(Y,\mathbf{T})$ are measurable spaces and if T is a transformation from X into Y, how should the concept of measurability be defined for T? Motivated by the special case in which Y is the real line, we shall say that T is a **measurable transformation** if the inverse image of every measurable set is measurable. We observe that this language is inconsistent with our earlier one concerning measurable functions; because of the special role of the real number 0, a measurable function is not necessarily a measurable transformation. This slight inconsistency is amply repaid by convenience in applications; confusion can always be avoided by use of the proper one of the terms "function" and "transformation." In the important case in which X itself belongs to $\mathbf{S}$ and Y is the real line, the two concepts, measurable transformation and measurable function, coincide.

If T is a measurable transformation from $(X,\mathbf{S})$ into $(Y,\mathbf{T})$, we shall denote by $T^{-1}(\mathbf{T})$ the class of all those subsets of X which have the form $T^{-1}(F)$ for some F in $\mathbf{T}$; it is clear that $T^{-1}(\mathbf{T})$ is a σ-ring contained in $\mathbf{S}$.

Theorem B. *If T is a measurable transformation from $(X,\mathbf{S})$ into $(Y,\mathbf{T})$, and if g is an extended real valued measurable function on Y, then gT is measurable with respect to the σ-ring $T^{-1}(\mathbf{T})$.*

Proof. Theorem A implies that, for every Borel set M on the real line,

$$N(gT) \cap (gT)^{-1}(M) = \{x\colon (gT)(x) \in M - \{0\}\} =$$
$$= T^{-1}(\{y\colon g(y) \in M - \{0\}\}) = T^{-1}(N(g) \cap g^{-1}(M));$$

it follows from the measurability of T that the set on the left belongs to $T^{-1}(\mathbf{T})$. ▮

A measurable transformation T from $(X,\mathbf{S})$ into $(Y,\mathbf{T})$ assigns in an obvious way a set function ν on $\mathbf{T}$ to every set function μ on $\mathbf{S}$; ν is defined for every F in $\mathbf{T}$ by $\nu(F) = \mu(T^{-1}(F))$. It is convenient and natural to write $\nu = \mu T^{-1}$.

Theorem C. *If T is a measurable transformation from a measure space $(X,\mathbf{S},\mu)$ into a measurable space $(Y,\mathbf{T})$, and if g is an extended real valued measurable function on Y, then*

$$\int g\,d(\mu T^{-1}) = \int (gT)\,d\mu,$$

in the sense that if either integral exists, then so does the other and the two are equal.

Proof. It is sufficient to treat non negative functions g. If g is the characteristic function of a measurable set F in Y, then it follows from Theorem A that gT is the characteristic function of $T^{-1}(F)$ and therefore

$$\int g\,d(\mu T^{-1}) = (\mu T^{-1})(F) = \mu(T^{-1}(F)) = \int (gT)\,d\mu.$$

It follows from this relation that the asserted equality is valid whenever g is a simple function. In the general case let $\{g_n\}$ be an increasing sequence of simple functions converging to g; then $\{g_nT\}$ is an increasing sequence of simple functions converging to gT and the desired conclusion follows by taking limits. ▮

If, in the notation of Theorem C, F is a measurable subset of Y, then an application of Theorem C to the function $\chi_F g$ yields the relation

$$\int_F g(y)\,d\mu T^{-1}(y) = \int_{T^{-1}(F)} g(T(x))\,d\mu(x).$$

We observe that either side of this equation may be obtained from the other by the formal substitution $y = T(x)$.

Theorem D. *If T is a measurable transformation from a measure space $(X,\mathbf{S},\mu)$ into a totally σ-finite measure space $(Y,\mathbf{T},\nu)$, such that μT^{-1} is absolutely continuous with respect to ν, then there exists a non negative measurable function ϕ on Y such that*

$$\int g(T(x))d\mu(x) = \int g(y)\phi(y)d\nu(y)$$

for every measurable function g, in the sense that if either integral exists, then so does the other and the two are equal.

The function ϕ plays the role of the Jacobian (or, rather, the absolute value of the Jacobian) in the theory of transformations of multiple integrals.

Proof. Write $\phi = \dfrac{d(\mu T^{-1})}{d\nu}$, (cf. §32), and apply 32.B to the result of Theorem C. ∎

If T is a one to one transformation from a measurable space $(X,\mathbf{S})$ *onto* a measurable space $(Y,\mathbf{T})$, and if both T and T^{-1} are measurable, we shall say that T is **measurability preserving.** A measurability preserving transformation T from a measure space $(X,\mathbf{S},\mu)$ onto a measure space $(Y,\mathbf{T},\nu)$ is **measure preserving** if $\mu T^{-1} = \nu$.

(1) The product of two measurable transformations is measurable.

(2) If T is a measurable transformation from $(X,\mathbf{S})$ into $(Y,\mathbf{T})$, and if a function f on X is measurable with respect to $T^{-1}(\mathbf{T})$, then $f(x_1) = f(x_2)$ whenever $T(x_1) = T(x_2)$. (Hint: if F_1 is a measurable set in Y containing $T(x_1)$, then there exists a measurable set F in Y such that

$$\{x: f(x) = f(x_1)\} \cap T^{-1}(F_1) = T^{-1}(F).$$

The fact that $x_1 \in T^{-1}(F)$ implies that $x_2 \in T^{-1}(F)$.)

(3) If T is a measurable transformation from $(X,\mathbf{S})$ *onto* $(Y,\mathbf{T})$, and if a real valued function f on X is measurable with respect to $T^{-1}(\mathbf{T})$, then there exists a unique measurable function g on Y such that $f = gT$. (Hint: in view of (2), g is unambiguously defined for every $y = T(x)$ by $g(y) = f(x)$. The fact that we have, for every Borel set M on the real line,

$$T^{-1}(\{y: g(y) \in M\}) = \{x: f(x) \in M\},$$

implies, since $T(X) = Y$, that $N(g) \cap \{y: g(y) \in M\} \in T$.) Does this result remain true if T maps X *into* Y?

(4) Suppose that $X = Y =$ the unit interval, $\mathbf{S} =$ the class of all Borel sets, and $\mathbf{T} =$ the class of all countable sets. If the transformation T is defined by $T(x) = x$, then T is a one to one, measurable transformation from X onto Y, but T is not measurability preserving. Is it possible to construct such an example for which $(X,\mathbf{S}) = (Y,\mathbf{T})$?

(5) If T is a measurable transformation from $(X,\mathbf{S})$ into $(Y,\mathbf{T})$, and if μ and ν are two measures on $\mathbf{S}$ such that $\nu \ll \mu$, then $\nu T^{-1} \ll \mu T^{-1}$.

§ 40. MEASURE RINGS

A **Boolean ring** is a ring in the usual algebraic sense, with the property that every element is idempotent. Equivalently, a Boolean ring is a set $\mathbf{R}$ and two algebraic operations (called addition and multiplication) defined for pairs of elements of $\mathbf{R}$, subject to the following restrictions. (a) Both addition and multiplication are commutative and associative, and multiplication is distributive with respect to addition. (b) There exists in $\mathbf{R}$ a unique element (denoted by 0) such that the result of adding 0 to any element E is E. (c) The result of adding any element to itself is 0. (d) The result of multiplying any element E by itself is E.

A typical example of a Boolean ring is a ring of subsets of a set X with $E \Delta F$ and $E \cap F$ playing the roles of the sum and the product of E and F, respectively. Since our introduction of Boolean rings is motivated exclusively by rings of sets, we shall adopt the mnemonic device of always denoting addition and multiplication in Boolean rings by Δ and $\cap$.

Most of the concepts we introduced and results we established for rings of sets carry over without change to Boolean rings in general. If, in particular, the formation of unions and differences is defined by

$$E \cup F = (E \Delta F) \Delta (E \cap F)$$

and

$$E - F = E \Delta (E \cap F),$$

then these operations are subject to the same formal identities as the corresponding operations on sets. A similar statement is true about the inclusion relations $E \subset F$ and $E \supset F$, defined by

$$E \cap F = E \quad \text{and} \quad E \cap F = F$$

respectively.

We recall that the union of any class of sets is the smallest set containing them all and their intersection is the largest set contained in them; similar statements are true about unions and intersections (as far as they can be formed) in every Boolean ring. If, for instance, E and F are elements of a Boolean ring $\mathbf{R}$, then $E \cup F$ is indeed the smallest element containing both E and F; i.e. $E \subset E \cup F$, $F \subset E \cup F$, and, if G is an element of $\mathbf{R}$ for which $E \subset G$ and $F \subset G$, then $E \cup F \subset G$. For an infinite set of elements in a Boolean ring, however, there need not be any element that contains them all, and, even if there is one, there need not be a smallest one. A **Boolean σ-ring** is a Boolean ring $\mathbf{S}$ with the property that every countable set of elements in $\mathbf{S}$ has a union; it is easy to verify that every countable set of elements in a Boolean σ-ring has an intersection. A typical example of a Boolean σ-ring is, of course, a σ-ring of subsets of a set X.

A **Boolean algebra** is a Boolean ring $\mathbf{R}$ in which there exists an element different from 0 (which, for obvious reasons, we shall denote by X), with the property that $E \subset X$ for every E in $\mathbf{R}$. A **Boolean σ-algebra** is a Boolean σ-ring which is a Boolean algebra.

The definitions of the concepts of additivity, measure, σ-finiteness, etc. for functions defined on a Boolean ring are the same as the corresponding definitions for set functions on a ring of sets. A measure μ on a Boolean ring is **positive** if it vanishes for the zero element only.

A measure μ on a σ-ring $\mathbf{S}$ of subsets of a set X is usually not positive. There are, however, several well known procedures which have the effect of making a positive measure out of μ. One such procedure is to consider the class $\mathbf{N}$ of measurable sets of measure zero and then, after observing that $\mathbf{N}$ is an ideal in the ring $\mathbf{S}$ (these words being used in their customary algebraic sense) to replace $\mathbf{S}$ by the quotient ring $\mathbf{S}/\mathbf{N}$. Another (equivalent) procedure is to write $E \sim F$ whenever $\mu(E \Delta F) = 0$ and then, after observing that the relation "$\sim$" is reflexive, symmetric, and transitive, to replace $\mathbf{S}$ by the set of all equivalence classes with respect to the relation $\sim$.

The most usual and most convenient procedure in measure theory (which is the one we shall adopt) is still another one.

We shall not replace **S** by another system—the elements of the Boolean σ-ring that we propose to consider are to be measurable sets. We shall, however, redefine the concept of equality; if two sets E and F in **S** are such that $\mu(E \Delta F) = 0$, then we shall consider them equal and we shall write $E = F\,[\mu]$. If $E_n = F_n\,[\mu]$, $n = 1, 2, \cdots$, then

$$E_1 - F_1 = E_2 - F_2 \quad \text{and} \quad \bigcup_{n=1}^{\infty} E_n = \bigcup_{n=1}^{\infty} F_n\,[\mu],$$

so that even with the altered concept of equality, **S** is a Boolean σ-ring with respect to the familiar set operations. If $E = F\,[\mu]$, then $\mu(E) = \mu(F)$, so that even with the altered concept of equality, the measure μ is unambiguously defined on **S**. Since the statements $\mu(E) = 0$ and $E = 0\,[\mu]$ are obviously equivalent, we see that, after the alteration of the concept of equality, μ becomes a positive measure.

If $(X,\mathbf{S},\mu)$ is a measure space, we shall use the symbol $\mathbf{S}(\mu)$ to denote the σ-ring **S** with equality interpreted modulo μ, as described above.

A **measure ring** $(\mathbf{S},\mu)$ is a Boolean σ-ring **S** and a positive measure μ on **S**. The preceding considerations show that if $(X,\mathbf{S},\mu)$ is a measure space, then $(\mathbf{S}(\mu),\mu)$ is a measure ring; we shall call it the measure ring **associated** with X or simply the measure ring of X. A **measure algebra** is a Boolean algebra which is at the same time a measure ring. The phrases [totally] finite and σ-finite are used for measure rings and measure algebras in the same way as for measure spaces.

An **isomorphism** between two measure rings $(\mathbf{S},\mu)$ and $(\mathbf{T},\nu)$ is a one to one transformation T from **S** onto **T** such that

$$T(E - F) = T(E) - T(F), \quad T\left(\bigcup_{n=1}^{\infty} E_n\right) = \bigcup_{n=1}^{\infty} T(E_n),$$

and

$$\mu(E) = \nu(T(E)),$$

whenever E, F, and E_n are elements of **S**, $n = 1, 2, \cdots$. Two measure rings are **isomorphic** if there exists an isomorphism between them. Two measure spaces $(X,\mathbf{S},\mu)$ and $(Y,\mathbf{T},\nu)$ are isomorphic if their associated measure rings $(\mathbf{S}(\mu),\mu)$ and $(\mathbf{T}(\nu),\nu)$ are isomorphic.

An **atom** of a measure ring $(\mathbf{S},\mu)$ [or of a measure μ] is an element E different from 0 such that if $F \subset E$, then either $F = 0$ or $F = E$; a measure ring with no atoms is **non atomic.** If $(X,\mathbf{S},\mu)$ is a measure space whose measure ring is non atomic, then both the measure space X and the measure μ are called non atomic.

If $(\mathbf{S},\mu)$ is a measure ring, we shall denote by $\mathcal{S}$ [or $\mathcal{S}(\mu)$] the set of elements of finite measure in $\mathbf{S}$ and, for any two elements E and F in $\mathcal{S}$, we shall write

$$\rho(E,F) = \mu(E \,\Delta\, F).$$

It is easy to verify that the function ρ is a metric for $\mathcal{S}$; we shall call $\mathcal{S}$ the metric space **associated** with $(\mathbf{S},\mu)$, or, simply, the metric space of $(\mathbf{S},\mu)$. We shall also use the symbol $\mathcal{S}(\mu)$ for the metric space associated with the measure ring $(\mathbf{S}(\mu),\mu)$ of a measure space $(X,\mathbf{S},\mu)$. A measure ring or a measure space is called **separable** if the associated metric space is separable.

Theorem A. *If $\mathcal{S}$ is the metric space of a measure ring $(\mathbf{S},\mu)$, and if*

$$f(E,F) = E \cup F \quad \textit{and} \quad g(E,F) = E \cap F,$$

then f, g, and also μ, are all uniformly continuous functions of their arguments.

Proof. The desired results are immediate consequences of the relations

$$\left.\begin{matrix}\mu((E_1 \cup F_1) - (E_2 \cup F_2)) + \mu((E_2 \cup F_2) - (E_1 \cup F_1)) \\ \mu((E_1 \cap F_1) - (E_2 \cap F_2)) + \mu((E_2 \cap F_2) - (E_1 \cap F_1))\end{matrix}\right\} \leqq$$

$$\leqq \mu(E_1 - E_2) + \mu(F_1 - F_2) + \mu(E_2 - E_1) + \mu(F_2 - F_1)$$

and

$$|\,\mu(E) - \mu(F)\,| = |\,\mu(E - F) - \mu(F - E)\,| \leqq$$

$$\leqq \mu(E - F) + \mu(F - E). \quad \blacksquare$$

Theorem B. *If $(X,\mathbf{S},\mu)$ is a σ-finite measure space such that the σ-ring $\mathbf{S}$ has a countable set of generators, then the metric space $\mathcal{S}(\mu)$ of measurable sets of finite measure is separable.*

Proof. Let $\{E_n\}$ be a sequence of sets in $\mathbf{S}$ such that $\mathbf{S} = \mathbf{S}(\{E_n\})$. Because of the σ-finiteness of μ, there is no loss of generality in assuming that $\mu(E_n) < \infty$ for every $n = 1, 2, \cdots$. Since (5.C) the ring generated by $\{E_n\}$ is also countable, we may assume that the class $\{E_n: n = 1, 2, \cdots\}$ is a ring. It follows from 13.D that, for every E in $\mathcal{S}(\mu)$ and for every positive number ϵ, there exists a positive integer n such that $\rho(E,E_n) < \epsilon$. Since this means that a countable set is dense in $\mathcal{S}(\mu)$, the proof of the theorem is complete. ▮

(1) The metric space $\mathcal{S}$ of a measure space $(X,\mathbf{S},\mu)$ is complete. (Hint: if $\{E_n\}$ is a fundamental sequence in $\mathcal{S}$, and if χ_n is the characteristic function of E_n, $n = 1, 2, \cdots$, then $\{\chi_n\}$ is fundamental in measure and therefore 22.E may be applied.)

(2) Is the metric space of a measure ring complete?

(3) There is a concept of completeness for Boolean rings which is related to but not identical with the concept of the same name for metric spaces. A Boolean ring $\mathbf{R}$ is **complete** if every subset $\mathbf{E}$ of $\mathbf{R}$ has a union. Clearly every complete Boolean ring is a Boolean σ-algebra; in the converse direction it is true that every totally finite measure algebra is complete. (Hint: let $\tilde{\mathbf{E}}$ be the set of all finite unions of elements of $\mathbf{E}$. Write $\alpha = \sup\{\mu(E): E \varepsilon \tilde{\mathbf{E}}\}$, find a sequence $\{E_n\}$ of elements of $\tilde{\mathbf{E}}$ such that $\lim_n \mu(E_n) = \alpha$, and set $E = \bigcup_{n=1}^{\infty} E_n$.)

(4) The result of (3) remains true for totally σ-finite measure algebras.

(5) If ρ is the metric on the metric space $\mathcal{S}$ of a measure ring $(\mathbf{S},\mu)$, then ρ is translation invariant in the sense that $\rho(E \,\Delta\, G, F \,\Delta\, G) = \rho(E,F)$ whenever E, F, and G are in $\mathcal{S}$.

(6) If a one to one transformation T from a measure ring $(\mathbf{S},\mu)$ onto a measure ring $(\mathbf{T},\nu)$ is such that $T(E - F) = T(E) - T(F)$, $T(E \cup F) = T(E) \cup T(F)$, and $\mu(E) = \nu(T(E))$, whenever E and F are in $\mathbf{S}$, then T is an isomorphism.

(7) If a one to one transformation T from a measure ring $(\mathbf{S},\mu)$ onto a measure ring $(\mathbf{T},\nu)$ is such that $\mu(E) = \nu(T(E))$ and $E \subset F$ if and only if $T(E) \subset T(F)$, then T is an isomorphism.

(8) A metric space $\mathcal{S}$ with metric ρ is **convex** if, for any two distinct elements E and F in $\mathcal{S}$, there exists an element G, different from both E and F and such that

$$\rho(E,F) = \rho(E,G) + \rho(G,F).$$

The metric space of a σ-finite measure ring is convex if and only if the measure ring is non atomic.

(9) An isomorphism between two measure rings is an isometry between their metric spaces.

(10) A totally σ-finite measure ring has (at most) countably many atoms.

(11) If $\mathcal{S}$ is the metric space of a measure space $(X,\mathbf{S},\mu)$ and if ν is a finite measure on $\mathbf{S}$ such that $\nu \ll \mu$, then the function ν is unambiguously defined and continuous on $\mathcal{S}$.

(12) If $(X,\mathbf{S},\mu)$ is a σ-finite measure space and $\{\nu_n\}$ is a sequence of finite signed measures on $\mathbf{S}$ such that each ν_n is absolutely continuous with respect to

μ and such that $\lim_n \nu_n(E)$ exists and is finite for every E in $\mathbf{S}$, then the set functions ν_n are uniformly absolutely continuous with respect to μ. (Hint: let $\mathcal{S}$ be the metric space of $(X,\mathbf{S},\mu)$ and write, for each fixed positive number ϵ,

$$\mathcal{E}_k = \bigcap_{n=k}^{\infty} \bigcap_{m=k}^{\infty} \left\{E: E \,\varepsilon\, \mathcal{S}, \, | \nu_n(E) - \nu_m(E) | \leqq \frac{\epsilon}{3}\right\}.$$

Since, by (11), each $\mathcal{E}_k$ is closed, and since, by (1), $\mathcal{S}$ is a complete metric space, the Baire category theorem implies that there exists a positive integer k_0, a positive number r_0, and a set E_0 in $\mathcal{S}$ such that $\{E: \rho(E,E_0) < r_0\} \subset \mathcal{E}_{k_0}$. Let δ be a positive number such that $\delta < r_0$ and such that $| \nu_n(E) | < \frac{\epsilon}{3}$ whenever $\mu(E) < \delta$ and $n = 1, \cdots, k_0$. Observe that if $\mu(E) < \delta$, then

$$\rho(E_0 - E, E_0) < r_0 \quad \text{and} \quad \rho(E_0 \cup E, E_0) < r_0,$$

and

$$| \nu_n(E) | \leqq$$

$$\leqq | \nu_{k_0}(E) | + | \nu_n(E_0 \cup E) - \nu_{k_0}(E_0 \cup E) | + | \nu_n(E_0 - E) - \nu_{k_0}(E_0 - E) |.)$$

(13) If, in the notation of (12), $\nu(E) = \lim_n \nu_n(E)$, then ν is a finite signed measure and $\nu \ll \mu$.

(14) If $\{\nu_n\}$ is a sequence of finite signed measures such that $\lim_n \nu_n(E) = \nu(E)$ exists and is finite for every measurable set E, then $\nu(E)$ is a signed measure. (Hint: if $| \nu_n(E) | \leqq c_n$, $n = 1, 2, \cdots$, write $\mu(E) = \sum_{n=1}^{\infty} \frac{1}{2^n c_n} | \nu_n |(E)$ and apply (13).)

(15a) Every Boolean ring $\mathbf{R}$ is isomorphic (in the customary algebraic sense of that word) to a ring of subsets of some set X. (Hint: consider the Boolean algebra $\mathbf{R}_0$ of two elements 0 and 1, and let X be the set of all homomorphisms of $\mathbf{R}$ into $\mathbf{R}_0$. If, for every E in $\mathbf{R}$,

$$T(E) = \{x: x \,\varepsilon\, X, x(E) = 1\},$$

then T is a homomorphism from $\mathbf{R}$ into the algebra of all subsets of X; all that remains to be proved is that if $E \,\varepsilon\, \mathbf{R}$ and $E \neq 0$, then there exists an x in X for which $x(E) = 1$. If $\mathbf{R}$ is finite, this result is easy. In the general case let X^* be the set of all functions from $\mathbf{R}$ into $\mathbf{R}_0$; in the customary product topology X^* is a compact Hausdorff space. If $\tilde{\mathbf{R}}$ is any finite subring of $\mathbf{R}$ such that $E \,\varepsilon\, \tilde{\mathbf{R}}$, and if $X^*(\tilde{\mathbf{R}})$ is the set of all those functions x^* in X^* which are homomorphisms on $\tilde{\mathbf{R}}$ and for which $x^*(E) = 1$, then the relation

$$\bigcap_{i=1}^{n} X^*(\tilde{\mathbf{R}}_i) \supset X^*(\tilde{\mathbf{R}})$$

(where $\tilde{\mathbf{R}}$ is the ring generated by $\tilde{\mathbf{R}}_1, \cdots, \tilde{\mathbf{R}}_n$) shows that the class $\{X^*(\tilde{\mathbf{R}})\}$ of sets has the finite intersection property.) This result is known as **Stone's theorem.**

(15b) The proof of Stone's theorem, outlined above, shows that $\mathbf{R}$ is isomorphic to a ring of open-closed sets in a compact Hausdorff space. If $\mathbf{R}$ is a Boolean algebra, then $\mathbf{R}$ is isomorphic to the ring of all open-closed sets in a compact Hausdorff space. (Hint: changing the notation of (15a) slightly, let X be the set of all those homomorphisms of **R** into $\mathbf{R}_0$ which map the maximal element of **R** on 1. Then the image of **R** under T contains a base for the topology of X. If a class of open-closed subsets of a compact Hausdorff space is a base and is closed under the formation of finite unions, then it contains every open-closed set.)

(15c) Every Boolean σ-algebra **S** is isomorphic to a σ-algebra of subsets of some set X modulo a σ-ideal. (Hint: map **S** by an algebraic isomorphism T on the algebra of all open-closed sets in a compact Hausdorff space X; let $\mathbf{S}_0$ be the σ-ring generated by the class of all open-closed subsets of X and let $\mathbf{N}_0$ be the class of all sets of the first category in $\mathbf{S}_0$. If $\{E_n\}$ is a sequence of open-closed sets, write $E = T(\bigcup_{n=1}^{\infty} T^{-1}(E_n))$; it follows that $E - \bigcup_{n=1}^{\infty} E_n$ is nowhere dense. In other words, the class of all open-closed sets is closed, modulo $\mathbf{N}_0$, under the formation of countable unions. The only essential thing that is still lacking is the fact, which ensures that T remains an isomorphism even after reduction modulo $\mathbf{N}_0$, that no non empty open-closed set belongs to $\mathbf{N}_0$; this result is, however, a special case of the Baire category theorem, which is just as valid for locally compact spaces as for complete metric spaces.)

§ 41. THE ISOMORPHISM THEOREM

The purpose of this section is to prove that the concept of a measure ring is not as general as it might appear. We shall show, in fact, that every measure ring, subject to certain not too restrictive conditions, is the measure ring of a measure space. Of the many theorems of this type we shall discuss only a rather special one, which we selected because it is important both historically and in current applications.

In what follows we restrict our attention to totally finite measure algebras. If $(\mathbf{S},\mu)$ is a totally finite measure algebra, then, unless we explicitly say otherwise, the symbol X will denote the maximal element of **S**; the algebra **S** and the measure μ are called **normalized** if $\mu(X) = 1$. A **partition** of an element E of **S** is a finite set **P** of disjoint elements of **S** whose union is E. The **norm** of a partition $\mathbf{P} = \{E_1, \cdots, E_k\}$, denoted by $|\mathbf{P}|$, is the maximum of the numbers $\mu(E_1), \cdots, \mu(E_k)$. If $\mathbf{P} = \{E_1, \cdots, E_k\}$ is a partition of E and if F is any element of **S** contained in E, we shall write $\mathbf{P} \cap F$ for the partition $\{E_1 \cap F, \cdots, E_k \cap F\}$ of F.

If $\mathbf{P}_1$ and $\mathbf{P}_2$ are partitions, we shall write $\mathbf{P}_1 \leqq \mathbf{P}_2$ if each element of $\mathbf{P}_1$ is contained in some element of $\mathbf{P}_2$; a sequence $\{\mathbf{P}_n\}$ of partitions is **decreasing** if $\mathbf{P}_{n+1} \leqq \mathbf{P}_n$ for $n = 1, 2, \cdots$. A sequence $\{\mathbf{P}_n\}$ of partitions is **dense** if to every element E of **S** and to every positive number ϵ there corresponds a positive integer n and an element E_0 of $\mathbf{S}$ which is equal to a union of elements of $\mathbf{P}_n$ and is such that $\rho(E,E_0) = \mu(E \Delta E_0) < \epsilon$.

Theorem A. *If* $(\mathbf{S},\mu)$ *is a totally finite, non atomic measure algebra, and if* $\{\mathbf{P}_n\}$ *is a dense, decreasing sequence of partitions of* X, *then* $\lim_n |\mathbf{P}_n| = 0$.

Proof. Since $\{|\mathbf{P}_n|\}$ is a decreasing sequence of positive numbers, it has a limit; we shall derive a contradiction from the assumption that this limit is a positive number δ.

If $\mathbf{P}_1 = \{E_1, \cdots, E_k\}$, then at least one of the elements E_i must be such that

$$|\mathbf{P}_n \cap E_i| \geqq \delta \quad \text{for} \quad n = 1, 2, \cdots.$$

Let F_1 be such an element and consider the sequence $\{\mathbf{P}_n \cap F_1\}$ of partitions of F_1. By a repetition of the argument just used we may find an element F_2 of the partition $\mathbf{P}_2$ such that $F_2 \subset F_1$ and

$$|\mathbf{P}_n \cap F_2| \geqq \delta \quad \text{for} \quad n = 1, 2, \cdots,$$

and we may proceed so on ad infinitum.

If $F = \bigcap_{n=1}^{\infty} F_n$, then $\mu(F) \geqq \delta > 0$, and therefore, since F is not an atom, there exists an element F_0 such that $F_0 \subset F$ and $0 < \mu(F_0) < \mu(F)$. We observe that the element F_0 is either contained in or disjoint from every element of each of the partitions $\mathbf{P}_n$, $n = 1, 2, \cdots$. It follows that if ϵ is a number smaller than either of the numbers $\mu(F_0)$ and $\mu(F) - \mu(F_0)$, then no element of $\mathbf{S}$ which is a union of such partition elements can have a distance less than ϵ from F_0. Since this contradicts the density of $\{\mathbf{P}_n\}$, the proof of the theorem is complete. ∎

Theorem B. *If* Y *is the unit interval,* $\mathbf{T}$ *is the class of all Borel subsets of* Y, *and* ν *is Lebesgue measure on* $\mathbf{T}$, *and if* $\{\mathbf{Q}_n\}$ *is a sequence of partitions into intervals of the maximal element* Y *of the measure algebra* $(\mathbf{T},\nu)$, *such that* $\lim_n |\mathbf{Q}_n| = 0$, *then* $\{\mathbf{Q}_n\}$ *is dense.*

Proof. To every positive number ϵ there corresponds a positive integer n such that $|\mathbf{Q}_n| < \frac{\epsilon}{2}$. If E is any subinterval of Y, let E_1 be the uniquely determined interval of the partition $\mathbf{Q}_n$ with the property that the left end point of E is contained in E_1. If E_1 does not contain the right end point of E, let E_2 be the interval of $\mathbf{Q}_n$ which is adjacent to E_1 on the right, and proceed so on a finite number of times until the process terminates with an interval E_k of $\mathbf{Q}_n$ which does contain the right end point of E. The union of the intervals $E_1, \cdots, E_k$ is at a distance less than ϵ from E; this proves that every subinterval of Y may be approximated by unions of elements of $\{\mathbf{Q}_n\}$. Since the class of all finite unions of intervals is dense, the proof of the theorem is complete. ∎

Theorem C. *Every separable, non atomic, normalized measure algebra* $(\mathbf{S},\mu)$ *is isomorphic to the measure algebra* $(\mathbf{T},\nu)$ *of the unit interval.*

Proof. Let $\{E_n\}$ be a dense sequence in the metric space $\mathcal{S}(\mu)$ of $(\mathbf{S},\mu)$. For each $n = 1, 2, \cdots$, the set of elements of the form $\bigcap_{i=1}^n A_i$, where, for each $i = 1, \cdots, n$, A_i is either E_i or $X - E_i$, is a partition $\mathbf{P}_n$ of X. It is clear that the sequence $\{\mathbf{P}_n\}$ of partitions is decreasing; the density of $\{E_n\}$ implies that the sequence $\{\mathbf{P}_n\}$ of partitions is dense. It follows from Theorem A that $\lim_n |\mathbf{P}_n| = 0$.

To each element E of the partition $\mathbf{P}_1$ we may make correspond a subinterval $T(E)$ of Y so that $\mu(E) = \nu(T(E))$ and so that these intervals constitute a partition of Y. Separately within each of these intervals we imitate similarly the behavior of $\mathbf{P}_2$, and we proceed so on by induction. We obtain in this manner a sequence $\{\mathbf{Q}_n\}$ of partitions of Y into intervals; the fact that the transformation T, from partition elements of $\{\mathbf{P}_n\}$ into intervals, is measure preserving implies that $\lim_n |\mathbf{Q}_n| = 0$, and therefore, by Theorem B, that $\{\mathbf{Q}_n\}$ is dense.

If we define T not only for partition elements occurring in $\{\mathbf{P}_n\}$ but also for finite unions of such elements by assigning to each such finite union the corresponding finite union of partition elements of $\{\mathbf{Q}_n\}$, then the transformation T is an isometry from a dense subset of the metric space $\mathcal{S}(\mu)$ onto a dense subset of $\mathfrak{I}(\nu)$. It follows that there is a unique isometric transformation $\bar{T}$ from $\mathcal{S}(\mu)$ onto $\mathfrak{I}(\nu)$ which coincides with T everywhere that the latter is defined. Since T preserves unions and differences, and since these operations are uniformly continuous functions of their arguments, it follows that $\bar{T}$ is an isomorphism. ∎

(1) If $(\mathbf{S},\mu)$ is a σ-finite, non atomic measure ring and if $E_0 \varepsilon \mathbf{S}$, $E_0 \neq 0$, then, for every positive number ϵ, there exists an element E of $\mathbf{S}$ such that $E \subset E_0$ and $0 < \mu(E) < \epsilon$. (Hint: if $\mu(E_0) < \infty$ and if E_1 is an element of $\mathbf{S}$ such that $E_1 \subset E_0$ and $0 < \mu(E_1) < \mu(E_0)$, then either $\mu(E_1) \leqq \frac{1}{2}\mu(E_0)$ or $\mu(E_0 - E_1) \leqq \frac{1}{2}\mu(E_0)$.)

(2) If $(\mathbf{S},\mu)$ is a σ-finite, non atomic measure ring, and if $E_0 \varepsilon \mathbf{S}$, then, for every extended real number α with $0 \leqq \alpha \leqq \mu(E_0)$, there exists an element E in $\mathbf{S}$ such that $E \subset E_0$ and $\mu(E) = \alpha$. (Hint: since the case $\alpha = \infty$ is trivial, there is no loss of generality in assuming that $\mu(E_0) < \infty$. The desired result follows by a transfinite exhaustion process. The method is similar to the one used in proving that any two points in a complete, convex metric space may be joined by a segment, and in fact the present assertion is a special case of this general theorem in metric geometry; (cf. 40.2 and 40.8).)

(3) If $(\mathbf{S},\mu)$ is a totally σ-finite, non atomic measure algebra, and if $E_0 \varepsilon \mathbf{S}$, then, for every extended real number α with $\mu(E_0) \leqq \alpha \leqq \mu(X)$, there exists an element E in $\mathbf{S}$ such that $E_0 \subset E$ and $\mu(E) = \alpha$. (Hint: if α is finite, apply (2) to $X - E_0$ and $\mu(X) - \alpha$.)

(4) If $(\mathbf{S},\mu)$ is a totally finite measure algebra, then the set of all values of μ is a closed set.

(5) If a σ-finite, non atomic measure ring $(\mathbf{S},\mu)$ contains at least one element different from 0, then its metric space $\mathcal{S}(\mu)$ has no isolated points. Is it true that, conversely, if $\mathcal{S}(\mu)$ has no isolated points, then $(\mathbf{S},\mu)$ is non atomic?

(6) Every separable, non atomic, totally σ-finite measure algebra $(\mathbf{S},\mu)$ such that $\mu(X) = \infty$, is isomorphic to the measure algebra $(\mathbf{T},\nu)$ of the real line. (Hint: it follows from (2) that there exists a sequence $\{X_n\}$ of elements in $\mathbf{S}$ such that $X = \bigcup_{n=1}^{\infty} X_n$ and $\mu(X_n) = 1$, $n = 1, 2, \cdots$, and hence such that Theorem C is applicable, for each n, to the algebra of subelements of X_n.)

(7) Every measure algebra is isomorphic to the measure algebra of a measure space; (cf. 40.15c).

§ 42. FUNCTION SPACES

There are certain metric spaces associated with an arbitrary measure space $(X,\mathbf{S},\mu)$ which are similar to the space $\mathcal{S}(\mu)$ of measurable sets of finite measure. The one lying nearest at hand is the class $\mathcal{L}_1$ (or $\mathcal{L}_1(\mu)$) of all (extended real valued) integrable functions. If for f in $\mathcal{L}_1$ we write

$$\|f\| = \int |f| d\mu,$$

and for f and g in $\mathcal{L}_1$ we write $\rho(f,g) = \|f - g\|$, (cf. § 23), then the function ρ has all but one of the usual properties of a metric. The missing property is, of course, the positiveness of ρ; if $\rho(f,g) = 0$, it does not necessarily follow that $f = g$. We know what does follow: by 25.B, $\rho(f,g) = 0$ is equivalent to $f = g\ [\mu]$. We adopt again the same attitude as in the case of the space of measurable sets of finite measure. Two elements (i.e. functions) in $\mathfrak{L}_1$ are to be regarded as equal if their distance is zero, or, equivalently, if they are equal almost everywhere $[\mu]$; with this understanding $\mathfrak{L}_1$ becomes a metric space which (cf. 26.B) is even known to be complete.

For some purposes in analysis it is desirable to generalize these considerations. If p is a real number, $p > 1$, we shall denote by $\mathfrak{L}_p$ (or $\mathfrak{L}_p(\mu)$) the class of all those measurable functions f for which $|f|^p$ is integrable. In analogy with the situation in $\mathfrak{L}_1$ we shall identify two elements of $\mathfrak{L}_p$ if they are equal almost everywhere $[\mu]$; up to a certain point the theory of $\mathfrak{L}_p$ imitates that of $\mathfrak{L}_1$ very closely. We define, for instance, for f in $\mathfrak{L}_p$,

$$\|f\|_p = \left(\int |f|^p d\mu\right)^{1/p},$$

and we write, if f and g are in $\mathfrak{L}_p$, $\rho_p(f,g) = \|f - g\|_p$. At this stage we run into difficulties. While it is clear that $\rho_p(f,g) = \rho_p(g,f) \geqq 0$, and while it is equally clear that $\rho_p(f,g) = 0$ if and only if $f = g\ [\mu]$, it is not clear that the triangle inequality is valid, nor yet, and this is much more serious, that ρ_p is always finite. In order to settle these difficulties we proceed now to present proofs of two classical results; the following one is known as **Hölder's inequality.**

Theorem A. *If p and q are real numbers greater than* 1 *such that* $\frac{1}{p} + \frac{1}{q} = 1$, *and if* $f \,\varepsilon\, \mathfrak{L}_p$ *and* $g \,\varepsilon\, \mathfrak{L}_q$, *then* $fg \,\varepsilon\, \mathfrak{L}_1$ *and* $\|fg\| \leqq \|f\|_p \cdot \|g\|_q$.

Proof. We consider an auxiliary function ϕ defined for all positive real numbers t by $\phi(t) = \frac{t^p}{p} + \frac{t^{-q}}{q}$. Differentiating we obtain

$$\phi'(t) = t^{p-1} - t^{-q-1},$$

so that 1 is (in the domain under consideration) the only critical value of ϕ. Since

$$\lim_{t\to 0}\phi(t) = \lim_{t\to\infty}\phi(t) = \infty,$$

it follows that the value of ϕ at 1 is a minimum, and therefore that

$$\frac{t^p}{p} + \frac{t^{-q}}{q} = \phi(t) \geqq \phi(1) = \frac{1}{p} + \frac{1}{q} = 1.$$

If a and b are any two positive numbers, and if we write $t = a^{1/q}/b^{1/p}$, it follows that

$$1 \leqq \frac{a^{p-1}}{bp} + \frac{b^{q-1}}{aq} \quad \text{or} \quad ab \leqq \frac{a^p}{p} + \frac{b^q}{q},$$

and it is clear that the latter inequality remains valid even if a and b are allowed to be 0.

We turn now to the proof of the theorem. If either $\| f \|_p = 0$ or $\| g \|_q = 0$, then the result is trivial; in all other cases we may write

$$a = \frac{|f|}{\| f \|_p} \quad \text{and} \quad b = \frac{|g|}{\| g \|_q}.$$

Applying the last written inequality we obtain

$$\frac{|fg|}{\| f \|_p \cdot \| g \|_q} \leqq \frac{1}{p}\frac{|f|^p}{\int |f|^p d\mu} + \frac{1}{q}\frac{|g|^q}{\int |g|^q d\mu}.$$

Since fg is measurable, this inequality shows already that $fg \,\varepsilon\, \mathfrak{L}_1$; by integrating it we obtain the desired result. ∎

Our next result is known as **Minkowski's inequality.**

Theorem B. *If p is a real number greater than* 1, *and if f and g are in $\mathfrak{L}_p$, then $f + g \,\varepsilon\, \mathfrak{L}_p$ and*

$$\| f + g \|_p \leqq \| f \|_p + \| g \|_p.$$

Proof. Hölder's inequality for a measure space containing two points, each of measure 1, yields the elementary inequality

$$| a_1 b_1 + a_2 b_2 | \leqq (| a_1 |^p + | a_2 |^p)^{1/p}(| b_1 |^q + | b_2 |^q)^{1/q},$$

where, as before, $\frac{1}{p} + \frac{1}{q} = 1$. It follows that

$$|f+g|^p \leq |f|\cdot|f+g|^{p-1} + |g|\cdot|f+g|^{p-1} \leq$$
$$\leq (|f|^p + |g|^p)^{1/p}\cdot(2|f+g|^{q(p-1)})^{1/q},$$

and hence that

$$|f+g|^p \leq 2^{p/q}(|f|^p + |g|^p).$$

This implies that $f + g \,\varepsilon\, \mathfrak{L}_p$; the desired inequality follows from the relations

$$(\|f+g\|_p)^p = \int |f+g|^p d\mu \leq$$

$$\leq \int |f|\cdot|f+g|^{p-1}d\mu + \int |g|\cdot|f+g|^{p-1}d\mu \leq$$

$$\leq \left(\int |f|^p d\mu\right)^{1/p}\left(\int |f+g|^p d\mu\right)^{1/q}$$

$$+ \left(\int |g|^p d\mu\right)^{1/p}\left(\int |f+g|^p d\mu\right)^{1/q} =$$

$$= (\|f\|_p + \|g\|_p)(\|f+g\|_p)^{p/q}. \quad ▮$$

Since it follows from Theorem B that if f, g, and h are in $\mathfrak{L}_p$, then

$$\rho_p(f,g) = \|f-g\|_p \leq \|f-h\|_p + \|h-g\|_p = \rho_p(f,h) + \rho_p(h,g),$$

we see that $\mathfrak{L}_p$ is indeed a metric space; the proof that serves to establish the completeness of $\mathfrak{L}_1$ carries over with only trivial changes and establishes the completeness of $\mathfrak{L}_p$.

(1) The metric space $\mathfrak{L}_p(\mu)$ on a measure space $(X,\mathbf{S},\mu)$ is separable if and only if the space $\mathcal{S}(\mu)$ of measurable sets of finite measure is separable. (Hint: if a class of sets is dense in $\mathcal{S}(\mu)$, then the set of all finite linear combinations with rational coefficients of the characteristic functions of these sets is dense in $\mathfrak{L}_p(\mu)$.)

(2) Another occasionally useful space is the set $\mathfrak{M}$ of all essentially bounded measurable functions. If we write, for any f in $\mathfrak{M}$,

$$\|f\|_\infty = \text{ess. sup. } \{|f(x)| : x \,\varepsilon\, X\}$$

and, for f and g in $\mathfrak{M}$, $\rho_\infty(f,g) = \|f - g\|_\infty$, then $\mathfrak{M}$ (with our by now familiar conventions as to what constitutes equality for two elements of a measure theoretically defined function space) becomes a complete metric space.

(3) The space $\mathfrak{L}_2$ is deservedly the most extensively studied of the function spaces we described; it is in a legitimate sense the most direct and fruitful generalization of ordinary, finite dimensional, Euclidean space. A **linear functional** on $\mathfrak{L}_2$ is a real valued function Λ on $\mathfrak{L}_2$ such that

$$\Lambda(\alpha f + \beta g) = \alpha\Lambda(f) + \beta\Lambda(g)$$

whenever α and β are real numbers and f and g are in $\mathfrak{L}_2$. A linear functional Λ is **bounded** if there exists a positive, finite constant c such that $|\Lambda(f)| \leqq c\|f\|_2$ for every f in $\mathfrak{L}_2$. It is an elementary geometric property of $\mathfrak{L}_2$ (whose proof depends on nothing deeper than that $\mathfrak{L}_2$ is complete) that corresponding to every bounded linear functional Λ there exists an element g of $\mathfrak{L}_2$ such that $\Lambda(f) = \int fg\,d\mu$ for every f in $\mathfrak{L}_2$. This fact may be used to prove the Radon–Nikodym theorem (of which, incidentally, it is in turn a reasonably easy consequence). For the sake of simplicity we shall restrict our outline of this proof to the case of finite measures. Suppose then that μ and ν are two finite measures such that $\nu \ll \mu$ and write $\lambda = \mu + \nu$.

(3a) If, for every f in $\mathfrak{L}_2(\lambda)$, $\Lambda(f) = \int f\,d\nu$, then Λ is a bounded linear functional on $\mathfrak{L}_2(\lambda)$.

(3b) If $\Lambda(f) = \int fg\,d\lambda$, then $0 \leqq g \leqq 1$ [λ]. (Hint: if f is the characteristic function of a measurable set E, then $\Lambda(f) = \nu(E) \leqq \lambda(E)$.)

(3c) If $E = \{x: g(x) = 1\}$, then $\lambda(E) = 0$. (Hint: $\lambda(E) = \nu(E)$.)

(3d) $\int f(1 - g)\,d\nu = \int fg\,d\mu$ for every non negative measurable function f.

(3e) If $g_0 = \dfrac{g}{1 - g}$, then, for every measurable set E, $\nu(E) = \int_E g_0\,d\mu$. (Hint: write $f = \dfrac{\chi_E}{1 - g}$.)

(4) Suppose that $(X,\mathbf{S},\mu)$ is a finite measure space and write, for any two real valued measurable functions f and g,

$$\rho_0(f,g) = \int \frac{|f - g|}{1 + |f - g|}\,d\mu.$$

The function ρ_0 is a metric; convergence with respect to ρ_0 is equivalent to convergence in measure.

§ 43. SET FUNCTIONS AND POINT FUNCTIONS

In this section we shall study the connection between certain functions of a real variable and finite measures on the real line. Throughout this section we shall assume that

X is the real line, $\mathbf{S}$ is the class of all Borel sets, and μ is Lebesgue measure on $\mathbf{S}$.

We shall consider monotone, non decreasing functions f on X, i.e. functions f for which $f(x) \leqq f(y)$ whenever $x \leqq y$; for brevity of expression we shall simply call such functions **monotone.** If f is a bounded monotone function, then it is easy to see that

$$\lim_{x \to -\infty} f(x) \quad \text{and} \quad \lim_{x \to +\infty} f(x)$$

always exist and are finite; it is customary to denote these limits by $f(-\infty)$ and $f(+\infty)$ respectively.

Theorem A. *If ν is a finite measure on $\mathbf{S}$ and if, for every real number x,*

$$f_\nu(x) = \nu(\{t: -\infty < t < x\}),$$

then f_ν is a bounded monotone function, continuous on the left and such that $f_\nu(-\infty) = 0$.

Proof. The boundedness and monotoneness of f_ν follow from the corresponding properties of ν. Since $f_\nu(-n) = \nu((-\infty, -n))$, $n = 1, 2, \cdots$, it follows that

$$f_\nu(-\infty) = \lim_n f_\nu(-n) = \nu(\textstyle\bigcap_{n=0}^{\infty} \{t: -\infty < t < -n\}) =$$

$$= \nu(0) = 0.$$

To prove that f_ν is continuous on the left at each x, suppose that $\{x_n\}$ is an increasing sequence of numbers such that $\lim_n x_n = x$; we have

$$0 = \nu(\textstyle\bigcap_{n=1}^{\infty} [x_n,x)) = \lim_n \nu([x_n,x)) = \lim_n (f_\nu(x) - f_\nu(x_n)). \quad \blacksquare$$

The following result goes in the converse direction.

Theorem B. *If f is a bounded monotone function, continuous on the left and such that $f(-\infty) = 0$, then there exists a unique finite measure ν on $\mathbf{S}$ such that $f = f_\nu$.*

Proof. In all details this proof parallels the construction of Lebesgue measure. If, in other words, we define ν for every semiclosed interval by $\nu([a,b)) = f(b) - f(a)$, then the results of § 8 are valid for ν in place of μ and hence the extension theorem

13.A may be applied. The only argument which needs modification is the one used to establish 8.C. We are to prove that if $[a_0,b_0)$ is a semiclosed interval contained in the union of a sequence $\{[a_i,b_i)\}$ of semiclosed intervals, then

$$\nu([a_0,b_0)) \leqq \sum_{i=1}^{\infty} \nu([a_i,b_i)).$$

If $a_0 = b_0$, the result is trivial; otherwise let ϵ be a positive number such that $\epsilon < b_0 - a_0$. Since f is continuous on the left at a_i, to every positive number δ and every positive integer i there corresponds a positive number ϵ_i such that

$$f(a_i) - f(a_i - \epsilon_i) < \frac{\delta}{2^i}, \quad i = 1, 2, \cdots.$$

If $F_0 = [a_0,b_0 - \epsilon]$ and $U_i = (a_i - \epsilon_i,b_i)$, $i = 1, 2, \cdots$, then $F_0 \subset \bigcup_{i=1}^{\infty} U_i$, and therefore, by the Heine–Borel theorem, there is a positive integer n such that

$$F_0 \subset \bigcup_{i=1}^{n} U_i.$$

From the analog of 8.B for ν we obtain

$$\begin{aligned} f(b_0 - \epsilon) - f(a_0) &\leqq \sum_{i=1}^{n} (f(b_i) - f(a_i - \epsilon_i)) = \\ &= \sum_{i=1}^{n} (f(b_i) - f(a_i)) + \sum_{i=1}^{n} (f(a_i) - f(a_i - \epsilon_i)) \leqq \\ &\leqq \sum_{i=1}^{\infty} (f(b_i) - f(a_i)) + \delta. \end{aligned}$$

Since ϵ and δ are arbitrary, the desired result follows from the fact that f is continuous on the left at b_0. ∎

Theorems A and B establish a one to one correspondence between all finite measures ν on **S** and some functions f_ν of a real variable; the following two theorems show how certain measure theoretic properties of ν may be characterized in terms of the corresponding function f_ν.

Theorem C. *If ν is a finite measure on* **S**, *then a necessary and sufficient condition that f_ν be continuous is that $\nu(\{x\}) = 0$ for every point x.*

Proof. If $\{x_n\}$ is a decreasing sequence of numbers such that $\lim_n x_n = x$, then

$$\nu(\{x\}) = \nu(\bigcap_{n=1}^{\infty} [x,x_n)) = \lim_n \nu([x,x_n)) = \lim_n (f_\nu(x_n) - f_\nu(x)).$$

The proof is completed by the observation that f_ν is continuous at x if and only if the last term of this relation vanishes. ∎

A real valued function f of a real variable is called **absolutely continuous** if to every positive number ϵ there corresponds a positive number δ such that

$$\sum_{i=1}^{n} |f(b_i) - f(a_i)| < \epsilon$$

for every finite, disjoint class $\{(a_i, b_i): i = 1, \cdots, n\}$ of bounded open intervals for which $\sum_{i=1}^{n} (b_i - a_i) < \delta$.

Theorem D. *If ν is a finite measure on* **S**, *then a necessary and sufficient condition that f_ν be absolutely continuous is that ν be absolutely continuous with respect to μ.*

Proof. If $\nu \ll \mu$, then to every positive number ϵ there corresponds a positive number δ such that $\nu(E) < \epsilon$ for every Borel set E for which $\mu(E) < \delta$. Hence if $\{(a_i, b_i): i = 1, \cdots, n\}$ is a finite, disjoint class of bounded open intervals for which

$$\mu(\bigcup_{i=1}^{n} [a_i, b_i)) = \sum_{i=1}^{n} (b_i - a_i) < \delta,$$

then

$$\sum_{i=1}^{n} |f_\nu(b_i) - f_\nu(a_i)| = \sum_{i=1}^{n} \nu([a_i, b_i)) =$$

$$= \nu(\bigcup_{i=1}^{n} [a_i, b_i)) < \epsilon.$$

Suppose, conversely, that f_ν is absolutely continuous. Let ϵ be any positive number and let δ be a positive number such that $\sum_{i=1}^{n} (b_i - a_i) < \delta$ implies $\sum_{i=1}^{n} |f_\nu(b_i) - f_\nu(a_i)| < \epsilon$. If E is a Borel set of Lebesgue measure zero, then there exists a disjoint sequence $\{[a_i, b_i)\}$ of semiclosed intervals such that

$$E \subset \bigcup_{i=1}^{\infty} [a_i, b_i) \quad \text{and} \quad \sum_{i=1}^{\infty} (b_i - a_i) < \delta.$$

Since it follows that $\sum_{i=1}^{n} |f_\nu(b_i) - f_\nu(a_i)| < \epsilon$ for every positive integer n, we have

$$\nu(E) \leqq \sum_{i=1}^{\infty} \nu([a_i, b_i)) = \sum_{i=1}^{\infty} |f_\nu(b_i) - f_\nu(a_i)| \leqq \epsilon.$$

Since ϵ is arbitrary, we must have $\nu(E) = 0$. ∎

For the statement of the next result (which is an easy but frequently useful consequence of the Lebesgue decomposition theorem) we need one more definition. We shall say that a finite measure ν on **S** is **purely atomic** if there exists a countable set C such that $\nu(X - C) = 0$.

Theorem E. *If ν is a finite measure on* **S**, *then there exist three uniquely determined measures ν_1, ν_2, and ν_3 on* **S** *whose sum is ν and which are such that ν_1 is absolutely continuous with respect to μ, ν_2 is purely atomic, and ν_3 is singular with respect to μ but $\nu_3(\{x\}) = 0$ for every point x.*

Proof. According to the Lebesgue decomposition theorem (32.C) there exist two measures ν_0 and ν_1 on **S** whose sum is ν and which are such that ν_0 is singular and ν_1 is absolutely continuous with respect to μ. Let C be the set of those points x for which $\nu_0(\{x\}) \neq 0$; the finiteness of ν implies that C is countable. If we write

$$\nu_2(E) = \nu_0(E \cap C) \quad \text{and} \quad \nu_3(E) = \nu_0(E - C),$$

then it is clear that the decomposition $\nu = \nu_1 + \nu_2 + \nu_3$ has all the desired properties. Uniqueness follows from the uniqueness of the Lebesgue decomposition and the easily verifiable uniqueness of C. ∎

(1) All the results of this section remain true for signed measures ν if the condition that f_ν be monotone is replaced by the condition that it be of bounded variation. (Hint: every function of bounded variation is the difference of two monotone functions.)

(2) Several of the well known properties of monotone functions and absolutely continuous functions may be proved by using the methods of this section; we indicate two examples.

(2a) A monotone function has (at most) countably many discontinuities. (Hint: for bounded monotone functions f which are continuous on the left and such that $f(-\infty) = 0$, apply Theorem B and the reasoning in the proof of Theorem C. The general case can be reduced to this special case by some obvious transformations.)

(2b) If a bounded monotone function f is absolutely continuous and such that $f(-\infty) = 0$, then there exists a non negative Lebesgue integrable function ϕ such that $f(x) = \int_{-\infty}^{x} \phi(t)d\mu(t)$. (Hint: apply Theorems B and D.)

(3) The purpose of the following considerations is to show that the results of 15.C and 15.1 can be extended to a very wide class of measures including the ones discussed in this section.

(3a) If two finite measures μ and ν on a σ-ring **S** of subsets of X agree on a lattice **L** of sets in **S**, then μ and ν agree on the σ-ring **S(L)** generated by **L**.

(Hint: if $E \varepsilon \mathbf{L}$, $F \varepsilon \mathbf{L}$, and $E \subset F$, then $\mu(F - E) = \nu(F - E)$. Apply 5.2, 8.5, and 13.A.)

(3b) If two finite measures μ and ν are defined on the class of Borel subsets of a metric space X and agree on the class $\mathbf{U}$ of open subsets of X, then they agree for all Borel sets.

(3c) If μ is a finite measure defined on the class of Borel subsets of a metric space X, and $\mathbf{U}$ is the class of open subsets of X, then $\mu(E) = \inf \{\mu(U) : E \subset U \varepsilon \mathbf{U}\}$ for every Borel set E. (Hint: the set function ν^* defined by $\nu^*(E) = \inf \{\mu(U) : E \subset U \varepsilon \mathbf{U}\}$ is a finite metric outer measure which defines a measure ν on the class of Borel sets, and ν agrees with μ on $\mathbf{U}$.)

(3d) If μ is a measure on the class of Borel subsets of a metric space X, and $\mathbf{C}$ is the class of closed subsets of X that have finite measure, then $\mu(E) = \sup \{\mu(C) : E \supset C \varepsilon \mathbf{C}\}$ for every Borel set E of σ-finite measure. (Hint: it is sufficient to consider sets E of finite measure. Write $\nu(F) = \mu(E \cap F)$ and apply (3c) to ν and $X - E$.)

(3e) If μ is a measure on the class of Borel subsets of a separable, complete, metric space X, and $\mathbf{C}_0$ is the class of compact subsets of X that have finite measure, then $\mu(E) = \sup \{\mu(C) : E \supset C \varepsilon \mathbf{C}_0\}$ for every Borel set E of σ-finite measure. (Hint: apply (3d) and 9.10.)

(4) If ν is a finite measure on $\mathbf{S}$ and if a Borel set E_0 is an atom of ν, then there exists a point x_0 in E_0 such that $\nu(E_0 - \{x_0\}) = 0$. (Hint: by means of (3) the general case may be reduced to the case in which E_0 is closed and bounded.)

(5) If ν is a finite measure on $\mathbf{S}$, then a necessary and sufficient condition that f_ν be continuous is that ν be non atomic.

(6) Most of the results of this section remain true for measures and signed measures ν which are not necessarily finite; what is essential is that $\nu(E)$ be finite whenever E is a bounded interval.

(7) In connection with (6) and for the purpose of constructing counter examples, it is interesting to observe that there exist σ-finite measures ν on $\mathbf{S}$ which are absolutely continuous with respect to μ, but for which $\nu(E) = \infty$ for every interval E with a non empty interior. (Hint: let f be a positive, Lebesgue integrable function such that $\int_{-\epsilon}^{+\epsilon} f^2 d\mu = \infty$ for every positive number ϵ; for example write $f(x) = (e^{|x|}\sqrt{|x|})^{-1}$. If $\{r_1, r_2, \cdots\}$ is an enumeration of the set of all rational numbers, if, for every x,

$$g(x) = \sum_{n=1}^{\infty} \frac{1}{2^n} f(x - r_n),$$

and if, for every Borel set E, $\nu(E) = \int_E g^2 d\mu$, then ν has all the desired properties. Observe that since $\int g d\mu = \sum_{n=1}^{\infty} \frac{1}{2^n} \int f d\mu$, the function g is finite valued a.e. $[\mu]$.)

Chapter IX

PROBABILITY

§ 44. HEURISTIC INTRODUCTION

The purpose of this section is to give an intuitive justification for the measure theoretic treatment of probability.

The principal undefined term in the theory of probability is "event." Intuitively speaking, an event is one of the possible outcomes of some physical experiment. To take a rather popular example, consider the experiment of rolling an ordinary six-sided die and observing the number x ($= 1, 2, 3, 4, 5$, or 6) showing on the top face of the die. "The number x is even"—"it is less than 4"—"it is equal to 6"—each such statement corresponds to a possible outcome of the experiment. From this point of view there are as many events associated with this particular experiment as there are combinations of the first six positive integers taken any number at a time. If, for the sake of aesthetic completeness and later convenience, we consider also the impossible event, "the number x is not equal to any of the first six positive integers," then there are altogether 2^6 admissible events associated with the experiment of the rolling die. For the purpose of studying this example in more detail let us introduce some notation. We write $\{2,4,6\}$ for the event "x is even," $\{1,2,3\}$ for "x is less than 4," and so on. The impossible event and the certain event ($= \{1,2,3,4,5,6\}$) deserve special names; we reserve for them the symbols 0 and X respectively.

Everyday language concerning events uses such phrases as these: "two events E and F are incompatible or mutually exclusive," "the event E is the opposite of the event F or complementary to F," "the event E consists of the simultaneous occurrence of F and G," and "the event E consists of the occurrence of at least one of the two events F and G." Such phrases suggest that there are relations between events and ways of making new events out of old that should certainly be a part of their mathematical theory.

The notion of complementary event is probably closest to the surface. If E is an event, we denote the complementary event by E': an experiment, one of whose outcomes is E, will be said to result in E' if and only if it does not result in E. Thus if $E = \{2,4,6\}$, then $E' = \{1,3,5\}$. We may also introduce combinations of events suggested by the logical concepts of "and" and "or." With any two events E and F we associate their "union" $E \cup F$ and their "intersection" $E \cap F$; here $E \cup F$ occurs if and only if at least one of the two events E and F occurs, while $E \cap F$ occurs if and only if both E and F occur. Thus if $E = \{2,4,6\}$ and $F = \{1,2,3\}$, then $E \cup F = \{1,2,3,4,6\}$ and $E \cap F = \{2\}$.

The considerations of the preceding paragraphs, and their obvious generalizations to more complicated experiments, justify the conclusion that probability theory consists of the study of Boolean algebras of sets. An event is a set, and its opposite event is the complementary set; mutually exclusive events are disjoint sets, and an event consisting of the simultaneous occurrence of two other events is a set obtained by intersecting two other sets—it is clear how this glossary, translating physical terminology into set theoretic terminology, may be continued.

For the traditional theory of probability, concerned with simple gambling games such as the rolling die, in which the total number of possible events is finite, the above heuristic reduction of the class of all pertinent events to a Boolean algebra of sets is adequate. For situations arising in modern theory and practice, and even for the more complicated gambling games, it is necessary to make an additional assumption. This assumption is that the system of events is closed under the formation of countably infinite unions, or, in the technical language we have already used, that the Boolean algebra is in fact a σ-algebra.

Perhaps an example, though a somewhat artificial one, might illustrate the need for the added assumption. Suppose that a player determines to roll a die repeatedly until the first time that the number showing on top is 6. Let E_n be the event that the first 6 appears only on the nth roll. The event $E = \bigcup_{n=1}^{\infty} E_n$ occurs if and only if the game ends in a finite number of rolls. The occurrence of the opposite event E' is at least logically (even if not practically) conceivable, and it seems reasonable to want to include a discussion of it in a general theory of probability. Numerous examples of this kind, together with some rather deep lying technical reasons, justify the statement that the mathematical theory of probability consists of the study of Boolean σ–algebras of sets.

This is not to say that all Boolean σ–algebras of sets are within the domain of probability theory. In general, statements concerning such algebras and the relations between their elements are merely qualitative; probability theory differs from the general theory in that it studies also the quantitative aspects of Boolean algebras. We proceed now to describe and motivate the introduction of numerical probabilities.

When we ask "what is the probability of a certain event?", we expect the answer to be a number, a number associated with the event. In other words, probability is a numerically valued function μ of events E, that is of sets of a Boolean σ–algebra. On intuitive and practical grounds we demand that the number $\mu(E)$ should give information about the occurrence habits of the event E. If, in a large number of repetitions of the experiment which may result in the event E, we observe that E actually occurs only a quarter of the time (the remaining three quarters of the experiments resulting therefore in E'), we may attempt to summarize this fact by saying that $\mu(E) = \frac{1}{4}$. Even this very rough first approximation to what is desired yields some suggestive clues concerning the nature of the function μ.

If, to begin with, $\mu(E)$ is to represent the proportion of times that E is expected to occur, then $\mu(E)$ must be a non negative real number, in fact a number in the unit interval [0,1]. If E and F are mutually exclusive events—say $E = \{1\}$ and $F = \{2,4,6\}$ in the example of the die—then the proportion of times that the union $E \cup F$ ($= \{1,2,4,6\}$ in the example) occurs is clearly the sum of the proportions associated with E and F separately. If an ace shows up one sixth of the time and an even number half the time, then the proportion of times in which the top face is either an ace or an even number is $\frac{1}{6} + \frac{1}{2}$. It follows therefore that the function μ cannot be completely arbitrary; it is necessary to subject it to the condition of additivity, that is to require that if $E \cap F = 0$, then $\mu(E \cup F)$ should be equal to $\mu(E) + \mu(F)$. Since the certain event X occurs every time, we should also require that $\mu(X) = 1$.

We are now separated from the final definition of probability theory only by a seemingly petty (but in fact very important) technicality. If μ is an additive set function on a Boolean σ-algebra of sets, and if $\{E_n\}$ is an infinite disjoint sequence of sets in the algebra, then it may or may not be true that $\mu(\bigcup_{n=1}^{\infty} E_n) = \sum_{n=1}^{\infty} \mu(E_n)$. The general condition of countable additivity is a further restriction on μ—a restriction without which modern probability theory could not function. It is a tenable point of view that our intuition demands infinite additivity just as much as finite additivity. At any rate, however, infinite additivity does not contradict our intuitive ideas, and the theory built on it is sufficiently far developed to assert that the assumption is justified by its success. To sum up:

numerical probability is a measure μ on a Boolean σ-algebra $\mathbf{S}$ of subsets of a set X, such that $\mu(X) = 1$.

In our development of measure theory in the preceding chapters, the concepts "measurable function," "integral," and "product space" played important roles; in the immediately following paragraphs we shall introduce the probability meaning of these concepts.

We begin with the frequently used term "random variable." "A random variable is a quantity whose values are determined by chance." What does that mean? The word "quantity" suggests magnitude—numerical magnitude. Ever since rigor has come to be demanded in mathematical definitions, it has been recognized that the word "variable," particularly a variable whose values are "determined" somehow or other, means in precise language a function. Accordingly a random variable is a function: a function whose numerical values are determined by chance. This means, in other words, that a random variable is a function attached to an experiment—once the experiment has been performed the value of the function is known. We have seen that the analytic correspondent of an experiment is a measure space X; it follows that a function of outcomes is a function of the points x of X. A random variable is a real valued function on the measure space X.

The preceding sentence does not yet fully describe the customary usage of "random variable." A function f on the measure space X is called a random variable only if probability questions concerning the values of f can be answered. An example of such a question is "what is the probability that f lies between α and β?" In measure theoretic language: "what is the measure of the set of those points x for which the inequality $\alpha \leqq f(x) \leqq \beta$ is satisfied?" In order for all such questions to be answerable, it is necessary and sufficient that the sets that occur in them belong to the basic σ-algebra $\mathbf{S}$ of X; in other words a random variable is a measurable function.

Let us consider in detail the random variable f associated with an honest die by the definition $f(x) = x$. The possible values of f are the first six positive integers. The arithmetic mean of these values, that is the number $\frac{1}{6}(1 + 2 + 3 + 4 + 5 + 6)$, is of considerable interest in probability theory; it is called the average, or mean value, or expectation of the random variable f. If the die is loaded, so that the probability p_x associated with x is not necessarily $\frac{1}{6}$, then the arithmetic mean is replaced by a weighted average; in this case the expectation of f is $1 \cdot p_1 + \cdots + 6 \cdot p_6$. The analogs of such weighted sums, in cases where the number of values of the function need not be finite, are given by integrals; if the measurable function is integrable, then its expectation is by definition the value of its integral.

We see thus that measurable functions and their integrals have their probability interpretations; in order to find such an interpretation of product spaces, we continue to study the example of the die. For simplicity we make again the classical assumption that any two faces are equally likely to turn up and that consequently the probability of any particular face showing is $\frac{1}{6}$. Consider the events $E = \{2,4,6\}$ and $F = \{1,2\}$. The first notion we want to introduce, the notion of conditional probability, can be used to answer such questions as these: "what is the probability of E when F is known to have occurred?" In the case of the example: if we know that x is less than 3, what can we say about the probability that x is even? The adjective "conditional" is clearly called for in the answer to a question of this type: we are evaluating probabilities subject to certain preassigned conditions.

To get a clue to the answer, consider first the event $G = \{2\}$ and ask for the conditional probability of E, given that G has already occurred. The intuitive answer is perfectly clear in this case, and is independent as it happens of any such numerical assumptions as the equal likelihood of the faces. If x is known to be 2, then x is certainly even, and the probability must be 1. What made the answer easy was the fact that G was contained in E. The general question of conditional probability asks us to evaluate the extent (measured by a numerical probability or proportion) to which the given event F is contained in the unknown event E. Phrased in this way, the question almost suggests its own answer: the extent to which F is contained in E can be measured by the extent to which E and F are likely to occur simultaneously, that is by $\mu(E \cap F)$. Almost—not quite. The trouble is that $\mu(E \cap F)$ may be very small for two reasons: one is that not much of F is contained in E, and the other is that there is not very much of E altogether. In other words it is not merely the absolute size of $E \cap F$ that matters: it is the relation or proportion of this size to the size of F that is relevant.

We are led therefore to define the conditional probability of E, given that F has already occurred, in symbols $\mu_F(E)$, as the ratio $\mu(E \cap F)/\mu(F)$. For $E = \{2,4,6\}$ and $G = \{2\}$, this gives the answer we derived above, $\mu_G(E) = 1$; for $E = \{2,4,6\}$ and $F = \{1,2\}$, we get the rather reasonable figure $\mu_F(E) = \frac{1}{2}$. In other words if it is known that x is either 1 or 2, then x is odd or even (i.e. equal to 1 or equal to 2) each with probability $\frac{1}{2}$.

Consider now the following two questions: "F happened, what is the chance of E?" and simply "what is the chance of E?"

The answers of course are $\mu_F(E)$ and $\mu(E)$ respectively. It might happen, and it does happen in the example given above, that the two answers are the same, that, in other words, knowledge of F contributes nothing to our knowledge of the probability of E. It seems natural in this situation to use the word "independent": the probability distribution of E is independent of the knowledge of F. This motivates the precise definition: two events E and F are independent if $\mu_F(E) = \mu(E)$. The definition is transformed into its more usual form and at the same time gains in symmetry if we recall the definition of $\mu_F(E)$. In symmetric form, E and F are independent in the sense of probability (statistically or stochastically independent) if and only if $\mu(E \cap F) = \mu(E)\mu(F)$.

Suppose now that we wish to make two independent trials of the same experiment—say, for example, to roll an honest die twice in succession. In a compound experiment such as this one, we do not expect the reported outcome of the experiment to be a number, but rather a pair of numbers (x_1,x_2). The points of the measure space associated with the compound experiment are, in other words, the points of the Cartesian product of the original measure space with itself; the problem is to determine how the probability is distributed among these points. For a clue to the answer, consider the events $E =$ "$x_1 < 3$" and $F =$ "$x_2 < 4$." We have $\mu(E) = \frac{1}{3}$ and $\mu(F) = \frac{1}{2}$; if we interpret the independence of trials to mean the independence of any two events such as E and F, we should have $\mu(E \cap F) = \frac{1}{6}$.

On the basis of the preceding paragraph we shall say that, if the analytic description of an experiment is given by a measure space $(X,\mathbf{S},\mu)$, then the analytic description of the experiment consisting of two independent trials of the given one is the Cartesian product of $(X,\mathbf{S},\mu)$ with itself.

What we can do once we can do again. Just as two repetitions of an experiment give rise to two dimensional Cartesian products, similarly any finite number of repetitions (say n) give rise to n-dimensional Cartesian products. The procedure can be extended also to infinity: the analytic model of an infinite sequence of independent repetitions of an experiment is an infinite dimensional Cartesian product space. Even if an actually infinite sequence of repetitions of an experiment is practically unthinkable, there is a point in considering infinite dimensional product spaces. The point is that many probability statements are assertions concerning what happens in the long run—assertions which can be made precise only by carefully formulated theorems concerning limits. Hence even if practice yields only approximations to infinity, it is the infinite sequence space that is the touchstone whereby the mathematical theory of probability can be tested against our intuitive ideas.

We leave now these heuristic considerations and, in the next section, turn to the detailed investigation of the basic concepts and results of probability theory.

§ 45. INDEPENDENCE

A **probability space** is a totally finite measure space $(X,\mathbf{S},\mu)$ for which $\mu(X) = 1$; the measure μ on a probability space is called a **probability measure.**

If $\mathbf{E}$ is a finite or infinite class of measurable sets in a probability space $(X,\mathbf{S},\mu)$, the sets of the class $\mathbf{E}$ are (stochastically) **independent** if

$$\mu(\bigcap_{i=1}^n E_i) = \prod_{i=1}^n \mu(E_i)$$

for every finite class $\{E_i: i = 1, \cdots, n\}$ of distinct sets in $\mathbf{E}$. In case the class $\mathbf{E}$ contains only two sets E and F, the condition of independence is expressed by the equation

$$\mu(E \cap F) = \mu(E)\mu(F).$$

An illuminating example of two independent sets E and F is obtained by taking for X the unit square with Lebesgue measure, $X = \{(x,y): 0 \leqq x \leqq 1, 0 \leqq y \leqq 1\}$, and writing $E = \{(x,y): 0 \leqq x \leqq 1, a \leqq y \leqq b\}$ and $F = \{(x,y): c \leqq x \leqq d, 0 \leqq y \leqq 1\}$, where a, b, c, and d are arbitrary numbers in the closed unit interval. We remark that it is *not* sufficient for the independence of the sets of a class $\mathbf{E}$ (even if $\mathbf{E}$ is a finite class) that any two distinct sets of $\mathbf{E}$ be independent.

If $\mathcal{E}$ is a finite or infinite set of real valued measurable functions on a probability space $(X,\mathbf{S},\mu)$, the functions of the set $\mathcal{E}$ are (stochastically) **independent** if

$$\mu(\bigcap_{i=1}^{n} \{x: f_i(x) \varepsilon M_i\}) = \prod_{i=1}^{n} \mu(\{x: f_i(x) \varepsilon M_i\})$$

for every finite subset $\{f_i: i = 1, \cdots, n\}$ of distinct functions in $\mathcal{E}$ and every finite class $\{M_i: i = 1, \cdots, n\}$ of Borel sets on the real line. An equivalent way of expressing this condition is to say that if, for each f in $\mathcal{E}$, M_f is a Borel set on the real line, then, for every possible choice of the Borel sets M_f, the sets of the class $\mathbf{E} = \{f^{-1}(M_f): f \varepsilon \mathcal{E}\}$ are independent. An illuminating example of two independent functions f and g is obtained by taking for X the unit square, as in the preceding paragraph, and defining f and g by $f(x,y) = x$ and $g(x,y) = y$.

As our examples of independent sets and functions might indicate, there is a very close connection between the concepts of stochastic independence and Cartesian product. Suppose, in fact, that f_1 and f_2 are two independent functions on a probability space $(X,\mathbf{S},\mu)$ and consider the transformation T from X into the Euclidean plane, defined by

$$T(x) = (f_1(x), f_2(x)).$$

If measurability in the plane is interpreted in the sense of Borel, then the facts that X is a measurable set and f_1 and f_2 are measurable functions imply that T is a measurable transformation; similarly, of course, the functions f_1 and f_2 are themselves measurable transformations from X into the real line. A direct comparison with the definition of independence shows that the fact that f_1 and f_2 are independent can be very simply expressed by the equation

$$\mu T^{-1} = \mu f_1^{-1} \times \mu f_2^{-1}.$$

(If the transformation T is denoted, as it may well be, by the symbol $f_1 \times f_2$, then the last written equation takes the form of an easily remembered distributive law.) If the functions g_1 and g_2 on the plane are defined by

$$g_1(y_1,y_2) = y_1 \quad \text{and} \quad g_2(y_1,y_2) = y_2,$$

then it is easy to verify that $f_1 = g_1 T$ and $f_2 = g_2 T$. From these very simple considerations we may already draw a non trivial conclusion.

Theorem A. *If f_1 and f_2 are independent functions, neither of which vanishes a.e., then a necessary and sufficient condition that both f_1 and f_2 be integrable is that their product f_1f_2 be integrable; if this condition is satisfied, then*

$$\int f_1f_2d\mu = \int f_1d\mu \cdot \int f_2d\mu.$$

Proof. Using the notation established above, we see (by 39.C) that the integrability of $|f_i|$ is equivalent to the integrability of g_i, $i = 1, 2$, and, by Fubini's theorem, the integrability of $|g_1|$ and $|g_2|$ implies that of $|g_1g_2|$. Conversely, of course, if $|g_1g_2|$ is integrable, then almost every section of $|g_1g_2|$ is integrable. Since each such section is a constant multiple of $|g_1|$ or of $|g_2|$ and since the assumption that f_1 and f_2 do not vanish a.e. implies that these constant factors may be selected to be different from zero, it follows that the integrability of $|g_1g_2|$ implies that of $|g_1|$ and $|g_2|$. Since, finally, another application of 39.C shows that $|g_1g_2|$ is integrable if and only if $|f_1f_2|$ is integrable, the assertion concerning integrability is proved. The multiplicative relation follows from Fubini's theorem. ▮

The use of product spaces in the study of independent functions extends far beyond the simple case indicated by the reasoning above. Suppose, for instance, that $\{f_n\}$ is a sequence of independent functions and let Y be the Cartesian product of a sequence of real lines in each of which measurability is interpreted in the sense of Borel. If, for every x,

$$T(x) = (f_1(x), f_2(x), \cdots),$$

then T is a measurable transformation from X into Y; a necessary and sufficient condition that the functions f_n be independent is that $\mu T^{-1} = \mu f_1^{-1} \times \mu f_2^{-1} \times \cdots$. If the functions g_n on Y are defined to be the coordinate functions, $g_n(y_1, y_2, \cdots) = y_n$, $n = 1, 2, \cdots$, then $f_n = g_nT$, $n = 1, 2, \cdots$. Similar results are true of course for arbitrary (finite, countable, or uncountable) sets of functions.

Theorem B. *If* $\{f_{ij}: i = 1, \cdots, k; j = 1, \cdots, n_i\}$ *is a set of independent functions, if* ϕ_i *is a real valued, Borel measurable function of* n_i *real variables,* $i = 1, \cdots, k$, *and if*

$$f_i(x) = \phi_i(f_{i1}(x), \cdots, f_{in_i}(x)),$$

then the functions $f_1, \cdots, f_k$ *are independent.*

Proof. The theorem is an easy application of the relations established above between product spaces and independence. Suppose that Y_{ij} is the real line, $i = 1, \cdots, k, j = 1, \cdots, n_i$, and $Y = \times_{ij} Y_{ij}$. If we write

$$T(x) = (f_{11}(x), \cdots, f_{1n_1}(x), \cdots, f_{k1}(x), \cdots, f_{kn_k}(x)),$$

$$g_{ij}(y_{11}, \cdots, y_{1n_1}, \cdots, y_{k1}, \cdots, y_{kn_k}) = y_{ij}$$

and

$$g_i = \phi_i(g_{i1}, \cdots, g_{in_i}),$$

then $f_i = g_i T$, $i = 1, \cdots, k$. Since the independence of the g_i's is obvious, the independence of the f_i's follows. ∎

We conclude this section by introducing a frequently used notation of probability theory. If f is a real valued measurable function on a probability space $(X, \mathbf{S}, \mu)$, such that f^2 is integrable, then it follows from Schwarz's inequality (i.e. Hölder's inequality with $p = 2$, cf. 42.A) that f itself is integrable and that, in fact,

$$\left(\int f d\mu\right)^2 \leqq \int f^2 d\mu.$$

If $\int f d\mu = \alpha$, then the **variance** of f, denoted by $\sigma^2(f)$, is defined by $\sigma^2(f) = \int (f - \alpha)^2 d\mu$. Since the integral of a constant function over a probability space is equal to the value of the constant, it follows from the definition of α, by multiplying out the last written integrand, that

$$\sigma^2(f) = \left(\int f^2 d\mu\right) - \left(\int f d\mu\right)^2;$$

it is clear that, for any real constant c, $\sigma^2(cf) = c^2\sigma^2(f)$.

Theorem C. *If f and g are independent functions with a finite variance, then*

$$\sigma^2(f+g) = \sigma^2(f) + \sigma^2(g).$$

Proof. We have

$$\sigma^2(f+g) = \int (f+g)^2 d\mu - \left(\int (f+g) d\mu\right)^2 =$$

$$= \int f^2 d\mu + 2\int fg d\mu + \int g^2 d\mu - \left(\int f d\mu\right)^2$$

$$- 2\left(\int f d\mu\right)\left(\int g d\mu\right) - \left(\int g d\mu\right)^2;$$

the desired result follows from Theorem A. ▮

(1) If F is a measurable set of positive measure in a probability space $(X,\mathbf{S},\mu)$, and if, for every measurable set E, $\mu_F(E) = \mu(F \cap E)/\mu(F)$, then μ_F is a probability measure on $\mathbf{S}$ such that $\mu_F(F) = 1$; the sets E and F are independent if and only if $\mu_F(E) = \mu(E)$. The number $\mu_F(E)$ is called the **conditional probability of E given F.**

(2) If $\{E_i: i = 1, \cdots, n\}$ is a finite class of measurable sets of positive measure, then

$$\mu(E_1 \cap \cdots \cap E_n) = \mu(E_1)\mu_{E_1}(E_2) \cdots \mu_{E_1 \cap \cdots \cap E_{n-1}}(E_n).$$

This result is known as the **multiplication theorem** for conditional probabilities.

(3) If $\{E_i: i = 1, \cdots, n\}$ is a finite, disjoint class of measurable sets of positive measure whose union is X (i.e. if $\{E_i\}$ is a partition of X), then, for every measurable set F, $\mu(F) = \sum_{i=1}^n \mu(E_i)\mu_{E_i}(F)$, and, if F has positive measure,

$$\mu_F(E_j) = \mu(E_j)\mu_{E_j}(F)/\textstyle\sum_{i=1}^n \mu(E_i)\mu_{E_i}(F).$$

This result is known as **Bayes' theorem.**

(4) Two partitions of X, say $\{E_i: i = 1, \cdots, n\}$ and $\{F_j: j = 1, \cdots, m\}$, are called independent if $\mu(E_i \cap F_j) = \mu(E_i)\mu(F_j)$ for $i = 1, \cdots, n$ and $j = 1, \cdots, m$. Two sets E and F are independent if and only if the partitions $\{E,E'\}$ and $\{F,F'\}$ are independent.

(5) Let $X = \{x: 0 \leqq x \leqq 1\}$ be the unit interval with Lebesgue measure. For every positive integer n define a function f_n on X by setting $f_n(x) = +1$ or -1 according as the integer i for which $\frac{i-1}{2^n} \leqq x < \frac{i}{2^n}$ is odd or even. The functions f_n are called the **Rademacher functions.** Any two of the functions f_1, f_2, and $f_1 f_2$ are independent, but the three together are not.

(6) If f and g are independent integrable functions, if M is a Borel set on the real line, and if $E = f^{-1}(M)$, then $\int_E fg d\mu = \int_E f d\mu \cdot \int g d\mu$. (Hint: observe that

$\chi_E(x) = \chi_M(f(x))$ and apply Theorem B to show that the product of f and $\chi_M(f)$ is independent of g.)

(7) If f and g are measurable functions with finite variances such that $\sigma(f)\sigma(g) \neq 0$, their **coefficient of correlation** is defined by

$$r(f,g) = \frac{\int fg d\mu - \int f d\mu \cdot \int g d\mu}{\sigma(f)\sigma(g)},$$

where $\sigma(f) = \sqrt{\sigma^2(f)}$ is the **standard deviation** of f. The functions f and g are **uncorrelated** if $r(f,g) = 0$. If f and g are independent, then they are uncorrelated. A necessary and sufficient condition that $\sigma^2(f+g) = \sigma^2(f) + \sigma^2(g)$ is that f and g be uncorrelated.

(8) Is it true that if f and g are uncorrelated, then they are independent? (Hint: let X be the unit interval and write $f(x) = \sin 2\pi x$, $g(x) = \cos 2\pi x$.)

(9) If f and g are independent integrable functions such that $(f+g)^2$ is integrable, then f^2 and g^2 are integrable.

§ 46. SERIES OF INDEPENDENT FUNCTIONS

Throughout this section we shall work with a fixed probability space $(X,\mathbf{S},\mu)$. Our first result is known as **Kolmogoroff's inequality.**

Theorem A. *If f_i, $i = 1, \cdots, n$ are independent functions such that $\int f_i d\mu = 0$, and $\int f_i^2 d\mu < \infty$, $i = 1, \cdots, n$, and if $f(x) = \bigcup_{k=1}^n \left| \sum_{i=1}^k f_i(x) \right|$ (i.e. f is the maximum of the absolute values of the partial sums of the f_i's), then, for every positive number ϵ,*

$$\mu(\{x: |f(x)| \geqq \epsilon\}) \leqq \frac{1}{\epsilon^2} \sum_{k=1}^n \sigma^2(f_k).$$

Proof. We write

$$E = \{x: |f(x)| \geqq \epsilon\}, \quad s_k = \sum_{i=1}^k f_i$$

and

$$E_k = \{x: |s_k(x)| \geqq \epsilon\} \cap \bigcap_{1 \leqq i < k} \{x: |s_i(x)| < \epsilon\}.$$

We have

$$\int_{E_k} s_n^2 d\mu = \int_{E_k} s_k^2 d\mu + \mu(E_k) \sum_{k<i\leqq n} \int f_i^2 d\mu \geqq$$

$$\geqq \int_{E_k} s_k^2 d\mu \geqq \mu(E_k)\epsilon^2.$$

Since $E = \bigcup_{k=1}^{n} E_k$ and since the sets E_k are disjoint, it follows that

$$\sum_{k=1}^{n} \sigma^2(f_k) = \int (f_1 + \cdots + f_n)^2 d\mu \geqq \int_E s_n^2 d\mu =$$
$$= \sum_{k=1}^{n} \int_{E_k} s_n^2 d\mu \geqq \sum_{k=1}^{n} \mu(E_k)\epsilon^2 = \mu(E)\epsilon^2. \quad \blacksquare$$

Theorem B. *If $\{f_n\}$ is a sequence of independent functions such that $\int f_n d\mu = 0$ and $\sum_{n=1}^{\infty} \sigma^2(f_n) < \infty$, then the series $\sum_{n=1}^{\infty} f_n(x)$ converges a.e.*

Proof. If we write

$$s_n(x) = \sum_{i=1}^{n} f_i(x), \quad n = 1, 2, \cdots,$$
$$a_m(x) = \sup \{| s_{m+k}(x) - s_m(x) |: k = 1, 2, \cdots\},$$
$$a(x) = \inf \{a_m(x): m = 1, 2, \cdots\},$$

then a necessary and sufficient condition for the convergence of $\sum_{n=1}^{\infty} f_n(x)$ at x is that $a(x) = 0$. By Kolmogoroff's inequality we have, for every positive number ϵ and every pair of positive integers m and n,

$$\mu(\{x: \bigcup_{k=1}^{n} | s_{m+k}(x) - s_m(x) | \geqq \epsilon\}) \leqq \frac{1}{\epsilon^2} \sum_{k=m+1}^{m+n} \sigma^2(f_k),$$

and therefore

$$\mu(\{x: a_m(x) \geqq \epsilon\}) \leqq \frac{1}{\epsilon^2} \sum_{k=m+1}^{\infty} \sigma^2(f_k).$$

It follows that

$$\mu(\{x: a(x) \geqq \epsilon\}) \leqq \frac{1}{\epsilon^2} \sum_{k=m+1}^{\infty} \sigma^2(f_k)$$

and therefore, using the convergence of $\sum_{n=1}^{\infty} \sigma^2(f_n)$, that $\mu(\{x: a(x) \geqq \epsilon\}) = 0$. The desired result is implied by the arbitrariness of ϵ. $\blacksquare$

The next result goes in the converse direction.

Theorem C. *If $\{f_n\}$ is a sequence of independent functions and c is a positive constant such that $\int f_n d\mu = 0$ and $| f_n(x) | \leqq c$*

a.e., $n = 1, 2, \cdots$, and if $\sum_{n=1}^{\infty} f_n(x)$ converges on a set of positive measure, then

$$\sum_{n=1}^{\infty} \sigma^2(f_n) < \infty.$$

Proof. If $s_0(x) = 0$ and $s_n(x) = \sum_{i=1}^{n} f_i(x)$, $n = 1, 2, \cdots$, then Egoroff's theorem implies (cf. 21.2) that there exists a positive number d such that the set

$$E = \bigcap_{n=0}^{\infty} \{x: | s_n(x) | \leqq d\}$$

has positive measure. If we write

$$E_n = \bigcap_{i=1}^{n} \{x: | s_i(x) | \leqq d\}, \quad n = 0, 1, 2, \cdots,$$

then $\{E_n\}$ is a decreasing sequence of sets whose intersection is E. If $F_n = E_{n-1} - E_n$, $n = 1, 2, \cdots$, and $\alpha_n = \int_{E_n} s_n^2 d\mu$, $n = 0, 1, 2, \cdots$, then

$$\alpha_n - \alpha_{n-1} = \int_{E_{n-1}} s_n^2 d\mu - \int_{F_n} s_n^2 d\mu - \int_{E_{n-1}} s_{n-1}^2 d\mu =$$

$$= \int_{E_{n-1}} f_n^2 d\mu + 2\int_{E_{n-1}} f_n s_{n-1} d\mu - \int_{F_n} s_n^2 d\mu, \quad n = 1, 2, \cdots.$$

Since

$$\int_{E_{n-1}} f_n^2 d\mu = \mu(E_{n-1})\sigma^2(f_n) \quad \text{and} \quad \int_{E_{n-1}} f_n s_{n-1} d\mu = 0,$$

and since $\mu(E_{n-1}) \geqq \mu(E)$ and $| s_n(x) | \leqq c + d$ for x in F_n, $n = 1, 2, \cdots$, it follows that

$$\alpha_n - \alpha_{n-1} \geqq \mu(E)\sigma^2(f_n) - (c + d)^2\mu(F_n), \quad n = 1, 2, \cdots.$$

Summing over n from 1 to k we obtain

$$d^2 \geqq \mu(E_k)d^2 \geqq \alpha_k \geqq \mu(E) \sum_{n=1}^{k} \sigma^2(f_n) - (c + d)^2. \quad ∎$$

We remark that Theorems B and C imply that if $\{f_n\}$ is a uniformly bounded sequence of independent functions such that $\int f_n d\mu = 0$ for $n = 1, 2, \cdots$, then the series $\sum_{n=1}^{\infty} f_n(x)$ either converges a.e. or diverges a.e.; the measure of the convergence set is always one of the extreme values, 0 or 1.

Theorem D. *If $\{f_n\}$ is a sequence of independent functions and c is a positive constant such that $|f_n(x)| \leqq c$ a.e., $n = 1, 2, \cdots$, then $\sum_{n=1}^{\infty} f_n(x)$ converges a.e. if and only if both the series $\sum_{n=1}^{\infty} \int f_n d\mu$ and $\sum_{n=1}^{\infty} \sigma^2(f_n)$ are convergent.*

Proof. The "if" follows by applying Theorem B to the sequence $\{g_n\}$ defined by $g_n(x) = f_n(x) - \int f_n d\mu$, $n = 1, 2, \cdots$. To prove the converse, we consider the Cartesian product of the space X with itself and on it the functions h_n defined by $h_n(x,y) = f_n(x) - f_n(y)$, $n = 1, 2, \cdots$. Since the convergence a.e. of $\sum_{n=1}^{\infty} f_n(x)$ implies that $\sum_{n=1}^{\infty} h_n(x,y)$ converges a.e., and since

$$\int h_n d(\mu \times \mu) = 0,$$

it follows from Theorem C that $\sum_{n=1}^{\infty} \sigma^2(h_n) < \infty$. Since, however, $\sigma^2(h_n) = 2\sigma^2(f_n)$, we see that $\sum_{n=1}^{\infty} \sigma^2(f_n) < \infty$. Since $\sigma^2(g_n) = \sigma^2(f_n)$, it follows from Theorem B that $\sum_{n=1}^{\infty} g_n(x)$ converges a.e. and therefore the relation

$$\int f_n d\mu = f_n(x) - g_n(x), \quad n = 1, 2, \cdots,$$

implies the convergence of $\sum_{n=1}^{\infty} \int f_n d\mu$. ∎

All our preceding results on series are included in the following very general assertion, known as Kolmogoroff's **three series theorem.**

Theorem E. *If $\{f_n\}$ is a sequence of independent functions and c is a positive constant, and if $E_n = \{x: |f_n(x)| \leqq c\}$, $n = 1, 2, \cdots$, then a necessary and sufficient condition for the convergence a.e. of $\sum_{n=1}^{\infty} f_n(x)$ is the convergence of all three series*

(a) $$\sum_{n=1}^{\infty} \mu(E_n'),$$

(b) $$\sum_{n=1}^{\infty} \int_{E_n} f_n d\mu,$$

(c) $$\sum_{n=1}^{\infty}\left(\int_{E_n} f_n^2 d\mu - \left(\int_{E_n} f_n d\mu\right)^2\right).$$

Proof. If we write

$$g_n(x) = \begin{cases} f_n(x) \\ c \end{cases} \quad \text{and} \quad h_n(x) = \begin{cases} f_n(x) \\ -c \end{cases} \quad \text{if} \quad \begin{cases} |f_n(x)| \leqq c, \\ |f_n(x)| > c, \end{cases}$$

then it is clear that the series

$$\sum_{n=1}^{\infty} f_n(x), \quad \sum_{n=1}^{\infty} g_n(x), \quad \text{and} \quad \sum_{n=1}^{\infty} h_n(x)$$

converge at the same points. It follows from Theorem D (applied separately to $\{g_n\}$ and $\{h_n\}$) that $\sum_{n=1}^{\infty} f_n(x)$ converges a.e. if and only if all four series

(d) $$\sum_{n=1}^{\infty}\left(\int_{E_n} f_n d\mu \pm c\mu(E_n')\right),$$

(e) $$\sum_{n=1}^{\infty}\left(\int_{E_n} f_n^2 d\mu - \left(\int_{E_n} f_n d\mu\right)^2 + c^2\mu(E_n)\mu(E_n') \mp 2c\mu(E_n')\int_{E_n} f_n d\mu\right)$$

are convergent. It is readily verified that the convergence of (d) and (e) (with all choices of the ambiguous signs) is equivalent to the convergence of (a), (b), and (c). All that the verification requires, in addition to the obvious additions and subtractions, is the remark that, since the terms of a convergent series are bounded, the termwise product of two convergent series one of which has non negative terms is convergent. ▮

(1) The following result, which is implicitly contained in our earlier discussion of the relation between mean convergence and convergence in measure, is known as **Tchebycheff's inequality.** If f is a measurable function with finite variance, then, for every positive number ϵ,

$$\mu(\{x: |f(x) - \int f d\mu| \geqq \epsilon\}) \leqq \frac{1}{\epsilon^2}\sigma^2(f).$$

Kolmogoroff's inequality for $n = 1$ reduces to Tchebycheff's. Since, in the notation of Theorem A,

$$\{x: |f(x)| \geqq \epsilon\} = \bigcup_{k=1}^{n} \{x: |\sum_{i=1}^{k} f_i(x)| \geqq \epsilon\},$$

an application of Tchebycheff's inequality separately to each partial sum yields

$$\mu(\{x: |f(x)| \geqq \epsilon\}) \leqq \frac{1}{\epsilon^2}\sum_{k=1}^{n} (n - k + 1)\sigma^2(f_k).$$

(2) An interesting special case of Theorem D is obtained by considering the sequence $\{f_n\}$ of Rademacher functions; (cf. 45.5). If $\{c_n\}$ is a sequence of real numbers, then $\sum_{n=1}^{\infty} c_n f_n(x)$ converges or diverges a.e. according as the series $\sum_{n=1}^{\infty} c_n^2$ converges or diverges. In the language of probability: a necessary and sufficient condition for the convergence with probability 1 of the series $\sum_{n=1}^{\infty} \pm c_n$ is the convergence of $\sum_{n=1}^{\infty} c_n^2$, it being understood that $+$ and $-$ are equally likely and that the ambiguous signs are determined independently of each other.

(3) The fact that the convergence set of a series of independent functions must have measure 0 or 1 is a consequence of the following very general principle, known as the **zero-one law**. Suppose that the probability space X is the Cartesian product of a sequence $\{X_n\}$ of probability spaces. If, for each positive integer n, $J_n = \{n+1, n+2, \cdots\}$, and if a measurable set E in X is a J_n-cylinder for every n, then $\mu(E) = 0$ or 1. (Hint: write, for every measurable set F, $\nu(F) = \mu(E \cap F)$. If F is a J-cylinder for a finite set J, then $\nu(F) = \mu(E)\mu(F)$; since a finite measure on the class of all measurable subsets of X is uniquely determined by its values on such cylinders, this relation remains valid for E in place of F.)

(4) If $\{E_n\}$ is a sequence of independent sets, then $\mu(\limsup_n E_n) = 0$ if and only if $\sum_{n=1}^{\infty} \mu(E_n) < \infty$; (cf. 9.6). (Hint: let χ_n be the characteristic function of E_n, and apply Theorem D to the sequence $\{\chi_n\}$.) This result is known as the **Borel-Cantelli lemma.**

(5) Two sequences $\{f_n\}$ and $\{g_n\}$ are **equivalent** in the sense of Khintchine if

$$\sum_{n=1}^{\infty} \mu(\{x : f_n(x) \neq g_n(x)\}) < \infty.$$

If $\{f_n\}$ is a sequence of independent functions, then a necessary and sufficient condition for the convergence a.e. of the series $\sum_{n=1}^{\infty} f_n(x)$ is the existence of an equivalent sequence $\{g_n\}$ of independent functions with finite **variances** such that the series $\sum_{n=1}^{\infty} \int g_n d\mu$ and $\sum_{n=1}^{\infty} \sigma^2(g_n)$ are convergent.

(6) If $\{f_n\}$ is a sequence of integrable functions and if f is a measurable function with finite variance such that, for every positive integer n, the functions

$$f_1, \cdots, f_n, f - (f_1 + \cdots + f_n)$$

are independent, then each f_n has finite variance and the series

$$\sum_{n=1}^{\infty} \left(f_n(x) - \int f_n d\mu\right)$$

converges a.e. (Hint: apply 45.9 and the three series theorem.)

§ 47. THE LAW OF LARGE NUMBERS

There are several limit theorems in the theory of probability which are collectively known as the law of large numbers; in this section we present two typical ones. The first of these is known as **Bernoulli's theorem** or the **weak law of large numbers.**

Theorem A. *If $\{f_n\}$ is a sequence of independent functions with finite variances, such that $\int f_n d\mu = 0$, $n = 1, 2, \cdots$, and $\lim_n \frac{1}{n^2} \sum_{i=1}^n \sigma^2(f_i) = 0$, then the sequence $\left\{\frac{1}{n} \sum_{i=1}^n f_i\right\}$ of averages converges to* 0 *in measure.*

Proof. Since σ^2 is homogeneous of degree 2 and, for independent functions, additive, we have

$$\int \left(\frac{1}{n} \sum_{i=1}^n f_i\right)^2 d\mu = \sigma^2\left(\frac{1}{n} \sum_{i=1}^n f_i\right) = \frac{1}{n^2} \sum_{i=1}^n \sigma^2(f_i).$$

In other words, the principal assumption of the theorem is equivalent to the assumption that the sequence of averages converges to zero in the mean of order two (i.e. converges to zero in the space $\mathfrak{L}_2$), and this implies convergence in measure. ∎

Two real valued measurable functions f and g on a probability space $(X,\mathbf{S},\mu)$ have the **same distribution** if $\mu(f^{-1}(M)) = \mu(g^{-1}(M))$ for all real Borel sets M. It is easy to verify that if f and g are two integrable functions with the same distribution and if $F = f^{-1}(M)$ and $G = g^{-1}(M)$, for some real Borel set M, then $\int_F f d\mu = \int_G g d\mu$. An interesting special case of Bernoulli's theorem is the one in which every two terms of the sequence $\{f_n\}$ have the same distribution. In this case $\sigma^2(f_n) = \sigma^2(f_1)$ for every positive integer n, and hence $\frac{1}{n^2} \sum_{i=1}^n \sigma^2(f_i) = \frac{1}{n} \sigma^2(f_1)$, so that the assumption on the asymptotic behavior of $\{\sigma^2(f_n)\}$ is automatically satisfied.

As auxiliary propositions for the proof of a sharper form of the law of large numbers we need the following two results from elementary analysis.

Theorem B. *If $\{y_n\}$ is a sequence of real numbers which converges to a finite limit y, then* $\lim_n \frac{1}{n} \sum_{i=1}^n y_i = y$.

Proof. Corresponding to every positive number ϵ, there exists a positive integer n_0 such that if $n > n_0$, then $|y_n - y| < \frac{\epsilon}{2}$. Let

n_1 be a positive integer greater than n_0 and such that

$$\frac{1}{n_1}\sum_{i=1}^{n_0}|y_i - y| < \frac{\epsilon}{2}.$$

If $n > n_1$, then

$$\left|\left(\frac{1}{n}\sum_{i=1}^{n} y_i\right) - y\right| = \left|\frac{1}{n}\sum_{i=1}^{n}(y_i - y)\right| \leqq$$

$$\leqq \left|\frac{1}{n}\sum_{i=1}^{n_0}(y_i - y)\right| + \left|\frac{1}{n}\sum_{i=n_0+1}^{n}(y_i - y)\right| <$$

$$< \frac{1}{n_1}\sum_{i=1}^{n_0}|y_i - y| + \frac{n - n_0}{n}\cdot\frac{\epsilon}{2} < \epsilon. \quad ▮$$

Theorem C. *If $\{y_n\}$ is a sequence of real numbers such that the series $\sum_{n=1}^{\infty}\frac{1}{n}y_n$ is convergent, then* $\lim_n \frac{1}{n}\sum_{i=1}^{n} y_i = 0.$

Proof. We write

$$s_0 = 0, \quad s_n = \sum_{i=1}^{n}\frac{1}{i}y_i, \quad t_n = \sum_{i=1}^{n} y_i, \quad n = 1, 2, \cdots.$$

Since $y_i = i(s_i - s_{i-1})$, $i = 1, 2, \cdots$, and

$$t_{n+1} = \sum_{i=1}^{n+1} i s_i - \sum_{i=1}^{n+1} i s_{i-1} = -\sum_{i=1}^{n} s_i + (n+1)s_{n+1},$$

$$n = 1, 2, \cdots,$$

it follows that

$$\frac{t_{n+1}}{n+1} = -\frac{n}{n+1}\cdot\frac{1}{n}\sum_{i=1}^{n} s_i + s_{n+1}.$$

Since the sequence $\{s_n\}$ converges to a finite limit, and since, by Theorem B, the sequence $\left\{\frac{1}{n}\sum_{i=1}^{n} s_i\right\}$ converges to the same limit, we have

$$\lim_n \frac{t_{n+1}}{n+1} = 0. \quad ▮$$

Theorem D. *If* $\{f_n\}$ *is a sequence of independent functions with finite variances, such that* $\int f_n d\mu = 0$, $n = 1, 2, \cdots$, *and* $\sum_{n=1}^{\infty} \frac{\sigma^2(f_n)}{n^2} < \infty$, *then the sequence* $\left\{\frac{1}{n}\sum_{i=1}^{n} f_i\right\}$ *converges to* 0 *almost everywhere.*

We remark that the hypothesis and, correspondingly, the conclusion of Theorem D are stronger than those of Theorem A. The present theorem is one form of the **strong law of large numbers.**

Proof. We write $g_n(x) = \frac{1}{n} f_n(x)$, $n = 1, 2, \cdots$, and apply 46.B to the sequence $\{g_n\}$. Since $\int g_n d\mu = 0$, $n = 1, 2, \cdots$, and

$$\sum_{n=1}^{\infty} \sigma^2(g_n) = \sum_{n=1}^{\infty} \frac{\sigma^2(f_n)}{n^2} < \infty,$$

it follows that the series

$$\sum_{n=1}^{\infty} \frac{1}{n} f_n(x)$$

converges a.e.; the desired result follows from Theorem C. ∎

(1) Two measurable functions have the same distribution if and only if they have the same distribution function; (cf. 18.11).

(2) If $\{\sigma_i^2\}$ is a sequence of non negative real numbers and if m and n are positive integers such that $m < n$, then

$$\frac{\sigma_1^2 + \cdots + \sigma_n^2}{n^2} \leqq \frac{\sigma_1^2 + \cdots + \sigma_m^2}{n^2} + \frac{\sigma_{m+1}^2}{(m+1)^2} + \cdots + \frac{\sigma_n^2}{n^2}.$$

This inequality can be used to show that the assumptions of Theorem D are not weaker than those of Theorem A. That they are properly stronger may be shown by constructing a sequence $\{f_n\}$ of independent functions for which $\sigma^2(f_n) = \frac{n+1}{\log(n+1)}$.

(3) Theorem D is the best possible result of its kind (involving restrictions on $\sigma^2(f_n)$ only) in the following sense: if $\{\sigma_n^2\}$ is a sequence of non negative real numbers such that $\sum_{n=1}^{\infty} \frac{\sigma_n^2}{n^2} = \infty$, then there exists a sequence $\{f_n\}$ of independent functions such that $\int f_n d\mu = 0$, $\sigma^2(f_n) = \sigma_n^2$, $n = 1, 2, \cdots$, and

$\left\{\frac{1}{n}\sum_{i=1}^{n} f_i\right\}$ does not converge to 0 a.e. (Hint: construct f_n so that if $\sigma_n^2 \leq n^2$, then

$$\mu(\{x: f_n(x) = n\}) = \mu(\{x: f_n(x) = -n\}) = \frac{\sigma_n^2}{2n^2},$$

$$\mu(\{x: f_n(x) = 0\}) = 1 - \frac{\sigma_n^2}{n^2},$$

and if $\sigma_n^2 > n^2$, then

$$\mu(\{x: f_n(x) = \sigma_n\}) = \mu(\{x: f_n(x) = -\sigma_n\}) = \tfrac{1}{2}.$$

Observe that if $\lim_n \frac{1}{n}\sum_{i=1}^{n} y_i = 0$, then $\lim_n \frac{1}{n} y_n = 0$, and apply the Borel–Cantelli lemma to $\{x: |f_n(x)| \geq n\}$.)

(4) If $\{f_n\}$ is a sequence of independent functions satisfying the conditions of Theorem D, then there exists an equivalent sequence $\{g_n\}$ of independent functions such that

$$\sum_{n=1}^{\infty} \frac{\sigma^2(g_n)}{n^2} = \infty;$$

in other words, the converse of the strong law of large numbers is not true.

(5) The following weak converse of the strong law of large numbers is true. If $\{f_n\}$ is a sequence of independent functions and c is a positive constant, such that $\int f_n d\mu = 0$ and $\left|\frac{1}{n} f_n(x)\right| \leq c$ a.e., $n = 1, 2, \cdots$, and if $\left\{\frac{1}{n}\sum_{i=1}^{n} f_i\right\}$ converges to 0 a.e. then

$$\sum_{n=1}^{\infty} \frac{\sigma^2(f_n)}{n^{2+\epsilon}} < \infty$$

for every positive number ϵ. (Hint: if $\{y_n\}$ is a sequence of real numbers such that $\lim_n \frac{1}{n}\sum_{i=1}^{n} y_i = 0$, or even such that the sequence $\left\{\frac{1}{n}\sum_{i=1}^{n} y_i\right\}$ is bounded, then the series $\sum_{n=1}^{\infty} \frac{y_n}{n^{1+\epsilon}}$ is convergent for every positive number ϵ.)

(6) The conclusion of Theorem D remains true if the assumption $\int f_n d\mu = 0$, $n = 1, 2, \cdots$, is replaced by $\lim_n \frac{1}{n}\sum_{i=1}^{n} \int f_i d\mu = 0$.

(7) The following is another theorem which is sometimes known as the strong law of large numbers. If $\{f_n\}$ is a sequence of independent integrable functions with the same distribution, such that $\int f_n d\mu = 0$, then $\lim_n \frac{1}{n}\sum_{i=1}^{n} f_i = 0$ a.e. The sequence of assertions below is designed to lead up to a proof of this result.

(7a) If $E_n = \{x: |f_1(x)| \leq n\}$, then $\sum_{n=1}^{\infty} \frac{1}{n^2}\int_{E_n} f_1^2 d\mu < \infty$. (Hint: let χ_n be the characteristic function of E_n, and write $g = \sum_{n=1}^{\infty} \frac{1}{n^2}\chi_n f_1^2$. If $k - 1 < |f_1(x)| \leq k$, then $\chi_n(x) = 0$ whenever $n < k$, and this implies by an elementary computation that $|g(x)| < 2|f_1(x)|$ and hence that g is integrable.)

(7b) If $F_n = \{x: |f_n(x)| \leqq n\}$ and if $g_n = \chi_{F_n} f_n$, then the sequence $\{g_n\}$ of independent functions is equivalent to $\{f_n\}$.

(7c) $\lim_n \frac{1}{n} \sum_{i=1}^n \int g_i d\mu = 0$. (Hint: $\int g_i d\mu = \int_{F_i} f_i d\mu = \int_{E_i} f_1 d\mu$ and $\{E_i\}$ is an increasing sequence of measurable sets whose union is X; (cf. Theorem B).)

(7d) $\sum_{n=1}^\infty \frac{1}{n^2} \sigma^2(g_n) < \infty$. (Hint: observe that $\int g_n^2 d\mu = \int_{F_n} f_n^2 d\mu = \int_{E_n} f_1^2 d\mu$ and apply (7a). This establishes the convergence of $\sum_{n=1}^\infty \frac{1}{n^2} \int g_n^2 d\mu$; the convergence of $\sum_{n=1}^\infty \frac{1}{n^2} \left(\int g_n d\mu\right)^2$ follows from the relation

$$\left(\int g_n d\mu\right)^2 \leqq \left(\int |f_1| d\mu\right)^2 .)$$

(8) The following converse of the version of the strong law of large numbers stated in (7) is true. If $\{f_n\}$ is a sequence of independent functions with the same distribution such that $\lim_n \frac{1}{n} \sum_{i=1}^n f_i = 0$ a.e., then f_n is integrable. (Hint: the relation $\lim_n \frac{1}{n} f_n = 0$ a.e., together with the Borel–Cantelli lemma, implies the convergence of the series $\sum_{n=1}^\infty \mu(\{x: |f_n(x)| > n\})$. Observe that $\mu(\{x: |f_n(x)| > n\}) = \mu(\{x: |f_1(x)| > n\})$ and apply 27.4.)

(9) Applying the strong law of large numbers to the Rademacher functions we obtain the celebrated theorem of Borel on **normal numbers**: almost every number in the unit interval has in its binary expansion an equal number of 0's and 1's. Similar considerations are valid with respect to any other radix r in place of 2 ($r \geqq 3$), and yield the theorem concerning **absolutely normal numbers**: almost every number is normal with respect to every radix simultaneously.

§ 48. CONDITIONAL PROBABILITIES AND EXPECTATIONS

If E and F are measurable subsets of a probability space $(X,\mathbf{S},\mu)$ such that $\mu(F) \neq 0$, we have defined the conditional probability of E given F by the equation

$$\mu_F(E) = \mu(E \cap F)/\mu(F)$$

(cf. § 44 and 45.1), and we have investigated slightly its dependence on E. We are now interested in the way in which $\mu_F(E)$ depends on F. If F is such that both $\mu(F)$ and $\mu(F')$ are different from 0, we introduce a measurable space Y consisting of exactly two points y_1 and y_2 (with the understanding that every subset of Y is measurable), and a measurable transformation T from X

into Y, defined by $T(x) = y_1$ or y_2 according as $x \varepsilon F$ or $x \varepsilon F'$. If for every subset A of Y we write

$$\nu_E(A) = \mu(E \cap T^{-1}(A)) \quad \text{and} \quad \nu(A) = \nu_X(A) = \mu(T^{-1}(A)),$$

then, clearly,

$$\mu_F(E) = \frac{\nu_E(\{y_1\})}{\nu(\{y_1\})} \quad \text{and} \quad \mu_{F'}(E) = \frac{\nu_E(\{y_2\})}{\nu(\{y_2\})}.$$

In other words conditional probability may be viewed as a measurable function on Y—that function which is, roughly speaking, the ratio of the two measures ν_E and ν.

Generalizing the considerations of the preceding paragraph, we may consider a finite, disjoint class $\{F_1, \cdots, F_n\}$ of measurable sets of positive measure such that $\bigcup_{i=1}^n F_i = X$, and correspondingly we may introduce a measurable space Y of n points $y_1, \cdots, y_n$. If $T(x) = y_i$ whenever $x \varepsilon F_i$, $i = 1, \cdots, n$, then T is a measurable transformation from X into Y, and once more we may represent conditional probabilities as ratios of two measures on Y. These considerations motivate the following general definition. If T is any measurable transformation from the probability space $(X,\mathbf{S},\mu)$ into a measurable space $(Y,\mathbf{T})$, and if we write $\nu_E(F) = \mu(E \cap T^{-1}(F))$ whenever E and F are measurable subsets of X and Y respectively, then it is clear that ν_E and μT^{-1} $(= \nu_X)$ are measures on $\mathbf{T}$ such that $\nu_E \ll \mu T^{-1}$. It follows from the Radon–Nikodym theorem that there exists an integrable function p_E on Y such that

$$\mu(E \cap T^{-1}(F)) = \int_F p_E(y) d\mu T^{-1}(y)$$

for every F in $\mathbf{T}$; the function p_E is uniquely determined modulo μT^{-1}. We shall call $p_E(y)$ the **conditional probability of E given y** or the conditional probability of E given that $T(x) = y$. Sometimes we shall use the phrase "the conditional probability of E for a given value of $T(x)$" to refer to the number $p_E(T(x))$. We shall generally write $p(E,y)$ for $p_E(y)$; on the occasions when it is necessary to consider p as a function of its first argument we shall write $p^y(E) = p(E,y)$.

If F is such that $\mu(T^{-1}(F)) \neq 0$, we may divide the equation which defines p by $\mu(T^{-1}(F))$ and obtain the relation

$$\mu_{T^{-1}(F)}(E) = \frac{\mu(E \cap T^{-1}(F))}{\mu(T^{-1}(F))} = \frac{1}{\mu(T^{-1}(F))} \int_F p(E,y) d\mu T^{-1}(y).$$

Since the extreme left term of this relation is the conditional probability of E given $T^{-1}(F)$, it is formally plausible that as "F shrinks to y," the left term should tend to the conditional probability of E given y and the right term should tend to the integrand $p(E,y)$. The use of the Radon–Nikodym theorem is a rigorous substitute for this rather shaky "difference quotient" approach.

Theorem A. *For each fixed measurable set E in X,*

$$0 \leqq p(E,y) \leqq 1 \ [\mu T^{-1}];$$

for each fixed disjoint sequence $\{E_n\}$ of measurable sets in X,

$$p(\bigcup_{n=1}^{\infty} E_n, y) = \sum_{n=1}^{\infty} p(E_n, y) \ [\mu T^{-1}].$$

Proof. The inequality is an immediate consequence of the fact that $0 \leqq \mu(E \cap T^{-1}(F)) \leqq \mu(E)$ for every measurable subset F of Y. To prove the equation, observe that

$$\int_F p(\bigcup_{n=1}^{\infty} E_n, y) d\mu T^{-1}(y) = \mu((\bigcup_{n=1}^{\infty} E_n) \cap T^{-1}(F)) =$$

$$= \sum_{n=1}^{\infty} \mu(E_n \cap T^{-1}(F)) = \sum_{n=1}^{\infty} \int_F p(E_n, y) d\mu T^{-1}(y) =$$

$$= \int_F (\sum_{n=1}^{\infty} p(E_n, y)) d\mu T^{-1}(y),$$

and apply the uniqueness assertion of the Radon–Nikodym theorem. ∎

Theorem A asserts that p^y behaves in certain respects like a measure. It is easy to obtain more evidence in this direction and to prove, for instance, that $p(X,y) = 1 \ [\mu T^{-1}]$, that if $E_1 \subset E_2$, then $p(E_1,y) \leqq p(E_2,y) \ [\mu T^{-1}]$, and that if $\{E_n\}$ is a decreasing sequence of measurable sets in X, then

$$p(\bigcap_{n=1}^{\infty} E_n, y) = \lim_n p(E_n, y) \ [\mu T^{-1}].$$

It is important, however, to remember that the exceptional sets of measure zero depend in each of these cases on the particular sets E_i under consideration, and it is in general incorrect to conclude that p^y is a measure for almost all values of y.

The defining equation of $p(E,y)$ may also be written in the form $\int_{T^{-1}(F)} \chi_E(x)d\mu(x) = \int_F p(E,y)d\mu T^{-1}(y)$. If, more generally, f is any integrable function on X, then we may consider its indefinite integral ν, defined by

$$\nu(F) = \int_{T^{-1}(F)} f(x)d\mu(x),$$

for all measurable sets F in Y, as a signed measure on $\mathbf{T}$. Since clearly $\nu \ll \mu T^{-1}$, it follows from the Radon–Nikodym theorem that there exists an integrable function e_f on Y such that

$$\int_{T^{-1}(F)} f(x)d\mu(x) = \int_F e_f(y)d\mu T^{-1}(y)$$

for every F in $\mathbf{T}$; the function e_f is uniquely determined modulo μT^{-1}. We shall call $e_f(y)$ the **conditional expectation of f given y**; we shall also write $e(f,y)$ instead of $e_f(y)$.

Since the relation between p and e is similar to the relation between a measure and an indefinite integral, it might seem that some such equation as

$$e(f,y) = \int f(x)dp^y(x)$$

ought to hold. Since, however, p^y is not in general a measure, the right term of this equation is undefined; the misbehavior of p is reflected, slightly enlarged, in the misbehavior of e.

Theorem B. *If f is an integrable function on Y, then fT is an integrable function on X and $e(fT,y) = f(y)$ $[\mu T^{-1}]$.*

Proof. It follows from 39.C that fT is integrable and that $\int_{T^{-1}(F)} f(T(x))d\mu(x) = \int_F f(y)d\mu T^{-1}(y)$ for every F in $\mathbf{T}$. ∎

(1) Suppose that $(X,\mathbf{S},\mu)$ and $(Y,\mathbf{T},\nu)$ are probability spaces and consider their Cartesian product $(X \times Y, \mathbf{S} \times \mathbf{T}, \mu \times \nu)$. If $T(x,y) = x$, then T is a measurable transformation from $X \times Y$ onto X. For every measurable set E

in $X \times Y$, $p(E,x) = \nu(E_x)$ $[\mu]$, and hence, in this case, p_E may indeed be defined for each E so that p^x is a measure for every x.

(2) Suppose that $(X,\mathbf{S},\mu)$ and $(Y,\mathbf{T},\nu)$ are probability spaces and let λ be a probability measure on $\mathbf{S} \times \mathbf{T}$ such that $\lambda \ll \mu \times \nu$, say $\lambda(E) = \int_E f d(\mu \times \nu)$. If $T(x,y) = x$, then, for every measurable set E in $X \times Y$,

$$p(E,x) = \int \chi_E(x,y) f(x,y) d\nu(y) \ [\mu].$$

(3) If T is a measurable transformation from a probability space $(X,\mathbf{S},\mu)$ into a measurable space $(Y,\mathbf{T})$, then $p(T^{-1}(F),y) = \chi_F(y)$ $[\mu T^{-1}]$ for every measurable set F in Y.

(4) The purpose of the following considerations is the construction of an example for which the conditional probabilities $p(E,y)$ cannot be determined so that p^y is a measure for almost every y. Let Y be the closed unit interval, let $\mathbf{T}$ be the class of all Borel subsets of Y, and let ν be Lebesgue measure on $\mathbf{T}$. Write $X = Y$ and let $\mathbf{S}$ be the σ-ring generated by $\mathbf{T}$ and a set M such that both M and M' are thick in Y. A probability measure μ is unambiguously defined on $\mathbf{S}$ by writing

$$\mu((A \cap M) \cup (B \cap M')) = \nu(A)$$

whenever A and B are in $\mathbf{T}$; we consider the transformation T from X onto Y defined by $T(x) = x$. Suppose that there exists a set C_0 of measure zero in $\mathbf{T}$ such that p^y is a measure on $\mathbf{S}$ whenever $y \, \varepsilon' \, C_0$.

(4a) If $D_0 = \{y: p(M,y) \neq 1\}$, then $\nu(D_0) = 0$.

(4b) If E_0 is the set of those points y for which it is not true that $p(T^{-1}(F),y) = \chi_F(y)$ identically for all F in $\mathbf{T}$, then $\nu(E_0) = 0$. (Hint: let $\mathbf{R}$ be a countable ring such that $\mathbf{S}(\mathbf{R}) = \mathbf{T}$. If, for each F in $\mathbf{R}$,

$$E_0(F) = \{y: p(T^{-1}(F),y) \neq \chi_F(y)\},$$

then $\nu(E_0(F)) = 0$. Make use of the fact that if two probability measures agree on $\mathbf{R}$, then they agree on $\mathbf{T}$ also.)

(4c) If $y \, \varepsilon' \, C_0 \cup D_0 \cup E_0$, then $y \, \varepsilon \, M$. (Hint: the relations $p(M,y) = 1$ and $p(T^{-1}(\{y\}),y) = 1$, together with the fact that p^y is a measure, imply that

$$p(M \cap T^{-1}(\{y\}),y) = 1.)$$

Since (4c) implies that the Borel set $C_0' \cap D_0' \cap E_0'$ of measure 1 is contained in the set M, we have derived a contradiction with the assumption that M' is thick.

(5) If X is the real line and μ is a probability measure on the class $\mathbf{S}$ of all Borel sets in X, and if T is a measurable transformation from X into a measurable space $(Y,\mathbf{T})$, then the conditional probabilities $p(E,y)$ may be determined so that p^y is a measure for almost every y. (Hint: write $q(x,y) = p((-\infty,x),y)$. There exists a measurable set C_0 in Y such that $\mu T^{-1}(C_0) = 0$ and such that if $y \, \varepsilon' \, C_0$, then q^y is a monotone function on the set of all rational numbers in X and, moreover, $\lim_n q^y\left(x - \frac{1}{n}\right) = q^y(x)$ for every rational number x. Let $\bar{q}^y$ be a left continuous monotone function on X which agrees with q^y for rational

values of x and let $\bar{p}^y$ be the measure on $\mathbf{S}$ determined by the conditions $\bar{p}^y((-\infty,x)) = \bar{q}^y(x)$; write $\bar{p}(E,y) = \bar{p}^y(E)$.)

(6) If T is a measurable transformation from a probability space $(X,\mathbf{S},\mu)$ into a measurable space $(Y,\mathbf{T})$, and if it is possible to determine the conditional probabilities $p(E,y)$ so that p^y is a measure for almost every y, then

$$e(f,y) = \int f(x)dp^y(x)\ [\mu T^{-1}]$$

for every integrable function f on X. (Hint: the relation is true if f is the characteristic function of a measurable set.)

(7) If T is a measurable transformation from a probability space $(X,\mathbf{S},\mu)$ into a measurable space $(Y,\mathbf{T})$, and if f and g are integrable functions with respect to μ and μT^{-1} respectively, such that the function h defined by $h(x) = f(x)g(T(x))$ is integrable on X, then

$$e(h,y) = e(f,y)g(y)\ [\mu T^{-1}].$$

§ 49. MEASURES ON PRODUCT SPACES

Does there exist a sequence of independent random variables with prescribed distributions? More precisely, if $\{\mu_n\}$ is a sequence of probability measures on the Borel sets of the real line, does there exist a probability space $(X,\mathbf{S},\mu)$ and a sequence $\{f_n\}$ of independent functions on X such that $\mu(f_n^{-1}(E)) = \mu_n(E)$ for every Borel set E and every positive integer n? More generally, if $\{(X_n,\mathbf{S}_n,\mu_n)\}$ is a sequence of probability spaces, does there exist a probability space $(X,\mathbf{S},\mu)$ and, for each positive integer n, a measurable transformation T_n from X into $X_1 \times \cdots \times X_n$ such that $\mu T_n^{-1} = \mu_1 \times \cdots \times \mu_n$? The affirmative answers to these questions are given by 38.B.

It is important for the purposes of probability theory to introduce the concept of independence, and, at the same time, to emphasize that it is not the general case. The main purpose of this section is to formulate and prove a theorem which does for dependent random variables what 38.B did for independent ones—a theorem, in other words, which asserts that there always exists a sequence of random variables with prescribed joint distributions. Unlike 38.B, however, the theorem of this section will apply to the case of uniformly bounded, real valued functions only; in other words, the components of the product space which we shall treat are all unit intervals. The result and its proof extend to more general cases, which, however, all have in common the fact that they depend on topological concepts. This peculiar and somewhat undesirable circumstance appears to be unavoidable; it is known that the general measure theoretic analog of Theorem A below is not true.

Suppose that, for each positive integer n, X_n is the closed unit interval and $\mathbf{S}_n$ is the class of all Borel sets in X_n, and write $(X,\mathbf{S}) = \times_{n=1}^{\infty} (X_n,\mathbf{S}_n)$. Let $\mathbf{F}_n$ be the σ-ring of all measurable $\{1, \cdots, n\}$-cylinders in X and let $\mathbf{F}$ $(= \bigcup_{n=1}^{\infty} \mathbf{F}_n)$ be the ring of all measurable, finite dimensional subsets of X; (cf. § 38).

Theorem A. *If μ is a set function on $\mathbf{F}$ such that, for each positive integer n, μ is a probability measure on $\mathbf{F}_n$, then μ has a unique extension to a probability measure on $\mathbf{S}$.*

Proof. We define a measurable transformation T_n, from X onto the measurable space $Y_n = \times_{i=1}^{n} X_i$, by

$$T_n(x_1, \cdots, x_n, x_{n+1}, \cdots) = (x_1, \cdots, x_n), \quad n = 1, 2, \cdots,$$

and we write, for every measurable subset A of Y_n, $\nu_n(A) = \mu(T_n^{-1}(A))$. If $\{E_i\}$ is a decreasing sequence of sets in $\mathbf{F}$ such that $0 < \epsilon \leqq \mu(E_i)$, $i = 1, 2, \cdots$, then, for each fixed i, there is a positive integer n and a Borel set A_i in Y_n such that $E_i = T_n^{-1}(A_i)$. Let B_i be a closed subset of A_i such that $\nu_n(A_i - B_i) \leqq \frac{\epsilon}{2^{i+1}}$. If $F_i = T_n^{-1}(B_i)$, then F_i is a compact subset of the product space X (in its product topology) and $\mu(E_i - F_i) \leqq \frac{\epsilon}{2^{i+1}}$. If $G_k = \bigcap_{i=1}^{k} F_i$, then $\{G_k\}$ is a decreasing sequence of compact subsets of X. Since

$$\mu(E_k - G_k) = \mu(\bigcup_{i=1}^{k} (E_k - F_i)) \leqq \mu(\bigcup_{i=1}^{k} (E_i - F_i)) \leqq \frac{\epsilon}{2},$$

it follows that

$$\mu(G_k) = \mu(E_k) - \mu(E_k - G_k) \geqq \frac{\epsilon}{2},$$

and hence that $G_k \neq 0$, $k = 1, 2, \cdots$. Since a decreasing sequence of non empty compact sets has a non empty intersection, it follows that μ is continuous from above at 0 and hence countably additive; the desired result follows from 13.A. ▌

Retaining the notation established above we proceed to the proof of an interesting property of product spaces of the type discussed in Theorem A.

Theorem B. *For every measurable set E in X,*

$$\lim_n p(E,T_n(x)) = \chi_E(x)\ [\mu];$$

in other words, the conditional probabilities of E, for given values of the first n coordinates of a point x, converge (except perhaps on a set of x's of measure zero) to 0 or 1 according as $x \varepsilon E$ or $x \varepsilon' E$.

Proof. It is convenient to prove almost uniform convergence instead of almost everywhere convergence—it follows, of course, from 21.A and 21.B that the two are equivalent. Let ϵ and δ be any two positive numbers and suppose that $\delta < 1$. By 13.D there exists a positive integer n_0 and a measurable $\{1, \cdots, n_0\}$–cylinder E_0 such that $\mu(E \Delta E_0) < \frac{\epsilon\delta}{2}$. We write $B = E \Delta E_0$ and we observe that if $x \varepsilon' B$, then

$$\chi_E(x) = \chi_{E_0}(x).$$

If $C_n = \{x\colon p(B,T_n(x)) \geqq \delta\}$, $D_n = C_n - \bigcup_{1 \leqq i < n} C_i$, $n = 1, 2, \cdots$, and $C = \bigcup_{n=1}^{\infty} C_n = \bigcup_{n=1}^{\infty} D_n$, then, for each n, C_n and D_n are measurable $\{1, \cdots, n\}$–cylinders. It follows that

$$\mu(B \cap D_n) = \int_{D_n} p(B,T_n(x))d\mu(x) \geqq \delta\mu(D_n),$$

and hence that

$$\begin{aligned}\frac{\epsilon\delta}{2} > \mu(B) &\geqq \mu(B \cap C) = \mu(B \cap \textstyle\bigcup_{n=1}^{\infty} D_n) = \\ &= \textstyle\sum_{n=1}^{\infty} \mu(B \cap D_n) \geqq \delta \sum_{n=1}^{\infty} \mu(D_n) = \\ &= \delta\mu(\textstyle\bigcup_{n=1}^{\infty} D_n) = \delta\mu(C).\end{aligned}$$

If we write $A = B \cup C$, then $\mu(A) \leqq \frac{\epsilon\delta}{2} + \frac{\epsilon}{2} < \epsilon$. Since

$$|p(E,T_n(x)) - p(E_0,T_n(x))| \leqq p(E \Delta E_0,T_n(x))\ [\mu], \qquad n = 1, 2, \cdots,$$

we may assume that these relations are valid for every x in X. If $n \geqq n_0$, then it follows from 38.A and 48.B that

$$| p(E,T_n(x)) - \chi_{E_0}(x) | \leqq p(B,T_n(x)).$$

If, in addition, $x \varepsilon' A$, then, in the first place $\chi_{E_0}(x) = \chi_E(x)$ and, in the second place $p(B,T_n(x)) < \delta$, so that $| p(E,T_n(x)) - \chi_E(x) | < \delta$. ∎

(1) Suppose that $\{(X_n,\mathbf{S}_n,\mu_n)\}$ is a sequence of probability spaces, $(X,\mathbf{S}) = \times_{n=1}^{\infty} (X_n,\mathbf{S}_n)$, and μ is a set function on $\mathbf{F}$ such that, for each positive integer n, μ is a probability measure on $\mathbf{F}_n$. If, on each $\mathbf{F}_n$, μ is absolutely continuous with respect to the product measure $\times_{i=1}^{\infty} \mu_i$, then μ has a unique extension to a probability measure on $\mathbf{S}$. (Hint: cf. the proof of 38.B.) The result and the method of proof extend to all cases in which the conditional probabilities $p(E,T_n(x))$ may be determined so that for almost every fixed x they define probability measures on each $\mathbf{F}_k$.

(2) The statement and the proof of Theorem A remain correct if the spaces X_n are compact metric spaces. It follows, by a trivial compactification, that Theorem A is true if each X_n is the real line. Does it remain true for arbitrary compact spaces?

(3) Retaining the notation of (1), we proceed to give an example to show that Theorem A is not necessarily true if the spaces X_n are not intervals. Let Y be the unit interval, $\mathbf{T}$ the class of all Borel subsets of Y, and ν Lebesgue measure on $\mathbf{T}$. Let $\{X_n\}$ be a decreasing sequence of thick subsets of Y such that $\bigcap_{n=1}^{\infty} X_n = 0$. Write $\mathbf{S}_n = \mathbf{T} \cap X_n$; if $E \varepsilon \mathbf{S}_n$, so that $E = F \cap X_n$ with F in $\mathbf{T}$, then write $\mu_n(E) = \nu(F)$. Form the product space $(X,\mathbf{S}) = \times_{n=1}^{\infty} (X_n,\mathbf{S}_n)$, and, for each positive integer n, let S_n be the measurable transformation from X_n into $X_1 \times \cdots \times X_n$ defined by $S_n(x_n) = (z_1, \cdots, z_n)$, $z_i = x_n$, $i = 1, \cdots, n$.

(3a) For each measurable $\{1, \cdots, n\}$-cylinder E in X,

$$(E = A \times X_{n+1} \times X_{n+2} \times \cdots, \quad A \varepsilon \mathbf{S}_1 \times \cdots \times \mathbf{S}_n),$$

write $\mu(E) = \mu_n(S_n^{-1}(A))$. The set function μ is thereby unambiguously defined on $\mathbf{F}$ and, for each fixed positive integer n, μ is a probability measure on $\mathbf{F}_n$.

(3b) If E_i is the set of all those points $(x_1, x_2, \cdots)$ in X whose first i coordinates are all equal to each other, $i = 1, 2, \cdots$, then $E_i \varepsilon \mathbf{F}_i$. (Hint: if

$$D_i = \{(y_1, \cdots, y_i): y_1 = \cdots = y_i\},$$

then D_i is a measurable subset of the i-dimensional Cartesian product of Y with itself, and

$$E_i = (D_i \cap (X_1 \times \cdots \times X_i)) \times X_{i+1} \times X_{i+2} \times \cdots.)$$

(3c) The set function μ on $\mathbf{F}$ is not continuous from above at 0. (Hint: consider the sets E_i, $i = 1, 2, \cdots$, defined in (3b), and observe that $\mu(E_i) = 1$ and $\bigcap_{i=1}^{\infty} E_i = 0$.)

(4) The zero-one law (46.3) is a special case of Theorem B. Indeed if E is a

J_n-cylinder and if F is a measurable subset of Y_n, then $T_n^{-1}(F)$ is a $\{1, \cdots, n\}$-cylinder and

$$\mu(E \cap T_n^{-1}(F)) = \mu(E)\mu T_n^{-1}(F) = \int_F \mu(E) d\mu_n,$$

and therefore $p(E, T_n(x))$ is a constant $(= \mu(E))$ almost everywhere $[\mu]$. It follows from Theorem B that $\chi_E(x) = \mu(E)$ $[\mu]$, and hence that $\mu(E)$ is either 0 or 1.

Chapter X

LOCALLY COMPACT SPACES

§ 50. TOPOLOGICAL LEMMAS

In this section we shall derive a few auxiliary topological results which, because of their special nature, are usually not discussed in topology books.

Throughout this chapter, unless in a special context we explicitly say otherwise, we shall assume that X is a locally compact Hausdorff space. We shall use the symbol $\mathfrak{F}$ for the class of all real valued, continuous functions f on X such that $0 \leqq f(x) \leqq 1$ for all x in X.

Theorem A. *If C is a compact set and U and V are open sets such that $C \subset U \cup V$, then there exist compact sets D and E such that $D \subset U$, $E \subset V$, and $C = D \cup E$.*

Proof. Since $C - U$ and $C - V$ are disjoint compact sets, there exist two disjoint open sets $\tilde{U}$ and $\tilde{V}$ such that $C - U \subset \tilde{U}$ and $C - V \subset \tilde{V}$; we write $D = C - \tilde{U}$ and $E = C - \tilde{V}$. It is easy to verify that $D \subset U$, $E \subset V$, and that D and E are compact; since $\tilde{U} \cap \tilde{V} = 0$, we have $D \cup E = (C - \tilde{U}) \cup (C - \tilde{V}) = C - (\tilde{U} \cap \tilde{V}) = C$. ▮

Theorem B. *If C is a compact set, F is a closed set, and $C \cap F = 0$, then there exists a function f in $\mathfrak{F}$ such that $f(x) = 0$ for x in C and $f(x) = 1$ for x in F.*

Proof. Since X is completely regular, corresponding to each point y in C there exists a function f_y in $\mathfrak{F}$ such that $f_y(y) = 0$ and $f_y(x) = 1$ for x in F. Since the class of all sets of the form

$\{x: f_y(x) < \frac{1}{2}\}$, y in C, is an open covering of C, and since C is compact, there exists a finite subset $\{y_1, \cdots, y_n\}$ of C such that

$$C \subset \bigcup_{i=1}^{n} \{x: f_{y_i}(x) < \tfrac{1}{2}\}.$$

If we write $g(x) = \prod_{i=1}^{n} f_{y_i}(x)$, then $g \in \mathfrak{F}$; since $0 \leqq f_y(x) \leqq 1$ for all x in X and all y in C, it follows that $g(x) < \frac{1}{2}$ for x in C and $g(x) = 1$ for x in F. It is easy to verify that if $f = (2g - 1) \cup 0$, then $f \in \mathfrak{F}$, $f(x) = 0$ for x in C, and $f(x) = 1$ for x in F. ∎

It is sometimes relevant to know not only whether or not a function f (in $\mathfrak{F}$) can be found which vanishes on C, as in Theorem B, but also whether or not it may be chosen so as not to vanish anywhere else. The answer is in general negative; the following theorem contains some of the pertinent details.

Theorem C. *If f is a real valued continuous function on X and c is a real number, then each of the three sets*

$$\{x: f(x) \geqq c\}, \quad \{x: f(x) \leqq c\}, \quad \textit{and} \quad \{x: f(x) = c\}$$

is a closed G_δ. If, conversely, C is a compact G_δ, then there exists a function f in $\mathfrak{F}$ such that $C = \{x: f(x) = 0\}$.

Proof. Since $\{x: f(x) \geqq c\} = \{x: -f(x) \leqq -c\}$ and since $\{x: f(x) = c\} = \{x: f(x) \geqq c\} \cap \{x: f(x) \leqq c\}$, it is sufficient to consider the set $\{x: f(x) \leqq c\}$. The fact that this set is closed (and that, for every $n = 1, 2, \cdots$, the set $\left\{x: f(x) < c + \frac{1}{n}\right\}$ is open) follows from the continuity of f; the fact that it is a G_δ is shown by the relation

$$\{x: f(x) \leqq c\} = \bigcap_{n=1}^{\infty} \left\{x: f(x) < c + \frac{1}{n}\right\}.$$

Suppose, conversely, that $C = \bigcap_{n=1}^{\infty} U_n$, where C is compact and $\{U_n\}$ is a sequence of open sets. For every $n = 1, 2, \cdots$, there exists a function f_n in $\mathfrak{F}$ (Theorem B) such that $f_n(x) = 0$ for x in C and $f_n(x) = 1$ for x in $X - U_n$. If we write $f(x) = \sum_{n=1}^{\infty} \frac{1}{2^n} f_n(x)$, then $f \in \mathfrak{F}$ and $f(x) = 0$ for x in C. For any x in $X - C$ there exists at least one positive integer n for which

$x \in X - U_n$; it follows that, for x in $X - C$, $f(x) \geqq \frac{1}{2^n} f_n(x) = \frac{1}{2^n} > 0$, and therefore that $C = \{x : f(x) = 0\}$. ▮

Theorem D. *If C is compact, U is open, and $C \subset U$, then there exist sets C_0 and U_0 such that C_0 is a compact G_δ, U_0 is a σ-compact open set, and*

$$C \subset U_0 \subset C_0 \subset U.$$

Proof. Since there exists a bounded open set V such that $C \subset V \subset U$, there is no loss of generality in assuming that U is bounded. Let f be a function in $\mathfrak{F}$ such that $f(x) = 0$ for x in C and $f(x) = 1$ for x in $X - U$, (Theorem B); write

$$U_0 = \{x : f(x) < \tfrac{1}{2}\} \quad \text{and} \quad C_0 = \{x : f(x) \leqq \tfrac{1}{2}\}.$$

Clearly $C \subset U_0 \subset C_0 \subset U$ and, by Theorem C, C_0 is a closed G_δ. The fact that C_0 is compact follows from the boundedness of U; the fact that U_0 is σ-compact is shown by the relation

$$U_0 = \bigcup_{n=1}^{\infty} \left\{x : f(x) \leqq \frac{1}{2} - \frac{1}{2^n}\right\}. \quad ▮$$

Theorem E. *If X is separable, then every compact subset C of X is a G_δ.*

Proof. If a point x of X is not in C, then there exist two disjoint open sets $U(x)$ and $V(x)$ such that $C \subset U(x)$ and $x \in V(x)$. Since X is separable and since the class $\{V(x) : x \in' C\}$ is an open covering of $X - C$, there exists a sequence $\{x_n\}$ of points in X such that

$$X - C \subset \bigcup_{n=1}^{\infty} V(x_n).$$

It follows that

$$\bigcap_{n=1}^{\infty} U(x_n) \supset C \supset \bigcap_{n=1}^{\infty} (X - V(x_n)) \supset \bigcap_{n=1}^{\infty} U(x_n). \quad ▮$$

(1) An alternative proof of Theorem B may be given by introducing the one-point compactification of X and using the known fact that every compact Hausdorff space is normal, and that therefore if C and D are two disjoint closed subsets of a compact Hausdorff space, then there exists a function f in $\mathfrak{F}$ such that $f(x) = 0$ for x in C and $f(x) = 1$ for x in D.

(2) Theorem C may be applied to prove the result, which is also easy to prove directly, that the class of all compact G_δ's is closed under the formation of finite unions and countable intersections.

(3) If X^* is the one-point compactification, by x^*, of an uncountable discrete space X, then the one-point set $\{x^*\}$ is a compact set which is not a G_δ.

(4) Let I be an arbitrary uncountable set; for each i in I, let X_i be the (compact Hausdorff) space consisting of the two real numbers 0 and 1, and let X be the Cartesian product $\times_i X_i$.

(4a) Every one-point set in X is a compact set which is not a G_δ.

(4b) We call a subset E of X an $\aleph_0$–set if there exists a countable set J in I such that E is a J–cylinder; (cf. 38.2). A compact set C in X is a G_δ if and only if it is an $\aleph_0$–set. (Hint: if C is compact, U is open, and $C \subset U$, then, by the definition of topology in X, there exist a finite subset J of I and an open set U_0 which is a J–cylinder such that $C \subset U_0 \subset U$.)

(4c) If f is any real valued continuous function on X and M is any Borel set on the real line, then $f^{-1}(M)$ is an $\aleph_0$–set.

(5) Let X^* and Y^* be the one-point compactifications (by x^* and y^*) of a countably infinite and an uncountable discrete space, respectively. The subsets

$$(\{x^*\} \times Y^*) - \{(x^*,y^*)\} \quad \text{and} \quad (X^* \times \{y^*\}) - \{(x^*,y^*)\}$$

of the locally compact Hausdorff space $(X^* \times Y^*) - \{(x^*,y^*)\}$ may be used to show that Theorem B is false if C is not required to be compact.

(6) The class of all σ–compact open sets is a base; (cf. Theorem D).

§ 51. BOREL SETS AND BAIRE SETS

The relations between measurability and continuity are most interesting, and have been studied most, in locally compact spaces. We continue with our study of a fixed locally compact Hausdorff space X; in the present section we shall introduce the basic concepts and results of a theory of measurability in X.

We shall denote by **C** the class of all compact subsets of X, by **S** the σ–ring generated by **C**, and by **U** the class of all open sets belonging to **S**. We shall call the sets of **S** the **Borel sets** of X, so that, for instance, **U** may be described as the class of all open Borel sets. A real valued function on X is **Borel measurable** (or simply a **Borel function**) if it is measurable with respect to the σ–ring **S**.

Theorem A. *Every Borel set is σ–bounded; every σ–bounded open set is a Borel set.*

Proof. Every compact set is trivially bounded and therefore σ–bounded. The class of all σ–bounded sets is a σ–ring; since this

σ-ring includes $\mathbf{C}$, it contains every set of the σ-ring generated by $\mathbf{C}$.

Suppose, conversely, that U is open and that $\{C_n\}$ is a sequence of compact sets such that

$$U \subset \bigcup_{n=1}^{\infty} C_n = K.$$

Since, for $n = 1, 2, \cdots, C_n - U$ is compact, it follows that

$$D = \bigcup_{n=1}^{\infty} (C_n - U) \in \mathbf{S};$$

since $D = K - U$, it follows that $U = K - (K - U) \in \mathbf{S}$. ▮

We shall denote by $\mathbf{C}_0$ the class of all those compact subsets of X which are G_δ's, by $\mathbf{S}_0$ the σ-ring generated by $\mathbf{C}_0$, and by $\mathbf{U}_0$ the class of all open sets belonging to $\mathbf{S}_0$. We shall call the sets of $\mathbf{S}_0$ the **Baire sets** of X, so that, for instance, $\mathbf{U}_0$ may be described as the class of all open Baire sets. A real valued function on X is **Baire measurable** (or simply a **Baire function**) if it is measurable with respect to the σ-ring $\mathbf{S}_0$.

On first glance it might appear that the Borel sets are the obvious objects of measure theoretic investigation in locally compact spaces. There are, however, several natural reasons for the introduction of the apparently artificial concept of Baire set. First: the theory of Baire sets is in some respects simpler than the theory of Borel sets, and knowledge about Baire sets frequently provides a successful tool for dealing with Borel sets; (cf. § 63). Second: the study of Baire sets is connected with the reasonable requirement that the concept of measurability in X should be so defined as to ensure that every continuous function (or at least every continuous function which vanishes outside some compact set) is measurable; (cf. Theorem B below). Third: the class of all Baire sets plays a distinguished role, in that it is the minimal σ-ring which contains sufficiently many sets to describe the topology of X; (cf. Theorem C below). Fourth: in all classical special cases of the theory of measure in topological spaces (e.g. in Euclidean spaces) the concepts of Borel set and Baire set coalesce; (cf. 50.E).

Theorem B. *If a real valued, continuous function f on X is such that the set $N(f) = \{x: f(x) \neq 0\}$ is σ-bounded, then f is Baire measurable.*

Proof. If a σ-bounded open set U is an F_σ, then there exists a sequence $\{C_n\}$ of compact sets such that $U = \bigcup_{n=1}^\infty C_n$. By 50.D, for each positive integer n there exists a compact Baire set D_n such that $C_n \subset D_n \subset U$. It follows that $U = \bigcup_{n=1}^\infty D_n$ and hence that U is a Baire set. The assumptions on f imply that, for every real number c, the set $N(f) \cap \{x: f(x) < c\}$ is a σ-bounded open set which is an F_σ. ▌

Theorem C. *If* **B** *is a subbase and if* $\hat{\mathbf{S}}$ *is a* σ*-ring containing* **B**, *then* $\hat{\mathbf{S}} \supset \mathbf{S}_0$.

Proof. If C is a compact set and U is an open set containing C, then there exists a set E which is a finite union of finite intersections of sets of **B** (and which therefore belongs to $\hat{\mathbf{S}}$) such that $C \subset E \subset U$. Hence, if $C = \bigcap_{n=1}^\infty U_n$, where each U_n is open, then, for every $n = 1, 2, \cdots$, there exists a set E_n in $\hat{\mathbf{S}}$ such that $C \subset E_n \subset U_n$; it follows that $C = \bigcap_{n=1}^\infty E_n \in \hat{\mathbf{S}}$. Since we have thus proved that $\mathbf{C}_0 \subset \hat{\mathbf{S}}$, the desired result follows from the definition of $\mathbf{S}_0$. ▌

The class of Baire sets was defined to be the σ-ring generated by the class of compact G_δ's; it appears conceivable (though upon reflection somewhat improbable) that a compact set may be a Baire set without being a G_δ, i.e. that compact sets other than the generating ones manage to get into $\mathbf{S}_0$. The purpose of the following theorem is to show that this does not happen.

Theorem D. *Every compact Baire set is a* G_δ.

Proof. Let C be a compact set in $\mathbf{S}_0$; by 5.D, there exists a sequence $\{C_n\}$ of sets in $\mathbf{C}_0$ such that C belongs to the σ-ring $\mathbf{S}(\{C_n\})$. By 50.C, for every $n = 1, 2, \cdots$, there exists a function f_n in $\mathfrak{F}$ such that $C_n = \{x: f_n(x) = 0\}$. If for each pair, x and y, of points in X we write

$$d(x,y) = \sum_{n=1}^\infty \frac{1}{2^n} |f_n(x) - f_n(y)|,$$

then $d(x,x) = 0$, $d(x,y) = d(y,x)$, and $0 \leqq d(x,y) \leqq d(x,z) + d(z,y)$. It follows that if we write $x \equiv y$ whenever $d(x,y) = 0$, then the relation "$\equiv$" is reflexive, symmetric, and transitive, and therefore an equivalence relation; we denote the set of all equivalence classes by Ξ. For every x in X we write $\xi = T(x)$ for the (uniquely determined) equivalence class which contains x.

If $T(x_1) = T(y_1)$ and $T(x_2) = T(y_2)$ (i.e. if $x_1 \equiv y_1$ and $x_2 \equiv y_2$), then

$$d(x_1,x_2) \leqq d(x_1,y_1) + d(y_1,y_2) + d(y_2,x_2) = d(y_1,y_2),$$

and, by symmetry, $d(y_1,y_2) \leqq d(x_1,x_2)$, so that $d(x_1,x_2) = d(y_1,y_2)$. This means that if $\xi_1 = T(x_1)$ and $\xi_2 = T(x_2)$ are two elements of Ξ, then the equation $\delta(\xi_1,\xi_2) = d(x_1,x_2)$ unambiguously defines the number $\delta(\xi_1,\xi_2)$. Since $\delta(\xi_1,\xi_2) = 0$ implies that $\xi_1 = \xi_2$, the function δ is a metric on Ξ. If $\xi_0 = T(x_0)$ is any point of the metric space Ξ, if r_0 is any positive number, and if $E = \{\xi: \delta(\xi_0,\xi) < r_0\}$, then $T^{-1}(E) = \{x: d(x_0,x) < r_0\}$; since $d(x_0,x)$ depends continuously on x, this proves that T is a continuous transformation from X onto Ξ.

A subset of X is the inverse image (under T) of a subset of Ξ if and only if it has the property that it contains, along with any of its points, all points equivalent to that one (i.e. if and only if it is a union of equivalence classes). Since each C_n has this property, since the class of all inverse image sets is a σ-ring, and since $C \varepsilon \mathbf{S}(\{C_n\})$, it follows that there exists a subset Γ of Ξ with $T^{-1}(\Gamma) = C$. Since $T(T^{-1}(\Gamma)) = \Gamma$, since T is continuous, and since C is compact, it follows that Γ is compact. Since every closed (and therefore every compact) subset of a metric space is a G_δ, there exists a sequence $\{\Delta_n\}$ of open subsets of Ξ with

$$\Gamma = \bigcap_{n=1}^{\infty} \Delta_n.$$

If we write $U_n = T^{-1}(\Delta_n)$, $n = 1, 2, \cdots$, then $C = \bigcap_{n=1}^{\infty} U_n$; since, by the continuity of T, U_n is open, it follows that $C \varepsilon \mathbf{C}_0$. ∎

Theorem E. *If X and Y are locally compact Hausdorff spaces, and if $\mathbf{A}_0$, $\mathbf{B}_0$, and $\mathbf{S}_0$ are the σ-rings of Baire sets in X, Y, and $X \times Y$ respectively, then $\mathbf{S}_0 = \mathbf{A}_0 \times \mathbf{B}_0$.*

Proof. If A and B are compact Baire sets in X and Y respectively, then $A \times B$ is a compact G_δ, and hence a compact Baire set in $X \times Y$. Since $\mathbf{A}_0 \times \mathbf{B}_0$ is the σ-ring generated by the class of all sets of the form $A \times B$, it follows that $\mathbf{A}_0 \times \mathbf{B}_0 \subset \mathbf{S}_0$. If U and V are open Baire sets in X and Y respectively, then

$U \times V \in \mathbf{A}_0 \times \mathbf{B}_0$. Since the class of all sets of the form $U \times V$ is a base for $X \times Y$, it follows from Theorem C that $\mathbf{A}_0 \times \mathbf{B}_0 \supset \mathbf{S}_0$. ∎

We conclude this section by stating, for the purpose of reference, an easily verified theorem related to the generation of Borel sets and Baire sets; (cf. 5.2 and 5.3).

Theorem F. *The class of all finite, disjoint unions of proper differences of sets of* $\mathbf{C}$ [*or of* $\mathbf{C}_0$] *is a ring; the* σ*–ring it generates coincides with* $\mathbf{S}$ [*or, respectively, with* $\mathbf{S}_0$].

(1) The definition of Borel set for the real line, when it is considered as a locally compact space, is equivalent with the definition in § 15.

(2) The entire space X is a Borel set if and only if it is σ–compact.

(3) The σ–ring generated by the class of all bounded open sets, or, equivalently, the σ–ring generated by $\mathbf{U}$, coincides with $\mathbf{S}$. (Hint: for every compact set C, let U be a bounded open set containing C, and consider $U - (U - C)$.)

(4) If X is the product space of 50.4, then the class of Baire sets coincides with the class of measurable sets, as defined in § 38.

(5) The σ–ring generated by the class of all bounded open Baire sets, or, equivalently, the σ–ring generated by $\mathbf{U}_0$, coincides with $\mathbf{S}_0$. (Hint: if C is compact, U is open, and $C \subset U$, then there exists a bounded open Baire set U_0 such that $C \subset U_0 \subset U$.)

(6) The term "Baire set" is suggested by the term "Baire function" as used in analysis. If $\mathfrak{B}$ is the smallest class of functions which contains all continuous functions and contains the limit of every pointwise (but not necessarily uniformly) convergent sequence of functions in it, then the functions of $\mathfrak{B}$ are called the **Baire functions** on X. A necessary and sufficient condition that a set be a Baire set is that it be a Borel set and that its characteristic function be a Baire function.

(7) Every Boolean σ–algebra is isomorphic to the class of all Baire sets, modulo Baire sets of the first category, in a totally disconnected, compact Hausdorff space. (Hint: cf. 40.15c and observe that the σ–ring generated by the class of all open-closed sets in a totally disconnected, compact Hausdorff space coincides with the class of all Baire sets.)

§ 52. REGULAR MEASURES

A **Borel measure** is a measure μ defined on the class $\mathbf{S}$ of all Borel sets and such that $\mu(C) < \infty$ for every C in $\mathbf{C}$; a **Baire measure** is a measure μ_0 defined on the class $\mathbf{S}_0$ of all Baire sets and such that $\mu_0(C_0) < \infty$ for every C_0 in $\mathbf{C}_0$.

Several aspects of the theories of Borel measures and Baire measures are so similar to each other that it is worth while to develop them simultaneously; for this purpose we adopt the following notational device. Throughout this section we shall use $\hat{\mathbf{C}}$, $\hat{\mathbf{U}}$, and $\hat{\mathbf{S}}$ to stand either for $\mathbf{C}$, $\mathbf{U}$, and $\mathbf{S}$ or else for $\mathbf{C}_0$, $\mathbf{U}_0$, and $\mathbf{S}_0$, respectively, and we shall study a measure $\hat{\mu}$ which is a Borel measure if $\hat{\mathbf{S}} = \mathbf{S}$ and a Baire measure if $\hat{\mathbf{S}} = \mathbf{S}_0$.

A set E in $\hat{\mathbf{S}}$ is **outer regular** (with respect to the measure $\hat{\mu}$) if

$$\hat{\mu}(E) = \inf \{\hat{\mu}(U) \colon E \subset U \in \hat{\mathbf{U}}\};$$

a set E in $\hat{\mathbf{S}}$ is **inner regular** (with respect to $\hat{\mu}$) if

$$\hat{\mu}(E) = \sup \{\hat{\mu}(C) \colon E \supset C \in \hat{\mathbf{C}}\}.$$

A set E in $\hat{\mathbf{S}}$ is **regular** if it is both inner regular and outer regular; a measure $\hat{\mu}$ is regular if every set E in $\hat{\mathbf{S}}$ is regular.

Loosely speaking, a measure is regular if all its values may be calculated from its values on the topologically important compact sets and open sets; if it is desired that the measure theoretic structure of X be not completely unrelated to its topological structure, the condition of regularity is a natural one to impose. The measure theoretic behavior of a non regular set is very pathological.

It is easy to verify that if $E \in \hat{\mathbf{S}}$ and $\hat{\mu}(E) = \infty$, or if $E \in \hat{\mathbf{U}}$, or if E is the intersection of a sequence of sets of finite measure in $\hat{\mathbf{U}}$, then E is outer regular. Dually, if $E \in \hat{\mathbf{S}}$ and $\hat{\mu}(E) = 0$, or if $E \in \hat{\mathbf{C}}$, or if E is the union of a sequence of sets in $\hat{\mathbf{C}}$, then E is inner regular. Our first purpose in the sequel is to show that the regularity of certain sets implies the regularity of many others. The motivation of the particular steps in the proof is furnished by 51.F; we progress from compact sets to their differences, and from differences to unions of differences. After that we shall show that the class of regular sets has sufficient closure properties to justify the application of the theorem on the monotone class generated by a ring, and thus we shall obtain the conclusion that certain measures are necessarily regular.

Theorem A. *If every set in $\hat{\mathbf{C}}$ is outer regular, then so is every proper difference of two sets of $\hat{\mathbf{C}}$; if every bounded set in $\hat{\mathbf{U}}$ is inner regular, then so is every proper difference of two sets of $\hat{\mathbf{C}}$.*

Proof. Let C and D be two sets in $\hat{\mathbf{C}}$ such that $C \supset D$. If C is outer regular, then, for every $\epsilon > 0$, there is a set U in $\hat{\mathbf{U}}$ such that $C \subset U$ and $\hat{\mu}(U) \leqq \hat{\mu}(C) + \epsilon$. Since $C - D \subset U - D \,\varepsilon\, \hat{\mathbf{U}}$, the relations

$$\hat{\mu}(U - D) - \hat{\mu}(C - D) = \hat{\mu}((U - D) - (C - D)) =$$
$$= \hat{\mu}(U - C) = \hat{\mu}(U) - \hat{\mu}(C) \leqq \epsilon$$

imply that $C - D$ is outer regular.

To prove the assertion concerning inner regularity, let U be a bounded set in $\hat{\mathbf{U}}$ such that $C \subset U$. If the bounded set $U - D$ (in $\hat{\mathbf{U}}$) is inner regular, then, for every $\epsilon > 0$, there is a set E in $\hat{\mathbf{C}}$ such that $E \subset U - D$ and $\hat{\mu}(U - D) \leqq \hat{\mu}(E) + \epsilon$. Since $C - D = C \cap (U - D) \supset C \cap E \,\varepsilon\, \hat{\mathbf{C}}$, the relations

$$\hat{\mu}(C - D) - \hat{\mu}(C \cap E) = \hat{\mu}((C - D) - (C \cap E)) =$$
$$= \hat{\mu}((C - D) - E) \leqq$$
$$\leqq \hat{\mu}((U - D) - E) =$$
$$= \hat{\mu}(U - D) - \hat{\mu}(E) \leqq \epsilon$$

imply that $C - D$ is inner regular. ▮

Theorem B. *A finite, disjoint union of inner regular sets of finite measure is inner regular.*

Proof. If $\{E_1, \cdots, E_n\}$ is a finite, disjoint class of inner regular sets of finite measure, then, for every $\epsilon > 0$ and for every $i = 1, \cdots, n$, there exists a set C_i in $\hat{\mathbf{C}}$ such that

$$C_i \subset E_i \quad \text{and} \quad \hat{\mu}(E_i) \leqq \hat{\mu}(C_i) + \frac{\epsilon}{n}.$$

If $C = \bigcup_{i=1}^{n} C_i$ and $E = \bigcup_{i=1}^{n} E_i$, then $E \supset C \,\varepsilon\, \hat{\mathbf{C}}$, and the relations

$$\hat{\mu}(E) = \sum_{i=1}^{n} \hat{\mu}(E_i) \leqq \sum_{i=1}^{n} \hat{\mu}(C_i) + \epsilon = \hat{\mu}(C) + \epsilon$$

imply that E is inner regular. ▮

It is easy, but unnecessary, to prove the analogous result for outer regular sets; the following theorem is much more inclusive.

Theorem C. *The union of a sequence of outer regular sets is outer regular; the union of an increasing sequence of inner regular sets is inner regular.*

Proof. If $\{E_i\}$ is a sequence of outer regular sets, then, for every $\epsilon > 0$ and for every $i = 1, 2, \cdots$, there exists a set U_i in $\hat{\mathbf{U}}$ such that

$$E_i \subset U_i \quad \text{ana} \quad \hat{\mu}(U_i) \leqq \hat{\mu}(E_i) + \frac{\epsilon}{2^i};$$

we write $U = \bigcup_{i=1}^{\infty} U_i$. If $E = \bigcup_{i=1}^{\infty} E_i$ and $\hat{\mu}(E) = \infty$, then E is trivially outer regular; if $\hat{\mu}(E) < \infty$, then

$$\hat{\mu}(U) - \hat{\mu}(E) = \hat{\mu}(U - E) \leqq \hat{\mu}(\bigcup_{i=1}^{\infty} (U_i - E_i)) \leqq$$
$$\leqq \sum_{i=1}^{\infty} \hat{\mu}(U_i - E_i) = \sum_{i=1}^{\infty} (\hat{\mu}(U_i) - \hat{\mu}(E_i)) \leqq \epsilon.$$

If $\{E_i\}$ is an increasing sequence of inner regular sets and $E = \bigcup_{i=1}^{\infty} E_i$, we make use of the relation

$$\hat{\mu}(E) = \lim_i \hat{\mu}(E_i).$$

We are to prove that, for every real number c with $c < \hat{\mu}(E)$, there is a set E in $\hat{\mathbf{C}}$ such that $C \subset E$ and $c < \hat{\mu}(C)$. To prove this, we need only select a value of i so that $c < \hat{\mu}(E_i)$, and then, using the inner regularity of E_i, find a set C in $\hat{\mathbf{C}}$ such that $C \subset E_i$ and $c < \hat{\mu}(C)$. ∎

Theorem D. *The intersection of a sequence of inner regular sets of finite measure is inner regular; the intersection of a decreasing sequence of outer regular sets of finite measure is outer regular.*

Proof. If $\{E_i\}$ is a sequence of inner regular sets of finite measure, then, for every $\epsilon > 0$ and for every $i = 1, 2, \cdots$, there exists a set C_i in $\hat{\mathbf{C}}$ such that

$$C_i \subset E_i \quad \text{and} \quad \hat{\mu}(E_i) \leqq \hat{\mu}(C_i) + \frac{\epsilon}{2^i};$$

we write $C = \bigcap_{i=1}^{\infty} C_i$. If $E = \bigcap_{i=1}^{\infty} E_i$, then $E \supset C \,\varepsilon\, \hat{\mathbf{C}}$ and

$$\hat{\mu}(E) - \hat{\mu}(C) = \hat{\mu}(E - C) \leqq \hat{\mu}(\bigcup_{i=1}^{\infty} (E_i - C_i)) \leqq$$
$$\leqq \sum_{i=1}^{\infty} \hat{\mu}(E_i - C_i) = \sum_{i=1}^{\infty} (\hat{\mu}(E_i) - \hat{\mu}(C_i)) \leqq \epsilon.$$

If $\{E_i\}$ is a decreasing sequence of outer regular sets of finite measure and $E = \bigcap_{i=1}^{\infty} E_i$, we make use of the relation

$$\hat{\mu}(E) = \lim_i \hat{\mu}(E_i).$$

We are to prove that, for every real number c with $c > \hat{\mu}(E)$, there is a set U in $\hat{\mathbf{U}}$ such that $E \subset U$ and $c > \hat{\mu}(U)$. To prove this we need only select a value of i so that $c > \hat{\mu}(E_i)$ and then, using the outer regularity of E_i, find a set U in $\hat{\mathbf{U}}$ such that $E_i \subset U$ and $\hat{\mu}(U) < c$. ∎

The duality between inner and outer regularity is even more thoroughgoing than is indicated by the similarities among the above proofs; we proceed to prove that the two kinds of regularity are essentially the same.

Theorem E. *A necessary and sufficient condition that every set in* $\hat{\mathbf{C}}$ *be outer regular is that every bounded set in* $\hat{\mathbf{U}}$ *be inner regular.*

Proof. Suppose that every set in $\hat{\mathbf{C}}$ is outer regular, let U be a bounded set in $\hat{\mathbf{U}}$, and let ϵ be a positive number. Let C be a set in $\hat{\mathbf{C}}$ such that $U \subset C$; since $C - U$ is compact and belongs to $\hat{\mathbf{S}}$, it follows from 51.D that $C - U \in \hat{\mathbf{C}}$, and therefore that there exists a set V in $\hat{\mathbf{U}}$ such that

$$C - U \subset V \quad \text{and} \quad \hat{\mu}(V) \leqq \hat{\mu}(C - U) + \epsilon.$$

Since $U = C - (C - U) \supset C - V \in \hat{\mathbf{C}}$, the relations

$$\hat{\mu}(U) - \hat{\mu}(C - V) = \hat{\mu}(U - (C - V)) = \hat{\mu}(U \cap V) \leqq$$
$$\leqq \hat{\mu}(V - (C - U)) = \hat{\mu}(V) - \hat{\mu}(C - U) \leqq \epsilon$$

imply that U is inner regular.

Suppose next that every bounded set in $\hat{\mathbf{U}}$ is inner regular, let C be a set in $\hat{\mathbf{C}}$, and let ϵ be a positive number. Let U be a bounded set in $\hat{\mathbf{U}}$ such that $C \subset U$; since $U - C$ is a bounded set in $\hat{\mathbf{U}}$, there exists a set D in $\hat{\mathbf{C}}$ such that

$$D \subset U - C \quad \text{and} \quad \hat{\mu}(U - C) \leqq \hat{\mu}(D) + \epsilon.$$

Since $C = U - (U - C) \subset U - D \in \hat{\mathbf{U}}$, the relations

$$\hat{\mu}(U - D) - \hat{\mu}(C) = \hat{\mu}((U - D) - C) = \hat{\mu}((U - C) - D) = \\ = \hat{\mu}(U - C) - \hat{\mu}(D) \leqq \epsilon$$

imply that C is outer regular. ▮

Theorem F. *Either the outer regularity of every set in* $\hat{\mathbf{C}}$ *or the inner regularity of every bounded set in* $\hat{\mathbf{U}}$ *is a necessary and sufficient condition for the regularity of the measure* $\hat{\mu}$.

Proof. The necessity of both conditions is trivial. To prove sufficiency, it is enough (Theorem C) to prove that every bounded set in $\hat{\mathbf{S}}$ is regular, since every set in $\hat{\mathbf{S}}$ is the union of an increasing sequence of bounded sets in $\hat{\mathbf{S}}$. Let E_0 be a bounded set in $\hat{\mathbf{S}}$ and let C_0 be a set in $\hat{\mathbf{C}}$ such that $E_0 \subset C_0$. By 5.E, the σ-ring $\hat{\mathbf{S}} \cap C_0$ is generated by the class of all sets of the form $C \cap C_0$, where $C \in \hat{\mathbf{C}}$. By 51.F (applied to the compact space C_0), this σ-ring is generated by the ring of all sets of the form $E \cap C_0$, where E is a finite, disjoint union of proper differences of sets of $\hat{\mathbf{C}}$. According as the condition on $\hat{\mathbf{C}}$ or on $\hat{\mathbf{U}}$ is assumed, it follows from Theorems A, B, and C that every set in this ring is outer or inner regular. Since, by Theorems C and D, the class of outer regular subsets of C_0 and the class of inner regular subsets of C_0 are both monotone classes, it follows from 6.B and Theorem E that, assuming either of the two conditions, if a subset of C_0 is in $\hat{\mathbf{S}}$, then it is regular, and hence, in particular, that E_0 is regular. ▮

Theorem G. *Every Baire measure* ν *is regular; if* $C \in \mathbf{C}$, *then*

$$\nu^*(C) = \inf\{\nu(U_0): C \subset U_0 \in \mathbf{U}_0\},$$

and, if $U \in \mathbf{U}$, *then*

$$\nu_*(U) = \sup\{\nu(C_0): U \supset C_0 \in \mathbf{C}_0\}.$$

Proof. Since every set in $\mathbf{C}_0$ may be written as the intersection of a decreasing sequence of sets of finite measure in $\mathbf{U}_0$, the regularity of ν follows from Theorem F. Since, by definition of outer measure,

$$\nu^*(C) = \inf\{\nu(E_0): C \subset E_0 \in \mathbf{S}_0\} \leqq \inf\{\nu(U_0): C \subset U_0 \in \mathbf{U}_0\},$$

for every $\epsilon > 0$, there exists a set E_0 in $\mathbf{S}_0$ such that $C \subset E_0$ and $\nu(E_0) \leqq \nu^*(C) + \frac{\epsilon}{2}$. The outer regularity of E_0 implies the existence of a set U_0 in $\mathbf{U}_0$ such that

$$E_0 \subset U_0 \quad \text{and} \quad \nu(U_0) \leqq \nu(E_0) + \frac{\epsilon}{2};$$

it follows that $C \subset U_0$ and $\nu(U_0) \leqq \nu^*(C) + \epsilon$. The proof of the assertion concerning inner measure exploits, in an entirely similar way, the inner regularity of every Baire set E_0. ∎

Theorem H. *Let μ be a Borel measure and let ν be the Baire contraction of μ (defined for every Baire set E by $\nu(E) = \mu(E)$). Either of the two conditions,*

$$\mu(C) = \nu^*(C) \quad \textit{for all } C \textit{ in } \mathbf{C},$$

$$\mu(U) = \nu_*(U) \quad \textit{for all bounded open } U \textit{ in } \mathbf{U},$$

is necessary and sufficient for the regularity of μ. If two regular Borel measures agree on all Baire sets, then they agree on all Borel sets.

Proof. If, for some C in $\mathbf{C}$, $\mu(C) = \nu^*(C)$, then according to Theorem G, for every $\epsilon > 0$ there exists a set U_0 in $\mathbf{U}_0$ such that

$$C \subset U_0 \quad \text{and} \quad \mu(U_0) = \nu(U_0) \leqq \nu^*(C) + \epsilon = \mu(C) + \epsilon;$$

this implies that C is outer regular and hence that μ is regular. The proof of the sufficiency of the condition involving ν_* exploits, in an entirely similar way, the last assertion of Theorem G.

Suppose next that μ is regular and let ϵ be an arbitrary positive number. For any C in $\mathbf{C}$, there exists a bounded set U in $\mathbf{U}$ such that $C \subset U$ and $\mu(U) \leqq \mu(C) + \epsilon$; similarly, for any bounded set U in $\mathbf{U}$, there exists a set C in $\mathbf{C}$ such that $C \subset U$ and $\mu(U) \leqq \mu(C) + \epsilon$. In either case, there exist sets C_0 in $\mathbf{C}_0$ and U_0 in $\mathbf{U}_0$ such that $C \subset U_0 \subset C_0 \subset U$, (50.D). It follows from Theorem G that

$$\nu^*(C) \leqq \nu(U_0) = \mu(U_0) \leqq \mu(U) \leqq \mu(C) + \epsilon,$$

and

$$\nu_*(U) \geqq \nu(C_0) = \mu(C_0) \geqq \mu(C) \geqq \mu(U) - \epsilon.$$

The arbitrariness of ϵ implies that

$$\nu^*(C) \leqq \mu(C) \quad \text{and} \quad \nu_*(U) \geqq \mu(U);$$

the reverse inequality is obvious in both cases. Since it has thus been shown that the values of a regular measure on the Baire sets uniquely determine its values on the compact sets, the last assertion of the theorem follows from 51.F. ∎

We conclude this section by introducing a concept which sometimes provides a useful tool for proving regularity. If μ is any Borel measure, its Baire contraction μ_0, defined for all E in $\mathbf{S}_0$ by $\mu_0(E) = \mu(E)$, is a Baire measure associated with μ in a natural way. If it happens that every set in $\mathbf{C}$ or every bounded set in $\mathbf{U}$, and therefore in either case, every set in $\mathbf{S}$ is $\mu_0{}^*$–measurable (i.e. if all compact sets, and therefore all Borel sets, belong to the domain of definition of the completion of μ_0), then we shall say that the Borel measure μ is **completion regular.** If μ is completion regular, then to every Borel set E there correspond two Baire sets A and B such that

$$A \subset E \subset B \quad \text{and} \quad \mu_0(B - A) = 0;$$

it follows from Theorem H that completion regularity implies regularity.

(1) Every Borel measure is σ–finite.

(2) If the space X is compact, then the class of all regular sets is a normal class; (cf. 6.2).

(3) If μ is a Borel measure and if there exists a countable set Y such that $\mu(E) = \mu(E \cap Y)$ for every Borel set E, then μ is regular.

(4) If X is the Euclidean plane and if μ is Lebesgue measure on the class of all Borel sets, then μ is a regular Borel measure in the sense of this section. If, however, for every Borel set E, $\mu(E)$ is defined to be the sum of the linear measures of all horizontal sections of E, then μ is not a Borel measure.

(5) Suppose that X is compact and x^* is a point such that $\{x^*\}$ is not a G_δ; (cf., for instance, 50.3). If, for every E in $\mathbf{S}$, $\mu(E) = \chi_E(x^*)$, then μ is a regular Borel measure which is not completion regular.

(6) If μ_1, μ_2, and μ are Borel measures such that $\mu = \mu_1 + \mu_2$, then the regularity of any two of them implies that of the third. (Hint: if $C \in \mathbf{C}$, $U \in \mathbf{U}$, $C \subset U$, and $\mu(U) \leqq \mu(C) + \epsilon$, then

$$\mu_1(C) + \mu_2(U) \leqq \mu(U) \leqq \mu_1(C) + \mu_2(C) + \epsilon.)$$

(7) Suppose that X and Y are compact Hausdorff spaces, T is a continuous transformation from X onto Y, and μ is a Borel measure on X. If $\nu = \mu T^{-1}$, and if D is a compact subset of Y, then D is regular with respect to ν if and only if $C = T^{-1}(D)$ is regular with respect to μ. (Hint: if $C \subset U \varepsilon \mathbf{U}$, then $T(X - U)$ and D are disjoint compact sets in Y. If V is a neighborhood of D which is disjoint from $T(X - U)$, then $C \subset T^{-1}(V) \subset U$.)

(8) If μ is a regular Borel measure, then, for every σ-bounded set E,

$$\mu^*(E) = \inf \{\mu(U): E \subset U \varepsilon \mathbf{U}\} \quad \text{and} \quad \mu_*(E) = \sup \{\mu(C): E \supset C \varepsilon \mathbf{C}\}.$$

(9) If μ and ν are Borel measures such that μ is regular and $\nu \ll \mu$, then ν is regular.

(10a) Let Ω be the first uncountable ordinal, and let $\bar{X}$ be the set of all ordinals less than or equal to Ω. Write $X = \bar{X} - \{\Omega\}$. If the class of all "intervals" of the form $\{x: \alpha < x \leq \beta\}$ together with the set $\{0\}$ is taken for a base, then $\bar{X}$ is compact.

(10b) The class of all unbounded, closed subsets of X is closed under the formation of countable intersections.

(10c) If, for every Borel set E in $\bar{X}$, $\mu(E) = 1$ or 0 according as E does or does not contain an unbounded, closed subset of X, then μ is a Borel measure.

(10d) The Borel measure μ is not regular. (Hint: every interval containing Ω has measure 1.)

§ 53. GENERATION OF BOREL MEASURES

The purpose of this section is to show how certain (regular) Borel measures may be obtained from more primitive set functions.

We define a **content** as a non negative, finite, monotone, additive, and subadditive set function on the class $\mathbf{C}$ of all compact sets. In other words, a content is a set function λ on $\mathbf{C}$ which is such that (a) $0 \leq \lambda(C) < \infty$ for all C in $\mathbf{C}$, (b) if C and D are compact sets for which $C \subset D$, then $\lambda(C) \leq \lambda(D)$, (c) if C and D are disjoint compact sets, then $\lambda(C \cup D) = \lambda(C) + \lambda(D)$, and (d) if C and D are any two compact sets, then $\lambda(C \cup D) \leq \lambda(C) + \lambda(D)$. We observe that, since $\lambda(0) + \lambda(0) = \lambda(0 \cup 0) = \lambda(0) < \infty$, a content must always vanish on the empty set.

The outline of our procedure from now on will be as follows. In terms of a given content λ we shall define a set function λ_* on the class of open Borel sets, and in terms of λ_* we shall define an outer measure μ^* on the class of all σ-bounded sets. Then we shall use the already established theory of μ^*-measurability to obtain from the outer measure μ^* a measure μ which will, in fact, turn out to be a regular Borel measure.

The **inner content** λ_*, **induced** by a content λ, is the set function defined for every U in $\mathbf{U}$ by

$$\lambda_*(U) = \sup\{\lambda(C): U \supset C \varepsilon \mathbf{C}\}.$$

Theorem A. *The inner content λ_* induced by a content λ vanishes at* 0, *and is monotone, countably subadditive, and countably additive.*

Proof. It is obvious that $\lambda_*(0) = 0$. If U and V are in $\mathbf{U}$, if $U \subset V$, and if C is a compact set contained in U, then $C \subset V$ and therefore $\lambda(C) \leqq \lambda_*(V)$. It follows that

$$\lambda_*(U) = \sup \lambda(C) \leqq \lambda_*(V).$$

If U and V are in $\mathbf{U}$ and if C is a compact set such that $C \subset U \cup V$, then (50.A) there exist compact sets D and E such that $D \subset U$, $E \subset V$, and $C = D \cup E$. Since $\lambda(C) \leqq \lambda(D) + \lambda(E) \leqq \lambda_*(U) + \lambda_*(V)$, it follows that

$$\lambda_*(U \cup V) = \sup \lambda(C) \leqq \lambda_*(U) + \lambda_*(V),$$

i.e. that λ_* is subadditive. It follows immediately, by mathematical induction, that λ_* is finitely subadditive. If $\{U_i\}$ is a sequence of sets in $\mathbf{U}$ and if C is a compact set such that $C \subset \bigcup_{i=1}^{\infty} U_i$, then, by the compactness of C, there is a positive integer n such that $C \subset \bigcup_{i=1}^{n} U_i$. It follows that

$$\lambda(C) \leqq \lambda_*(\bigcup_{i=1}^{n} U_i) \leqq \sum_{i=1}^{n} \lambda_*(U_i) \leqq \sum_{i=1}^{\infty} \lambda_*(U_i),$$

and therefore that

$$\lambda_*(\bigcup_{i=1}^{\infty} U_i) = \sup \lambda(C) \leqq \sum_{i=1}^{\infty} \lambda_*(U_i),$$

i.e. that λ_* is countably subadditive.

Suppose next that U and V are two disjoint sets in $\mathbf{U}$ and let C and D be compact sets such that $C \subset U$ and $D \subset V$. Since C and D are disjoint and since $C \cup D \subset U \cup V$, we have

$$\lambda(C) + \lambda(D) = \lambda(C \cup D) \leqq \lambda_*(U \cup V),$$

and therefore

$$\lambda_*(U) + \lambda_*(V) = \sup \lambda(C) + \sup \lambda(D) \leqq \lambda_*(U \cup V).$$

The subadditivity of λ_* implies now that λ_* is additive and hence, by mathematical induction, that λ_* is finitely additive. If $\{U_i\}$ is a disjoint sequence of sets in $\mathbf{U}$, then

$$\lambda_*(\bigcup_{i=1}^{\infty} U_i) \geqq \lambda_*(\bigcup_{i=1}^{n} U_i) = \sum_{i=1}^{n} \lambda_*(U_i);$$

since this is true for every $n = 1, 2, \cdots$, it follows that

$$\lambda_*(\bigcup_{i=1}^{\infty} U_i) \geqq \sum_{i=1}^{\infty} \lambda_*(U_i).$$

The countable additivity of λ_* follows from its already proved countable subadditivity. ▮

If λ is a content and λ_* is the inner content induced by λ, we define a set function μ^* on the hereditary σ-ring of all σ-bounded sets by

$$\mu^*(E) = \inf\{\lambda_*(U): E \subset U \in \mathbf{U}\}.$$

The set function μ^* is called the **outer measure induced** by λ; the terminology is justified by the following result.

Theorem B. *The outer measure μ^* induced by a content λ is an outer measure.*

Proof. The equation $\mu^*(0) = 0$ follows from the facts that $0 \subset 0 \in \mathbf{U}$ and $\lambda_*(0) = 0$. If E and F are two σ-bounded sets such that $E \subset F$, and if U is a set in $\mathbf{U}$ such that $F \subset U$, then $E \subset U$ and therefore $\mu^*(E) \leqq \lambda_*(U)$. It follows that

$$\mu^*(E) \leqq \inf \lambda_*(U) = \mu^*(F).$$

If $\{E_i\}$ is a sequence of σ-bounded sets, then, for every $\epsilon > 0$ and for every $i = 1, 2, \cdots$, there exists a set U_i in $\mathbf{U}$ such that

$$E_i \subset U_i \quad \text{and} \quad \lambda_*(U_i) \leqq \mu^*(E_i) + \frac{\epsilon}{2^i}.$$

It follows that

$$\mu^*(\bigcup_{i=1}^{\infty} E_i) \leqq \lambda_*(\bigcup_{i=1}^{\infty} U_i) \leqq \sum_{i=1}^{\infty} \lambda_*(U_i) \leqq \sum_{i=1}^{\infty} \mu^*(E_i) + \epsilon;$$

the arbitrariness of ϵ implies the countable subadditivity of μ^*. ▮

It might be conjectured that the procedures of Theorems A and B actually yield extensions of λ and λ_* respectively, i.e., for instance, that μ^* is such that $\mu^*(C) = \lambda(C)$ for every compact set C. This is not true in general; the best that can be said is contained in the following result.

Theorem C. *If λ_* is the inner content and μ^* is the outer measure induced by a content λ, then $\mu^*(U) = \lambda_*(U)$ for every U in $\mathbf{U}$ and $\mu^*(C^0) \leqq \lambda(C) \leqq \mu^*(C)$ for every C in $\mathbf{C}$.*

(We recall that C^0 denotes the interior of the set C.)

Proof. If $U \varepsilon \mathbf{U}$, then the relation $U \subset U \varepsilon \mathbf{U}$ implies that $\mu^*(U) \leqq \lambda_*(U)$. If $V \varepsilon \mathbf{U}$ and $U \subset V$, then $\lambda_*(U) \leqq \lambda_*(V)$ and therefore

$$\lambda_*(U) \leqq \inf \lambda_*(V) = \mu^*(U).$$

If $C \varepsilon \mathbf{C}$, $U \varepsilon \mathbf{U}$, and $C \subset U$, then $\lambda(C) \leqq \lambda_*(U)$, and therefore

$$\lambda(C) \leqq \inf \lambda_*(U) = \mu^*(C).$$

If $C \varepsilon \mathbf{C}$, $D \varepsilon \mathbf{C}$, and $D \subset C^0$ $(\subset C)$, then $\lambda(D) \leqq \lambda(C)$, and therefore

$$\mu^*(C^0) = \lambda_*(C^0) = \sup \lambda(D) \leqq \lambda(C). \quad ▮$$

Theorem D. *If μ^* is the outer measure induced by a content λ, then a σ–bounded set E is μ^*–measurable if and only if*

$$\mu^*(U) \geqq \mu^*(U \cap E) + \mu^*(U \cap E')$$

for every U in $\mathbf{U}$.

Proof. Let λ_* be the inner content induced by λ, let A be an arbitrary σ–bounded set, and let U be a set in $\mathbf{U}$ such that $A \subset U$. From the relations

$$\lambda_*(U) = \mu^*(U) \geqq \mu^*(U \cap E) + \mu^*(U \cap E') \geqq$$
$$\geqq \mu^*(A \cap E) + \mu^*(A \cap E')$$

it follows that

$$\mu^*(A) = \inf \lambda_*(U) \geqq \mu^*(A \cap E) + \mu^*(A \cap E');$$

the reverse inequality and the converse follow from the subadditivity of μ^* and the definition of μ^*–measurability. ▮

Theorem E. *If μ^* is the outer measure induced by a content λ, then the set function μ, defined for every Borel set E by $\mu(E) = \mu^*(E)$, is a regular Borel measure.*

We shall call μ the Borel measure **induced** by the content λ.

Proof. We shall prove first that every compact set C (and therefore every Borel set) is μ^*–measurable; it will then follow immediately that μ is a measure on the class of all Borel sets. In virtue of Theorem D it is sufficient to prove that

$$\mu^*(U) \geqq \mu^*(U \cap C) + \mu^*(U \cap C')$$

for all U in $\mathbf{U}$. Let D be a compact subset of $U \cap C'$ and let E be a compact subset of $U \cap D'$; we observe that both the sets $U \cap C'$ and $U \cap D'$ belong to $\mathbf{U}$. Since $D \cap E = 0$ and $D \cup E \subset U$, it follows that

$$\mu^*(U) = \lambda_*(U) \geqq \lambda(D \cup E) = \lambda(D) + \lambda(E),$$

where λ_* is, of course, the inner content induced by λ. Therefore

$$\begin{aligned} \mu^*(U) &\geqq \lambda(D) + \sup \lambda(E) = \lambda(D) + \lambda_*(U \cap D') = \\ &= \lambda(D) + \mu^*(U \cap D') \geqq \lambda(D) + \mu^*(U \cap C); \end{aligned}$$

this in turn implies that

$$\begin{aligned} \mu^*(U) \geqq \mu^*(U \cap C) + \sup \lambda(D) &= \mu^*(U \cap C) + \lambda_*(U \cap C') = \\ &= \mu^*(U \cap C) + \mu^*(U \cap C'). \end{aligned}$$

To prove that $\mu(C) < \infty$, we observe that there exists a compact set F such that $C \subset F^0$; it follows that

$$\mu(C) = \mu^*(C) \leqq \mu^*(F^0) \leqq \lambda(F) < \infty.$$

The fact that the measure μ is regular follows, finally, from the relations

$$\begin{aligned} \mu(C) &= \mu^*(C) = \inf\{\lambda_*(U) : C \subset U \in \mathbf{U}\} = \\ &= \inf\{\mu^*(U) : C \subset U \in \mathbf{U}\} = \inf\{\mu(U) : C \subset U \in \mathbf{U}\}. \quad \blacksquare \end{aligned}$$

We conclude with a result which we shall have opportunity to use later.

Theorem F. *Suppose that T is a homeomorphism of X onto itself and that λ is a content. If, for every C in $\mathbf{C}$, $\hat{\lambda}(C) = \lambda(T(C))$, and if μ and $\hat{\mu}$ are the Borel measures induced by λ and $\hat{\lambda}$ respectively, then $\hat{\mu}(E) = \mu(T(E))$ for every Borel set E. If, in particular, λ is invariant under T, then the same is true of μ.*

Proof. If λ_* and $\hat{\lambda}_*$ are the inner contents induced by λ and $\hat{\lambda}$ respectively, and if $U \varepsilon \mathbf{U}$, then the relations

$$\{\hat{\lambda}(C): U \supset C \varepsilon \mathbf{C}\} = \{\lambda(T(C)): U \supset C \varepsilon \mathbf{C}\} =$$

$$= \{\lambda(D): D = T(C), U \supset C \varepsilon \mathbf{C}\} =$$

$$= \{\lambda(D): U \supset T^{-1}(D) \varepsilon \mathbf{C}\} =$$

$$= \{\lambda(D): T(U) \supset D \varepsilon \mathbf{C}\}$$

imply that $\hat{\lambda}_*(U) = \lambda_*(T(U))$. If μ^* and $\hat{\mu}^*$ are the outer measures induced by λ and $\hat{\lambda}$ respectively, then a similar computation shows that, for every σ-bounded set E, $\hat{\mu}^*(E) = \mu^*(T(E))$, and hence that, for every Borel set E, $\hat{\mu}(E) = \mu(T(E))$. The last assertion of the theorem is an immediate consequence of the preceding ones. ∎

(1) The following are examples of non negative, finite set functions defined on the class $\mathbf{C}$ of all compact subsets of a locally compact Hausdorff space; some of them are contents, while others fail to possess exactly one of the principal defining properties (monotoneness, additivity, and subadditivity) of a content.

(1a) X^* is the one-point compactification of an infinite discrete space X; for every compact set C in X^*, $\lambda(C) = 0$ or 1 according as C is finite or infinite.

(1b) X is a discrete space consisting of a finite number of points; $\lambda(C) = 1$ for every compact set C.

(1c) X is the closed interval $[-1, +1]$; $\lambda(C) = 1$ or 0 according as $0 \varepsilon C^0$ or $0 \varepsilon' C^0$.

(1d) $X^* = \{X, x^*\}$ is, as in (1a), the one-point compactification of an infinite discrete space X; $\lambda(C) = 1$ or 0 according as $x^* \varepsilon C$ or $x^* \varepsilon' C$.

(1e) $X = \left\{0, \pm\frac{1}{n}: n = 1, 2, \cdots\right\}$. If C contains infinitely many negative numbers, then $\lambda(C) = 0$; otherwise $\lambda(C) = 1$ or 0 according as $0 \varepsilon C$ or $0 \varepsilon' C$.

(1f) Let μ_0 be a Baire measure on X, and, for every C in $\mathbf{C}$, write

$$\lambda(C) = \sup \{\mu_0(C_0): C \supset C_0 \varepsilon \mathbf{C}_0\}.$$

(1g) Let μ be a Borel measure on X, and, for every C in $\mathbf{C}$, write $\lambda(C) = \mu(C^0)$.

(2) If λ and $\hat{\lambda}$ are two contents inducing the outer measures μ^* and $\hat{\mu}^*$ respectively, and if, for every C in $\mathbf{C}$, $\lambda(C) \leq \hat{\lambda}(C) \leq \mu^*(C)$, then $\mu^* = \hat{\mu}^*$. (Hint: in view of the first part of Theorem C, it is sufficient to prove that $\mu^*(U) = \sup \{\hat{\lambda}(C): U \supset C \varepsilon \mathbf{C}\}$ for every U in $\mathbf{U}$.)

(3) The result of (2) may be strengthened to the following converse of Theorem C. If λ and $\hat{\lambda}$ are two contents, inducing the outer measures μ^* and $\hat{\mu}^*$ respec-

tively, and if, for every C in $\mathbf{C}$, $\mu^*(C^0) \leqq \hat{\lambda}(C) \leqq \mu^*(C)$, then $\mu^* = \hat{\mu}^*$. (Hint: Theorem E implies that

$$\mu^*(U) = \sup \{\mu^*(C): U \supset C \,\varepsilon\, \mathbf{C}\}$$

for every U in $\mathbf{U}$. It is to be proved that

$$\mu^*(U) = \sup \{\hat{\lambda}(C): U \supset C \,\varepsilon\, \mathbf{C}\}.$$

If $\epsilon > 0$ and $U \,\varepsilon\, \mathbf{U}$, then there exists a set C in $\mathbf{C}$ such that $C \subset U$ and $\mu^*(U) \leqq \mu^*(C) + \epsilon$, and there exists a set D in $\mathbf{C}$ such that $C \subset D^0 \subset D \subset U$.)

(4) If μ is the Borel measure induced by a content λ, and if $\lambda(C) > 0$ whenever $C^0 \neq 0$, then $\mu(U) > 0$ for every non empty U in $\mathbf{U}$.

(5) Independently of any content λ we might consider those outer measures μ^* on the class of all σ-bounded sets which have the property that

$$\mu^*(C) = \inf \{\mu^*(U): C \subset U \,\varepsilon\, \mathbf{U}\} < \infty$$

for every C in $\mathbf{C}$. Are Theorems D and E true for any such outer measure?

§ 54. REGULAR CONTENTS

We have remarked before on the fact that the values of a content need not coincide (on compact sets, of course) with the values of the Borel measure it induces. There is, however, an important class of contents which are such that the process of § 53 is actually an extension. In this section we shall study such contents and use our results to derive an important extension theorem which asserts, in fact, the existence of certain Borel measures whose uniqueness was established in 52.H.

A content λ is **regular** if, for every C in $\mathbf{C}$,

$$\lambda(C) = \inf \{\lambda(D): C \subset D^0 \subset D \,\varepsilon\, \mathbf{C}\}.$$

This definition of regularity for contents imitates the definition of (outer) regularity for measures as closely as possible in view of the restricted domain of definition of a content.

Theorem A. *If μ is the Borel measure induced by a regular content λ, then $\mu(C) = \lambda(C)$ for every C in $\mathbf{C}$.*

Proof. If $C \,\varepsilon\, \mathbf{C}$, then, because of the regularity of λ, for every $\epsilon > 0$ there exists a set D in $\mathbf{C}$ such that

$$C \subset D^0 \quad \text{and} \quad \lambda(D) \leqq \lambda(C) + \epsilon.$$

It follows from 53.C that

$$\lambda(C) \leqq \mu(C) \leqq \mu(D^0) \leqq \lambda(D) \leqq \lambda(C) + \epsilon;$$

the desired result follows from the arbitrariness of ϵ. ▌

The following result goes in the converse direction.

Theorem B. *If μ is a regular Borel measure and if, for every C in* **C**, *$\lambda(C) = \mu(C)$, then λ is a regular content and the Borel measure induced by λ coincides with μ.*

Proof. It is clear that λ is a content. The regularity of μ implies that, for every C in **C** and for every $\epsilon > 0$, there exists a set U in **U** such that

$$C \subset U \quad \text{and} \quad \mu(U) \leqq \mu(C) + \epsilon.$$

If D is a set in **C** such that $C \subset D^0 \subset D \subset U$, then

$$\lambda(D) = \mu(D) \leqq \mu(U) \leqq \mu(C) + \epsilon = \lambda(C) + \epsilon;$$

this proves the regularity of λ. If $\hat{\mu}$ is the Borel measure induced by λ, then, by Theorem A, $\hat{\mu}(C) = \lambda(C) = \mu(C)$ for every C in **C**, and therefore, indeed, $\hat{\mu} = \mu$. ▌

Theorem C. *If μ_0 is a Baire measure and if, for every C in* **C**,

$$\lambda(C) = \inf\{\mu_0(U_0): C \subset U_0 \in \mathbf{U}_0\},$$

then λ is a regular content.

Proof. It is easy to verify that λ is non negative, finite, and monotone.

If C and D are sets in **C** and U_0 and V_0 are sets in $\mathbf{U}_0$ such that $C \subset U_0$ and $D \subset V_0$, then $C \cup D \subset U_0 \cup V_0 \in \mathbf{U}_0$, and therefore

$$\lambda(C \cup D) \leqq \mu_0(U_0 \cup V_0) \leqq \mu_0(U_0) + \mu_0(V_0).$$

It follows that

$$\lambda(C \cup D) \leqq \inf \mu_0(U_0) + \inf \mu_0(V_0) = \lambda(C) + \lambda(D),$$

i.e. that λ is subadditive.

If C and D are disjoint sets in **C**, then there exist disjoint sets U_0 and V_0 in $\mathbf{U}_0$ such that $C \subset U_0$ and $D \subset V_0$. If $C \cup D \subset W_0 \in \mathbf{U}_0$, then

$$\lambda(C) + \lambda(D) \leqq \mu_0(U_0 \cap W_0) + \mu_0(V_0 \cap W_0) \leqq \mu_0(W_0),$$

and therefore

$$\lambda(C) + \lambda(D) \leqq \inf \mu_0(W_0) = \lambda(C \cup D).$$

The fact that λ is additive follows from the fact, proved above, that λ is subadditive.

To prove that λ is regular, let C be any compact set and let ϵ be any positive number. By the definition of λ, there exists a set U_0 in $\mathbf{U}_0$ such that

$$C \subset U_0 \quad \text{and} \quad \mu_0(U_0) \leqq \lambda(C) + \epsilon.$$

If D is a compact set such that $C \subset D^0 \subset D \subset U_0$, then

$$\lambda(D) \leqq \mu_0(U_0) \leqq \lambda(C) + \epsilon. \quad ▮$$

Theorem D. *If μ_0 is a Baire measure, then there exists a unique, regular Borel measure μ such that $\mu(E) = \mu_0(E)$ for every Baire set E.*

Proof. If, for every C in **C**,

$$\lambda(C) = \inf \{\mu_0(U_0): C \subset U_0 \in \mathbf{U}_0\},$$

then, by Theorem C, λ is a regular content; let μ be the regular Borel measure induced by λ. By Theorem A, $\mu(C) = \lambda(C)$ for every C in **C**. Since (52.G) every Baire measure is regular, we have $\lambda(C) = \mu_0(C)$, and consequently $\mu(C) = \mu_0(C)$ for every C in $\mathbf{C}_0$. This proves the existence of μ; uniqueness was explicitly stated and proved in 52.H. ▮

(1) Which of the set functions described in 53.1 are regular?

(2) If, in the notation of 53.F, λ is a regular content, then so is $\hat{\lambda}$.

(3) If μ is a Borel measure and if, for every C in **C**, $\lambda(C) = \sup \{\mu(C_0): C \supset C_0 \in \mathbf{C}_0\}$, then μ is completion regular if and only if λ is a regular content; (cf. 53.1f).

(4) A content λ is **inner regular** if, for every C in **C**, $\lambda(C) = \sup \{\lambda(D): C^0 \supset D \in \mathbf{C}\}$. The following analogs of Theorems A and B are true.

(4a) If μ is the Borel measure induced by an inner regular content λ, then $\mu(C^0) = \lambda(C)$ for every C in $\mathbf{C}$.

(4b) If μ is a regular Borel measure and if, for every C in $\mathbf{C}$, $\lambda(C) = \mu(C^0)$, then λ is an inner regular content and the Borel measure induced by λ coincides with μ.

§ 55. CLASSES OF CONTINUOUS FUNCTIONS

If X is, as usual, a locally compact Hausdorff space, we shall denote by $\mathcal{L}(X)$ or simply by $\mathcal{L}$ the class of all those real valued, continuous functions on X which vanish outside a compact set. In other words $\mathcal{L}$ is the class of all those continuous functions f on X for which the set

$$N(f) = \{x : f(x) \neq 0\}$$

is bounded. If X is not compact and if X^* is the one-point compactification of X by x^*, then the point x^* is frequently called the **point at infinity,** and consequently $\mathcal{L}$ may be described as the class of all those continuous functions which vanish in a neighborhood of infinity. We shall denote by $\mathcal{L}_+(X)$ or simply by $\mathcal{L}_+$ the subclass of all non negative functions in $\mathcal{L}$. The first of our results concerning these function spaces has been implicit in many of our preceding constructions.

Theorem A. *If C is any compact Baire set, then there exists a decreasing sequence $\{f_n\}$ of functions in $\mathcal{L}_+$ such that*

$$\lim_n f_n(x) = \chi_C(x)$$

for every x in X.

Proof. If $C = \bigcap_{n=1}^{\infty} U_n$, where each U_n is a bounded open set, then for each positive integer n there exists a function g_n in $\mathcal{F}$ (cf. § 50) such that

$$g_n(x) = \begin{cases} 1 & \text{if } x \in C, \\ 0 & \text{if } x \in' U_n. \end{cases}$$

If $f_n = g_1 \cap \cdots \cap g_n$, then $\{f_n\}$ is a decreasing sequence of non negative continuous functions such that

$$\lim_n f_n(x) = \chi_C(x)$$

for every x in X; the fact that U_n is bounded implies that $f_n \varepsilon \mathfrak{L}_+$, $n = 1, 2, \cdots$. ∎

If μ_0 is a Baire measure in X, if $f \varepsilon \mathfrak{L}$, and if $\{x: f(x) \neq 0\} \subset C \varepsilon \mathbf{C}_0$, then the facts that $\mu_0(C) < \infty$ and that f is bounded and Baire measurable (51.B) imply that f is integrable with respect to μ_0 and that

$$\int f d\mu_0 = \int_C f d\mu_0.$$

These statements are true, in particular, if μ is a Borel measure and μ_0 is its Baire contraction.

Theorem B. *If a Baire measure μ is such that the measure of every non empty Baire open set is positive, and if $f \varepsilon \mathfrak{L}_+$, then a necessary and sufficient condition that $\int f d\mu = 0$ is that $f(x) = 0$ for every x in X.*

Proof. The sufficiency of the condition is trivial. To prove necessity, suppose that $\int f d\mu = 0$ and let U be a bounded open Baire set such that $\{x: f(x) \neq 0\} \subset U$. If $E = \{x: f(x) = 0\}$, then, since

$$0 = \int f d\mu \geqq \int_{U-E} f d\mu,$$

it follows from the fact that f is non negative that $\mu(U - E) = 0$. Since $U - E$ is an open Baire set, we must have $U - E = 0$, or $U \subset E$. ∎

Theorem C. *If μ_0 is a Baire measure and $\epsilon > 0$, then, corresponding to every integrable simple Baire function f, there exists an integrable simple function g,*

$$g = \sum_{i=1}^n \alpha_i \chi_{C_i},$$

such that C_i is a compact Baire set, $i = 1, \cdots, n$, and

$$\int |f - g| d\mu_0 \leqq \epsilon.$$

Proof. Write $f = \sum_{i=1}^n \alpha_i \chi_{E_i}$ and let c be a positive number such that $|f(x)| \leqq c$ for every x in X (i.e. such that $|\alpha_i| \leqq c$

for $i = 1, \cdots, n$). The regularity of μ_0 implies that, for each $i = 1, \cdots, n$, there exists a compact Baire set C_i such that

$$C_i \subset E_i \quad \text{and} \quad \mu_0(E_i) \leqq \mu_0(C_i) + \frac{\epsilon}{nc}.$$

It follows that if $g = \sum_{i=1}^n \alpha_i \chi_{C_i}$, then

$$\int |f - g| d\mu_0 = \sum_{i=1}^n |\alpha_i| \mu_0(E_i - C_i) \leqq \epsilon. \quad \blacksquare$$

Theorem D. *If μ_0 is a Baire measure, if $\epsilon > 0$, and if $g = \sum_{i=1}^n \alpha_i \chi_{C_i}$ is a simple function such that C_i is a compact Baire set, $i = 1, \cdots, n$, then there exists a function h in $\mathfrak{L}$ such that*

$$\int | g - h | d\mu_0 \leqq \epsilon.$$

Proof. Since $\{C_1, \cdots, C_n\}$ is a finite, disjoint class of compact sets, there exists a finite, disjoint class $\{U_1, \cdots, U_n\}$ of bounded open Baire sets such that $C_i \subset U_i$, $i = 1, \cdots, n$. Because of the regularity of μ_0, there is no loss of generality in assuming that

$$\mu_0(U_i) \leqq \mu_0(C_i) + \frac{\epsilon}{nc}, \quad i = 1, \cdots, n,$$

where c is a positive number such that $| g(x) | \leqq c$ for every x in X. For each $i = 1, \cdots, n$, there exists a function h_i in $\mathfrak{F}$ such that $h_i(x) = 1$ for x in C_i and $h_i(x) = 0$ for x in $X - U_i$; we write $h = \sum_{i=1}^n \alpha_i h_i$. Since $h_i \,\varepsilon\, \mathfrak{L}_+$, $i = 1, \cdots, n$, it is clear that $h \,\varepsilon\, \mathfrak{L}$; the disjointness of the U_i implies that $| h(x) | \leqq c$ for all x in X. We have

$$\int | g - h | d\mu_0 = \sum_{i=1}^n \int_{U_i - C_i} | h | d\mu_0 \leqq \sum_{i=1}^n c \cdot \mu_0(U_i - C_i) \leqq \epsilon. \quad \blacksquare$$

(1) If μ is a regular Borel measure, then the class of all finite linear combinations of characteristic functions of compact sets is dense in $\mathfrak{L}_p(\mu)$, $1 \leqq p < \infty$.

(2) If μ is a regular Borel measure, then $\mathfrak{L}$ is dense in $\mathfrak{L}_p(\mu)$, $1 \leqq p < \infty$.

(3) If μ is a regular Borel measure, E is a Borel set of finite measure, and f is a Borel measurable function on E, then, for every $\epsilon > 0$, there exists a compact set C in E such that $\mu(E - C) \leqq \epsilon$ and such that f is continuous on C. (Hint: if f is a simple function, the result may be proved by the technique used in the

proof of Theorem C. In the general case, there exists a sequence $\{f_n\}$ of simple functions converging to f; by Egoroff's theorem and the regularity of μ, there exists a compact set C_0 in E such that $\mu(E) \leq \mu(C_0) + \frac{\epsilon}{2}$ and such that $\{f_n\}$ converges to f uniformly on C_0. Let C_n be a compact subset of E such that $\mu(E) \leq \mu(C_n) + \frac{\epsilon}{2^{n+1}}$ and such that f_n is continuous on C_n; the set

$$C = \bigcap_{n=0}^{\infty} C_n$$

satisfies the required conditions.) This result is known as **Lusin's theorem.**

§ 56. LINEAR FUNCTIONALS

A **linear functional** on $\mathcal{L}$ is a real valued function Λ of the functions in $\mathcal{L}$ such that

$$\Lambda(\alpha f + \beta g) = \alpha\Lambda(f) + \beta\Lambda(g)$$

for every pair, f and g, of functions in $\mathcal{L}$ and every pair, α and β, of real numbers. A linear functional Λ on $\mathcal{L}$ is **positive** if $\Lambda(f) \geqq 0$ for every f in $\mathcal{L}_+$. We observe that a positive linear functional Λ is monotone in the sense that if $f \in \mathcal{L}$, $g \in \mathcal{L}$, and $f \geqq g$, then $\Lambda(f) \geqq \Lambda(g)$. It is easy to verify that if μ_0 is a Baire measure in X and if $\Lambda(f) = \int f d\mu_0$ for every f in $\mathcal{L}$, then Λ is a positive linear functional; the main purpose of this section is to show that every positive linear functional may be obtained in this way.

We shall find it convenient to employ a somewhat unusual but very suggestive notation. If E is any subset of X and f is any real valued function on X, then we shall write $E \subset f$ [or $E \supset f$] if $\chi_E(x) \leqq f(x)$ [or $\chi_E(x) \geqq f(x)$] for every x in X.

Theorem A. *If Λ is a positive linear functional on $\mathcal{L}$ and if, for every C in* **C**,

$$\lambda(C) = \inf \{\Lambda(f): C \subset f \in \mathcal{L}_+\},$$

then λ is a regular content. If μ is the Borel measure induced by λ, then

$$\mu(U) \leqq \Lambda(f)$$

for every bounded open set U and for every f in $\mathcal{L}_+$ for which $U \subset f$.

Proof. The fact that Λ is positive implies that $\lambda(C) \geqq 0$ for every C in **C**. To prove that λ is finite, let C be any compact set and let U be any bounded open set containing C. Since there exists a function f in $\mathcal{L}_+$ such that $f(x) = 1$ for x in C and $f(x) = 0$ for x in $X - U$, it follows that $C \subset f \in \mathcal{L}_+$ and therefore

$$\lambda(C) \leqq \Lambda(f) < \infty.$$

If C and D are compact sets, $C \supset D$, and if $C \subset f \in \mathcal{L}_+$, then $D \subset f$, and, therefore, $\lambda(D) \leqq \Lambda(f)$. It follows that $\lambda(D) \leqq \inf \Lambda(f) = \lambda(C)$, i.e. that λ is monotone.

If C and D are compact sets, and if $C \subset f \in \mathcal{L}_+$ and $D \subset g \in \mathcal{L}_+$, then

$$C \cup D \subset f + g \in \mathcal{L}_+,$$

and therefore $\lambda(C \cup D) \leqq \Lambda(f + g) = \Lambda(f) + \Lambda(g)$. It follows that

$$\lambda(C \cup D) \leqq \inf \Lambda(f) + \inf \Lambda(g) = \lambda(C) + \lambda(D),$$

i.e. that λ is subadditive.

If C and D are disjoint compact sets, then there exist disjoint bounded open sets U and V such that $C \subset U$ and $D \subset V$. Let f and g be functions in $\mathcal{L}_+$ such that $f(x) = 1$ for x in C, $f(x) = 0$ for x in $X - U$, $g(x) = 1$ for x in D, and $g(x) = 0$ for x in $X - V$. If $C \cup D \subset h \in \mathcal{L}_+$, then

$$\lambda(C) + \lambda(D) \leqq \Lambda(hf) + \Lambda(hg) = \Lambda(h(f + g)) \leqq \Lambda(h).$$

It follows that

$$\lambda(C) + \lambda(D) \leqq \inf \Lambda(h) = \lambda(C \cup D);$$

the additivity of λ now follows from its subadditivity.

We have thus proved that λ is a content; it remains to prove that λ is regular. For every C in **C** and for every $\epsilon > 0$, there exists a function f in $\mathcal{L}_+$ such that

$$C \subset f \quad \text{and} \quad \Lambda(f) \leqq \lambda(C) + \frac{\epsilon}{2}.$$

If γ is a real number, $0 < \gamma < 1$, and if $D = \{x: f(x) \geqq \gamma\}$, then

$$C \subset \{x: f(x) \geqq 1\} \subset \{x: f(x) > \gamma\} \subset D^0 \subset D \in \mathbf{C}.$$

Since $D \subset \frac{1}{\gamma} f \in \mathfrak{L}_+$, it follows that

$$\lambda(D) \leqq \frac{1}{\gamma} \Lambda(f) \leqq \frac{1}{\gamma}\left(\lambda(C) + \frac{\epsilon}{2}\right).$$

Since γ may be chosen so that

$$\frac{1}{\gamma}\left(\lambda(C) + \frac{\epsilon}{2}\right) \leqq \lambda(C) + \epsilon,$$

it follows that $\lambda(D) \leqq \lambda(C) + \epsilon$; the arbitrariness of ϵ implies that λ is regular.

The last assertion of the theorem is an easy consequence of the regularity of μ. Indeed, if C is a compact set contained in U, then $C \subset f$ and therefore

$$\mu(C) = \lambda(C) \leqq \Lambda(f);$$

it follows that $\mu(U) = \sup \mu(C) \leqq \Lambda(f)$. ∎

Theorem B. *If Λ is a positive linear functional on $\mathfrak{L}$, if, for every C in* $\mathbf{C}$,

$$\lambda(C) = \inf \{\Lambda(f): C \subset f \in \mathfrak{L}_+\},$$

and if μ is the Borel measure induced by the content λ, then

$$\int f d\mu \leqq \Lambda(f)$$

for every f in $\mathfrak{L}_+$.

Proof. Since both $\int f d\mu$ and $\Lambda(f)$ depend linearly on f, it is sufficient to prove the inequality for functions f such that $0 \leqq f(x) \leqq 1$ for all x in X.

Let n be a fixed positive integer and write, for $i = 1, \cdots, n$,

$$f_i(x) = \begin{cases} 0 & \text{if } f(x) < \dfrac{i-1}{n}, \\ \dfrac{f(x) - \dfrac{i-1}{n}}{\dfrac{1}{n}} = nf(x) - (i-1) & \text{if } \dfrac{i-1}{n} \leqq f(x) \leqq \dfrac{i}{n}, \\ 1 & \text{if } \dfrac{i}{n} < f(x). \end{cases}$$

Since, for $i = 1, \cdots, n$,

$$f_i = ([nf - (i-1)] \cup 0) \cap 1 = ([nf - (i-1)] \cap 1) \cup 0,$$

the functions f_i all belong to $\mathcal{L}_+$. Since for any x for which $\dfrac{j-1}{n} \leqq f(x) \leqq \dfrac{j}{n}$, we have

$$f_i(x) = \begin{cases} 1 & \text{if } 1 \leqq i \leqq j-1, \\ 0 & \text{if } j+1 \leqq i \leqq n, \end{cases}$$

it follows that $f(x) = \dfrac{1}{n} \sum_{i=1}^n f_i(x)$ for every x in X.

If, for $i = 0, 1, \cdots, n$, $U_i = \left\{x \colon f(x) > \dfrac{i}{n}\right\}$, then U_i is a bounded open set such that, for $i = 1, \cdots, n$, $U_i \subset f_i$, and hence, by Theorem A, $\mu(U_i) \leqq \Lambda(f_i)$. Since $U_0 \supset U_1 \supset \cdots \supset U_n = 0$, we have

$$\Lambda(f) = \frac{1}{n} \sum_{i=1}^n \Lambda(f_i) \geqq \frac{1}{n} \sum_{i=1}^n \mu(U_i) =$$

$$= \sum_{i=1}^n \left(\frac{i}{n} - \frac{i-1}{n}\right) \mu(U_i) =$$

$$= \sum_{i=1}^{n-1} \frac{i}{n} [\mu(U_i) - \mu(U_{i+1})] =$$

$$= \sum_{i=1}^{n-1} \frac{i+1}{n} \mu(U_i - U_{i+1}) - \frac{1}{n} \mu(U_1) \geqq$$

$$\geqq \sum_{i=1}^{n-1} \int_{U_i - U_{i+1}} f d\mu - \frac{1}{n}\mu(U_1) =$$

$$= \int_{U_1} f d\mu - \frac{1}{n}\mu(U_1) \geqq \int f d\mu - \frac{1}{n}\mu(U_0).$$

The arbitrariness of n and the finiteness of $\mu(U_0)$ imply the desired result. ▌

Theorem C. *If Λ is a positive linear functional on $\mathfrak{L}$, if, for every C in $\mathbf{C}$,*

$$\lambda(C) = \inf \{\Lambda(f): C \subset f \in \mathfrak{L}_+\},$$

and if μ is the Borel measure induced by the content λ, then, corresponding to every compact set C and every positive number ϵ, there exists a function f_0 in $\mathfrak{L}_+$ such that $C \subset f_0$, $f_0 \leqq 1$, and

$$\Lambda(f_0) \leqq \int f_0 d\mu + \epsilon.$$

Proof. Let g_0 be a function in $\mathfrak{L}_+$ such that

$$C \subset g_0 \quad \text{and} \quad \Lambda(g_0) \leqq \lambda(C) + \epsilon.$$

If $f_0 = g_0 \cap 1$, then it follows that

$$\Lambda(f_0) \leqq \Lambda(g_0) \leqq \mu(C) + \epsilon \leqq \int f_0 d\mu + \epsilon. \quad ▌$$

Theorem D. *If Λ is a positive linear functional on $\mathfrak{L}$, then there exists a Borel measure μ such that, for every f in $\mathfrak{L}$,*

$$\Lambda(f) = \int f d\mu.$$

Proof. Write $\lambda(C) = \inf \{\Lambda(f): C \subset f \in \mathfrak{L}_+\}$ for every C in $\mathbf{C}$, let μ be the Borel measure induced by the content λ, and let f be any fixed function in $\mathfrak{L}$.

Let C be a compact set such that $\{x: f(x) \neq 0\} \subset C$, and let ϵ be a positive number. According to Theorem C, there exists a function f_0 in $\mathfrak{L}_+$ such that $C \subset f_0$, $f_0 \leqq 1$, and $\Lambda(f_0) \leqq \int f_0 d\mu + \epsilon$.

We observe that since $C \subset f_0$, it follows that $ff_0 = f$. If c is a positive number such that $|f(x)| \leqq c$ for all x in X, then the function $(f + c)f_0$ belongs to $\mathfrak{L}_+$ and hence, by Theorem B,

$$\Lambda(f) + c\Lambda(f_0) = \Lambda((f + c)f_0) \geqq \int (f + c)f_0 d\mu =$$

$$= \int f d\mu + c \int f_0 d\mu.$$

It follows that

$$\Lambda(f) \geqq \int f d\mu + c \left[\int f_0 d\mu - \Lambda(f_0) \right] \geqq \int f d\mu - c\epsilon;$$

the arbitrariness of ϵ implies that $\Lambda(f) \geqq \int f d\mu$, i.e. that Theorem B is true for all f in $\mathfrak{L}$. Applying this inequality to $-f$ yields its own reverse. ▮

Theorem E. *If μ is a regular Borel measure, if, for every f in $\mathfrak{L}$, $\Lambda(f) = \int f d\mu$, and if, for every C in* **C**,

$$\lambda(C) = \inf \{\Lambda(f) : C \subset f \, \varepsilon \, \mathfrak{L}_+\},$$

then $\mu(C) = \lambda(C)$ for every C in **C**. *Hence, in particular, the representation of a positive linear functional as an integral with respect to a regular Borel measure is unique.*

Proof. It is clear that $\mu(C) \leqq \lambda(C)$. If $C \, \varepsilon \, \mathbf{C}$ and $\epsilon > 0$, then, by the regularity of μ, there exists a bounded open set U containing C such that $\mu(U) \leqq \mu(C) + \epsilon$. Let f be a function in $\mathfrak{F}$ such that $f(x) = 1$ for x in C and $f(x) = 0$ for x in $X - U$; then $C \subset f \, \varepsilon \, \mathfrak{L}_+$ and

$$\lambda(C) \leqq \Lambda(f) = \int f d\mu \leqq \mu(U) \leqq \mu(C) + \epsilon.$$

The arbitrariness of ϵ implies the desired result. ▮

(1) If x_0 is a point of X and $\Lambda(f) = f(x_0)$ for every f in $\mathfrak{L}$, and if $\mu(E) = \chi_E(x_0)$ for every Borel set E, then $\Lambda(f) = \int f d\mu$.

(2) If μ_0 is a Baire measure and $\Lambda(f) = \int f d\mu_0$ for every f in $\mathfrak{L}$, and if μ is a

Borel measure such that $\Lambda(f) = \int f d\mu$, then $\mu(E) = \mu_0(E)$ for every Baire set E.

(3) If μ_0 is a Baire measure and $\Lambda(f) = \int f d\mu_0$ for every f in $\mathfrak{L}$, write

$$\lambda_*(U) = \sup \{\Lambda(f): U \supset f \in \mathfrak{L}_+\}$$

for every U in $\mathbf{U}$, and

$$\mu^*(E) = \inf \{\lambda_*(U): E \subset U \in \mathbf{U}\}$$

for every σ-bounded set E; then $\mu^*(E) = \mu_0(E)$ for every Baire set E.

(4) Let X be the one-point compactification, by ∞, of the countable discrete space of positive integers. A function f in $\mathfrak{L}$ is, in this case, a convergent sequence $\{f(n)\}$ of real numbers with $f(\infty) = \lim_n f(n)$; the most general positive linear functional Λ is defined by

$$\Lambda(f) = \sum_{1 \leq n \leq \infty} f(n)\Lambda_n,$$

where $\Sigma_n \Lambda_n$ is a convergent series of positive numbers.

(5) A linear functional Λ on $\mathfrak{L}$ is **bounded** if there exists a constant k such that $|\Lambda(f)| \leq k \sup \{|f(x)|: x \in X\}$ for every f in $\mathfrak{L}$. Every bounded (but not necessarily positive) linear functional is the difference of two bounded positive linear functionals. The proof of this assertion is not trivial; it may be achieved by imitating the derivation of the Jordan decomposition of a signed measure.

(6) If X is compact, then every positive linear functional on $\mathfrak{L}$ is bounded.

Chapter XI

HAAR MEASURE

§ 57. FULL SUBGROUPS

Before beginning our investigation of measure theory in topological groups, we shall devote this brief section to the proof of two topological results which have important measure theoretic applications. The results concern full subgroups; a subgroup Z of a topological group X is **full** if it has a non empty interior. We shall show that a full subgroup Z of a topological group X embraces the entire topological character of X—everything in X that goes beyond Z is described by the left coset structure of Z which is topologically discrete. We shall show also that a locally compact topological group always has sufficiently small full subgroups—i.e. full subgroups in which none of the measure theoretic pathology of the infinite can occur.

Theorem A. *If Z is a full subgroup of a topological group X, then every union of left cosets of Z is both open and closed in X.*

Proof. Since the complement of any union of left cosets is itself such a union and since a set whose complement is open is closed, it is sufficient to prove that every such union is open. Since a union of open sets is open, it is sufficient to prove that each left coset of Z is open, and for this, in turn, it is sufficient to prove that Z is open.

Since $Z^0 \neq 0$, there is an element z_0 in Z^0. If z is any element of Z, then $zz_0^{-1} \varepsilon Z$ and therefore $zz_0^{-1}Z = Z$. It follows that

$zz_0^{-1}Z^0 = Z^0$ and hence that

$$z = (zz_0^{-1})z_0 \, \varepsilon \, Z^0.$$

Since z is an arbitrary element of Z, we have thereby proved that $Z \subset Z^0$, i.e. that Z is open. ▮

Theorem B. *If E is any Borel set in a locally compact topological group X, then there exists a σ–compact full subgroup Z of X such that $E \subset Z$.*

Proof. It is sufficient to prove (cf. 51.A) that if $\{C_n\}$ is a sequence of compact sets in X, then there exists a σ–compact full subgroup Z of X such that $C_n \subset Z$ for $n = 1, 2, \cdots$.

Let D be a compact set which contains a neighborhood of e. We write $D_0 = D$ and, for $n = 0, 1, 2, \cdots$,

$$D_{n+1} = D_n^{-1}D_n \cup C_{n+1}.$$

If $Z = \bigcup_{n=0}^{\infty} D_n$, then Z is σ–compact, has a non empty interior, and contains each C_n; we shall complete the proof by showing that $Z^{-1}Z \subset Z$.

We show first that if, for any $n = 0, 1, 2, \cdots, e \, \varepsilon \, D_n$, then $D_n \subset D_{n+1}$. Indeed, if $e \, \varepsilon \, D_n$, then $e \, \varepsilon \, D_n^{-1}$; it follows that if $x \, \varepsilon \, D_n$, then

$$x \, \varepsilon \, (D_n^{-1})x \subset D_n^{-1}D_n \subset D_{n+1}.$$

Since $e \, \varepsilon \, D_0$, it follows by mathematical induction that $D_n \subset D_{n+1}$ for $n = 0, 1, 2, \cdots$.

If x and y are any two elements of Z, then, because of the result of the preceding paragraph, both x and y belong to D_n for some positive integer n, and therefore

$$x^{-1}y \, \varepsilon \, D_n^{-1}D_n \subset D_{n+1} \subset Z. \quad ▮$$

§ 58. EXISTENCE

A **Haar measure** is a Borel measure μ in a locally compact topological group X, such that $\mu(U) > 0$ for every non empty Borel open set U, and $\mu(xE) = \mu(E)$ for every Borel set E. The purpose of this section is to prove, for every locally compact topological group, the existence of at least one Haar measure.

The second defining property of a Haar measure may be called left invariance (or invariance under left translations); we observe that the first property is equivalent to the assertion that μ is not identically zero. Indeed, if $\mu(U) = 0$ for some non empty Borel open set U, and if C is any compact set, then the class $\{xU: x \varepsilon C\}$ is an open covering of C. Since C is compact, there exists a finite subset $\{x_1, \cdots, x_n\}$ of C such that $C \subset \bigcup_{i=1}^n x_iU$, and the left invariance of μ implies that $\mu(C) \leqq \sum_{i=1}^n \mu(x_iU) = n\mu(U) = 0$. Since the vanishing of μ on the class $\mathbf{C}$ of all compact sets implies its vanishing on the class $\mathbf{S}$ of all Borel sets, we obtain the desired result: a Haar measure is a left invariant Borel measure which is not identically zero.

Before exhibiting the construction of Haar measures, we remark on the asymmetry of their definition. Left translations and right translations play a perfectly symmetrical role in groups; there is something unfair about our emphasis on left invariance. The concept we defined should really be called "left Haar measure"; an analogous definition of "right Haar measure" should accompany it, and the relations between the two should be thoroughly investigated. Indeed in the sequel we shall occasionally make use of this modified (and thereby more precise) terminology. In most contexts, however, and specifically in connection with the problem of the existence of Haar measures, the perfect left–right symmetry justifies an asymmetric treatment; since the mapping which sends each x in X into x^{-1} interchanges left and right and preserves all other topological and group theoretic properties, every "left theorem" automatically implies and is implied by its corresponding "right theorem." It is, in particular, easy to verify that if μ is a left Haar measure, and if the set function ν is defined, for every Borel set E, by $\nu(E) = \mu(E^{-1})$, then ν is a right Haar measure, and conversely.

If E is any bounded set and F is any set with a non empty interior, we define the "ratio" $E:F$ as the least non negative integer n with the property that E may be covered by n left translations of F, i.e. that there exists a set $\{x_1, \cdots, x_n\}$ of n elements in X such that $E \subset \bigcup_{i=1}^n x_iF$. It is easy to verify that (since E is bounded and F^0 is not empty) $E:F$ is always finite, and that, if A has the properties of both E and F, i.e. if A is a bounded set with a non empty interior, then

$$E:F \leqq (E:A)(A:F).$$

Our construction of Haar measure is motivated by the following considerations. In order to construct a Borel measure in a locally compact Hausdorff space it is sufficient, in view of the results of the preceding chapter, to construct a content λ, i.e. a set function with certain additivity properties on **C**. If C is a compact set and U is a non empty open set, then the ratio $C:U$ serves as a comparison between the sizes of C and U. If we form the limit, in a certain sense, of the product of this ratio by a suitable factor depending on the size of U, as U becomes smaller and smaller, the resulting number should serve as the value of λ at C.

The outline in the preceding paragraph is not quite accurate. In order to illustrate the inaccuracy and make our procedure more intuitive, we mention an example. Suppose that X is the Euclidean plane, μ is Lebesgue measure, and C is an arbitrary compact set. If U_r is the interior of a circle of radius r, and if we write, for every $r > 0$, $n(r) = C:U_r$, then, clearly, $n(r)\pi r^2 \geqq \mu(C)$. It is known that $\lim_{r\to 0} n(r)\pi r^2$ exists and is equal not to $\mu(C)$ but to $\frac{2\pi\sqrt{3}}{9}\mu(C)$; in other words, starting with the usual notion of measure, which assigns the value πr^2 to U_r, our procedure yields a different measure which is a constant multiple of the original one. For this reason, in an attempt to eliminate such a factor of proportionality, we shall replace the ratio $C:U$ by the ratio of two ratios, i.e. by $(C:U)/(A:U)$, where A is a fixed compact set with a non empty interior.

Theorem A. *For each fixed, non empty open set U and compact set A with a non empty interior, the set function λ_U, defined for all compact sets C by*

$$\lambda_U(C) = \frac{C:U}{A:U}$$

is non negative, finite, monotone, subadditive, and left invariant; it is additive in the restricted sense that if C and D are compact sets for which $CU^{-1} \cap DU^{-1} = 0$, *then*

$$\lambda_U(C \cup D) = \lambda_U(C) + \lambda_U(D).$$

Proof. The verification of all parts of this theorem, except possibly the last, consists of a straightforward examination of the definition of ratios such as $C:U$. To prove the last assertion, let xU be a left translation of U and observe that if $C \cap xU \neq 0$, then $x \in CU^{-1}$ and if $D \cap xU \neq 0$, then $x \in DU^{-1}$. It follows that no left translation of U can have a non empty intersection with both C and D and hence that λ_U has the stated additivity property. ∎

Theorem B. *In every locally compact topological group X there exists at least one regular Haar measure.*

Proof. In view of 53.E and 53.F it is sufficient to construct a left invariant content which is not identically zero; 53.C implies that the induced measure is not identically zero and hence is a regular Haar measure.

Let A be a fixed compact set with a non empty interior and let **N** be the class of all neighborhoods of the identity. For each U in **N**, we construct the set function λ_U, defined for all compact sets C by $\lambda_U(C) = \dfrac{C:U}{A:U}$; since $C:U \leqq (C:A)(A:U)$, it follows that $0 \leqq \lambda_U(C) \leqq C:A$ for every C in **C**. Theorem A shows that each λ_U is almost a content; it fails to be a content only because it is not necessarily additive. We shall make use of the modern form of Cantor's diagonal process, i.e. of Tychonoff's theorem on the compactness of product spaces, to pick out a limit of the λ_U's which has all their properties and is in addition additive.

If to each set C in **C** we make correspond the closed interval $[0, C:A]$, and if we denote by Φ the Cartesian product (in the topological sense) of all these intervals, then Φ is a compact Hausdorff space whose points are real valued functions ϕ defined on **C**, such that, for each C in **C**, $0 \leqq \phi(C) \leqq C:A$. For each U in **N** the function λ_U is a point in this space.

For each U in $\mathbf{N}$ we denote by $\Lambda(U)$ the set of all those functions λ_V for which $V \subset U$; i.e.

$$\Lambda(U) = \{\lambda_V: U \supset V \,\varepsilon\, \mathbf{N}\}.$$

If $\{U_1, \cdots, U_n\}$ is any finite class of neighborhoods of the identity, i.e. any finite subclass of $\mathbf{N}$, then $\bigcap_{i=1}^n U_i$ is also a neighborhood of the identity and, moreover,

$$\bigcap_{i=1}^n U_i \subset U_j, \quad j = 1, \cdots, n.$$

It follows that

$$\Lambda(\bigcap_{i=1}^n U_i) \subset \bigcap_{i=1}^n \Lambda(U_i),$$

and hence, since $\Lambda(U)$ always contains λ_U and is therefore non empty, that the class of all sets of the form $\Lambda(U)$, $U \,\varepsilon\, \mathbf{N}$, has the finite intersection property. The compactness of Φ implies that there is a point λ in the intersection of the closures of all $\Lambda(U)$;

$$\lambda \,\varepsilon\, \bigcap \{\overline{\Lambda(U)}: U \,\varepsilon\, \mathbf{N}\}.$$

We shall prove that λ is the desired content.

It is clear that $0 \leqq \lambda(C) \leqq C:A < \infty$ for every C in $\mathbf{C}$. To prove that λ is monotone, we remark that if, for each fixed C in $\mathbf{C}$, $\xi_C(\phi) = \phi(C)$, then ξ_C is a continuous function on Φ, and hence, for any two compact sets C and D, the set

$$\Delta = \{\phi: \phi(C) \leqq \phi(D)\} \subset \Phi$$

is closed. If $C \subset D$ and $U \,\varepsilon\, \mathbf{N}$, then $\lambda_U \,\varepsilon\, \Delta$ and consequently $\Lambda(U) \subset \Delta$. The fact that Δ is closed implies that $\lambda \,\varepsilon\, \overline{\Lambda(U)} \subset \Delta$, i.e. that λ is monotone.

The proof of the subadditivity of λ is entirely similar to the above continuity argument; we omit it and turn instead to the proof that λ is additive. If C and D are compact sets such that $C \cap D = 0$, then there exists a neighborhood U of e such that $CU^{-1} \cap DU^{-1} = 0$. If $V \,\varepsilon\, \mathbf{N}$ and $V \subset U$, then $CV^{-1} \cap DV^{-1} = 0$ and hence (Theorem A)

$$\lambda_V(C \cup D) = \lambda_V(C) + \lambda_V(D).$$

This means that, whenever $V \subset U$, λ_V belongs to the closed set

$\Delta = \{\phi: \phi(C \cup D) = \phi(C) + \phi(D)\}$, and hence that $\Lambda(U) \subset \Delta$. It follows that $\lambda \,\varepsilon\, \Lambda(U) \subset \Delta$, i.e. that λ is additive.

Another application of the continuity argument shows that $\lambda(A) = 1$ (since $\lambda_U(A) = 1$ for every U in $\mathbf{N}$), and hence that the set function λ (which is already known to be a content) is not identically zero. The fact that λ is left invariant follows, again by continuity, from the left invariance of each λ_U. ∎

(1) The existence of a right Haar measure follows from the existence of a left Haar measure by consideration of the group $\hat{X}$ dual to X. The group $\hat{X}$ has, by definition, the same elements and the same topology as X; the product (in $\hat{X}$) of two elements x and y (in that order) is, however, defined to be the product (in X) of y and x (in that order).

(2) Haar measure is obviously not unique, since, for any Haar measure μ and any positive number c, the product $c\mu$ is also a Haar measure.

(3) If, for every U in $\mathbf{N}$, λ_U is the set function described in Theorem A, then, for every compact set C with $C^0 \neq 0$, $0 < \dfrac{1}{A:C} \leqq \lambda_U(C)$. It follows that $\lambda(C) > 0$ whenever $C^0 \neq 0$.

(4) The following is a well known example of a group in which the left and right Haar measures are essentially different. Let X be the set of all matrices of the form $\begin{pmatrix} x & y \\ 0 & 1 \end{pmatrix}$, where $0 < x < \infty$ and $-\infty < y < +\infty$; it is easy to verify that, with respect to ordinary matrix multiplication, X is a group. If X is topologized in the obvious way as a subset (half plane) of the Euclidean plane, then X becomes a locally compact topological group. If we write, for every Borel set E in X,

$$\mu(E) = \iint_E \frac{1}{x^2}\, dxdy \quad \text{and} \quad \nu(E) = \iint_E \frac{1}{x}\, dxdy$$

(where the integrals are with respect to Lebesgue measure in the half plane), then μ and ν are, respectively, left and right Haar measures in X. Since $\mu(E^{-1}) = \nu(E)$, this example shows also that there may exist measurable sets E for which $\mu(E) < \infty$ and $\mu(E^{-1}) = \infty$.

(5) If C and D are two compact sets such that $\mu(C) = \mu(D) = 0$, does it follow that $\mu(CD) = 0$?

(6) If μ is a Haar measure in X, then a necessary and sufficient condition that X be discrete is that $\mu(\{x\}) \neq 0$ for at least one x in X.

(7) Every locally compact topological group with Haar measure satisfies the condition of 31.10; (cf. § 57).

(8) If a Haar measure μ in X is finite, then X is compact.

(9) If μ is a Haar measure in X, then the following four assertions are mutually equivalent: (a) X is σ-compact; (b) μ is totally σ-finite; (c) every disjoint class of non empty open Borel sets is countable; and (d) for every non empty open Borel set U, there exists a sequence $\{x_n\}$ of elements in X such that

$$X = \bigcup_{n=1}^{\infty} x_n U.$$

§ 59. MEASURABLE GROUPS

A topological group is, by definition, a group X with a topology satisfying a suitable separation axiom and such that the transformation (from $X \times X$ onto X) which takes (x,y) into $x^{-1}y$ is continuous. For our present purposes it is convenient to replace this definition by an equivalent one which requires that the transformation S (from $X \times X$ onto itself), defined by $S(x,y) = (x,xy)$ be a homeomorphism. If, indeed, X is a topological group in the usual sense, then it follows that S is continuous; since S is clearly one to one and $S^{-1}(x,v) = (x,x^{-1}y)$, it follows similarly that S^{-1} is continuous, and hence that S is a homeomorphism. If, conversely, it is known that S is a homeomorphism, then S^{-1} is continuous and, therefore, so is the transformation S^{-1} followed by projection on the second coordinate. (In case X is the real line, the transformation S is easy to visualize; its effect is that of a shearing which moves every point in the plane vertically by an amount equal to its distance from the y–axis.)

Motivated by the preceding paragraph and the fact that every locally compact topological group has a Haar measure, we define the following measure theoretic analog of the concept of a topological group. A **measurable group** is a σ–finite measure space $(X,\mathbf{S},\mu)$ such that (a) μ is not identically zero, (b) X is a group, (c) the σ–ring $\mathbf{S}$ and the measure μ are invariant under left translations, and (d) the transformation S of $X \times X$ onto itself, defined by $S(x,y) = (x,xy)$, is measurability preserving. (To say that $\mathbf{S}$ is invariant under left translations means, of course, that $xE \in \mathbf{S}$ for every x in X and every E in $\mathbf{S}$; by a measurable subset of $X \times X$ we mean, as always, a set in the σ–ring $\mathbf{S} \times \mathbf{S}$.)

If X is a locally compact group, $\mathbf{S}$ is the class of all Baire sets in X, and μ is a Haar measure, then the fact that S is a homeomorphism (and therefore Baire measurability preserving), together with the fact (51.E) that the class of all Baire sets in $X \times X$ coincides with $\mathbf{S} \times \mathbf{S}$, implies that $(X,\mathbf{S},\mu)$ is a measurable group. The main purpose of the following discussion of measurable groups is to see how much one can say about a locally compact topological group by exploiting its measure theoretic structure only.

If X is any measurable space (and hence, in particular, if X is any measurable group), then the one to one transformation R of $X \times X$ onto itself, defined by $R(x,y) = (y,x)$, is measurability preserving—the reason for this is the immediately verifiable fact that if E is a measurable rectangle, then so also are $R(E)$ and $R^{-1}(E)(= R(E))$. Since the product of measurability preserving transformations is measurability preserving, this remark gives us a large stock of measurability preserving transformations in a measurable group—namely all transformations which may be obtained by multiplying powers of S and R. We shall in particular frequently use, in addition to the shearing transformation S, its reflected analog $T = R^{-1}SR$; we observe that $T(x,y) = (yx,y)$.

Throughout the remainder of this section we shall assume that

> μ and ν are two measures (possibly but not necessarily identical) such that $(X,\mathbf{S},\mu)$ and $(X,\mathbf{S},\nu)$ are measurable groups, and R, S, and T are the measurability preserving transformations described in the preceding paragraphs.

Theorem A. *If E is any subset of $X \times X$, then*

$$(S(E))_x = xE_x \quad \text{and} \quad (T(E))^y = yE^y$$

for every x and y in X.

Proof. The result for S follows from the relation

$$\chi_{S(E)}(x,y) = \chi_E(x,x^{-1}y),$$

together with the facts that $y \in (S(E))_x$ if and only if $\chi_{S(E)}(x,y) = 1$, and $x^{-1}y \in E_x$ if and only if $\chi_E(x,x^{-1}y) = 1$. The proof for T is similar. ▌

Theorem B. *The transformations S and T are measure preserving transformations of the measure space $(X \times X, \mathbf{S} \times \mathbf{S}, \mu \times \nu)$ onto itself.*

Proof. If E is a measurable subset of $X \times X$, then, by Fubini's theorem and Theorem A,

$$(\mu \times \nu)(S(E)) = \int \nu((S(E))_x)d\mu(x) = \int \nu(xE_x)d\mu(x) =$$

$$= \int \nu(E_x)d\mu(x) = (\mu \times \nu)(E);$$

this establishes the measure preserving character of S. The result for T follows similarly by considering the sections $(T(E))^y$. ∎

Theorem C. *If* $Q = S^{-1}RS$, *then*

$$(Q(A \times B))_{x^{-1}} = xA \cap B^{-1},$$

and

$$(Q(A \times B))^{y^{-1}} = \begin{cases} Ay & \text{if } y \varepsilon B, \\ 0 & \text{if } y \varepsilon' B. \end{cases}$$

Proof. We observe that $Q(x,y) = (xy,y^{-1})$ and that $Q^{-1} = Q$. The first conclusion follows from the relation

$$\chi_{Q(A\times B)}(x^{-1},y) = \chi_{A\times B}(x^{-1}y,y^{-1}) = \chi_{xA}(y)\chi_B(y^{-1}),$$

together with the facts that $y \varepsilon (Q(A \times B))_{x^{-1}}$ if and only if $\chi_{Q(A\times B)}(x^{-1},y) = 1$, and $y \varepsilon xA \cap B^{-1}$ if and only if $\chi_{xA}(y)\chi_B(y^{-1}) = 1$. The second conclusion follows from the relation

$$\chi_{Q(A\times B)}(x,y^{-1}) = \chi_{A\times B}(xy^{-1},y) = \chi_{Ay}(x)\chi_B(y),$$

together with the facts that $x \varepsilon (Q(A \times B))^{y^{-1}}$ if and only if $\chi_{Q(A\times B)}(x,y^{-1}) = 1$, and that $x \varepsilon Ay$ and $y \varepsilon B$ if and only if $\chi_{Ay}(x)\chi_B(y) = 1$. ∎

Theorem D. *If A is a measurable subset of X [of positive measure], and $y \varepsilon X$, then Ay is a measurable set [of positive measure] and A^{-1} is a measurable set [of positive measure]. If f is a measurable function, A is a measurable set of positive measure, and, for every x in X, $g(x) = f(x^{-1})/\mu(Ax)$, then g is measurable.*

Proof. The measurability of Ay follows by selecting any measurable set B containing y and observing that, according to Theorem C, Ay is a section of the measurable set $Q(A \times B)$ (where $Q = S^{-1}RS$). For the remainder of the proof we shall make use of the fact that Q is a measure preserving transformation of $(X \times X, \mathbf{S} \times \mathbf{S}, \mu \times \mu)$ onto itself. It follows that if $\mu(A) > 0$, then, by Theorem C,

$$0 < (\mu(A))^2 = (\mu \times \mu)(Q(A \times A)) = \int \mu(x^{-1}A \cap A^{-1})d\mu(x),$$

and hence, in particular, that $x^{-1}A \cap A^{-1}$ is a measurable set of positive measure for at least one value of x. We have proved, in other words, that if A is a measurable set of positive measure, then there exists a measurable set B of positive measure such that $B \subset A^{-1}$. (This implies, in particular, that as soon as we will have proved that A^{-1} is measurable, the result concerning the positiveness of $\mu(A^{-1})$ will follow automatically.) Since $B \subset A^{-1}$ implies that $y^{-1}B \subset y^{-1}A^{-1}$, and since $\mu(y^{-1}B) = \mu(B)$, another application of our result yields the existence of a measurable set C of positive measure such that $C \subset (y^{-1}B)^{-1} \subset (y^{-1}A^{-1})^{-1} = Ay$. This settles all our assertions about Ay. To prove the measurability of A^{-1} we observe that it follows from Theorem C and what we have just proved that, if $\mu(A) > 0$, then

$$\{y: \mu((Q(A \times A))^y) > 0\} = A^{-1}.$$

This proves that if $\mu(A) > 0$, then A^{-1} is measurable; if $\mu(A) = 0$, we may find a measurable set B of positive measure, disjoint from A, and deduce the measurability of A^{-1} from the relation $A^{-1} = (A \cup B)^{-1} - B^{-1}$.

What we have already proved implies that if $\hat{f}(x) = f(x^{-1})$, then $\hat{f}$ is measurable. If A and B are measurable sets, if $f_0(y) = \mu((Q(A \times B))^y)$, and if $\hat{f}_0(y) = f_0(y^{-1})$, then both f_0 and $\hat{f}_0$ are measurable, and we have, by Theorem C,

$$\hat{f}_0(y) = \mu(Ay)\chi_B(y).$$

We have proved, in other words, that if $h(y) = \mu(Ay)$, then h is measurable on every measurable set, and it follows that $\frac{1}{h}$ has the same property. ▌

Theorem E. *If A and B are measurable sets of positive measure, then there exist measurable sets C_1 and C_2 of positive measure and elements x_1, y_1, x_2, and y_2, such that*

$$x_1C_1 \subset A, \quad y_1C_1 \subset B, \quad C_2x_2 \subset A, \quad C_2y_2 \subset B.$$

Proof. Since $\mu(B) > 0$ implies that $\mu(B^{-1}) > 0$, it follows that $(\mu \times \mu)(A \times B^{-1}) = \mu(A)\mu(B^{-1}) > 0$. Theorem C implies that $x^{-1}A \cap B$ is measurable for every x in X and of positive measure for at least one x in X. If x_1 is such that $\mu(x_1{}^{-1}A \cap B)$

> 0, and if $y_1 = e$, then, for $C_1 = x_1^{-1}A \cap B$, we have $x_1C_1 \subset A$ and $y_1C_1 \subset B$.

Applying this result to A^{-1} and B^{-1} we may find C_0, x_0, y_0 so that $x_0C_0 \subset A^{-1}$ and $y_0C_0 \subset B^{-1}$, and we may write $C_2 = C_0^{-1}$, $x_2 = x_0^{-1}$, $y_2 = y_0^{-1}$. ▮

Theorem F. *If A and B are measurable sets and if $f(x) = \mu(x^{-1}A \cap B)$, then f is a measurable function and*

$$\int f d\mu = \mu(A)\mu(B^{-1}).$$

If $g(x) = \mu(xA \,\Delta\, B)$, and if $\epsilon < \mu(A) + \mu(B)$, then the set $\{x: g(x) < \epsilon\}$ is measurable.

The first half of this result is sometimes known as the **average theorem.**

Proof. The first assertion follows from the fact that if, as before, $Q = S^{-1}RS$, then Q is a measure preserving transformation of $(X \times X, \mathbf{S} \times \mathbf{S}, \mu \times \mu)$ onto itself and

$$f(x) = \mu((Q(A \times B^{-1}))_x).$$

If $\hat{f}(x) = f(x^{-1})$, then it follows that $\hat{f}$ is measurable. The second assertion is a consequence of this and the relation

$$\{x: g(x) < \epsilon\} = \{x: \hat{f}(x) > \tfrac{1}{2}(\mu(A) + \mu(B) - \epsilon)\}. \quad ▮$$

(1) Is the Cartesian product of two measurable groups a measurable group?

(2) If μ is a Haar measure in a compact group X of cardinal number greater than that of the continuum, then $(X,\mathbf{S},\mu)$ is not a measurable group. (Hint: let $D = \{(x,y): x = y\} = S(X \times \{e\})$. If D is in $\mathbf{S} \times \mathbf{S}$, then there exists a countable class $\mathbf{R}$ of rectangles such that $D \,\varepsilon\, \mathbf{S}(\mathbf{R})$. Let $\mathbf{E}$ be the [countable] class of sides of rectangles belonging to $\mathbf{R}$. Since $D \,\varepsilon\, \mathbf{S}(\mathbf{E}) \times \mathbf{S}(\mathbf{E})$, it follows that every section of D belongs to $\mathbf{S}(\mathbf{E})$. Since, however, by 5.9c, $\mathbf{S}(\mathbf{E})$ has cardinal number not greater than that of the continuum, this contradicts the assumption on the cardinal number of X.)

(3) If μ is a Haar measure on a locally compact group X, then, for every Baire set E and every x in X, the vanishing of any one of the numbers,

$$\mu(E), \quad \mu(xE), \quad \mu(Ex), \quad \text{and} \quad \mu(E^{-1}),$$

implies the vanishing of all others.

(4) If $(X,\mathbf{S},\mu)$ is a measurable group such that μ is totally finite, and if A is any measurable set such that $\mu(xA - A) = 0$ for every x in X, then either $\mu(A) = 0$ or $\mu(X - A) = 0$. (Hint: apply the average theorem to A and

$X - A$.) Properly formulated, this result remains valid even if μ is not necessarily finite; in the language of ergodic theory it asserts that a measurable group, considered as a group of measure preserving transformations on itself, is metrically transitive.

(5) If μ is a Haar measure on a compact group X, then, for every Baire set E and every x in X,

$$\mu(E) = \mu(xE) = \mu(Ex) = \mu(E^{-1}).$$

§ 60. UNIQUENESS

Our purpose in this section is to prove that the measure in a measurable group is essentially unique.

Theorem A. *If μ and ν are two measures such that $(X,\mathbf{S},\mu)$ and $(X,\mathbf{S},\nu)$ are measurable groups, and if E in $\mathbf{S}$ is such that $0 < \nu(E) < \infty$, then, for every non negative measurable function f on X,*

$$\int f(x)d\mu(x) = \mu(E)\int \frac{f(y^{-1})}{\nu(Ey)}\,d\nu(y).$$

The important part of this result, as far as it concerns the uniqueness proof, is its qualitative aspect, which asserts that every μ–integral may be expressed in terms of a ν–integral.

Proof. If $g(y) = f(y^{-1})/\nu(Ey)$, then our results in the preceding section imply that g is a non negative measurable function along with f. If, as before, we write

$$S(x,y) = (x,xy) \quad \text{and} \quad T(x,y) = (yx,y),$$

then in the measure space $(X \times X, \mathbf{S} \times \mathbf{S}, \mu \times \nu)$ both the transformations S and T are measure preserving and, therefore, so is the transformation $S^{-1}T$. Since $S^{-1}T(x,y) = (yx,x^{-1})$, it follows from Fubini's theorem that

$$\begin{aligned}\mu(E)\int g(y)d\nu(y) &= \int \chi_E(x)d\mu(x)\int g(y)d\nu(y) = \\ &= \int \chi_E(x)g(y)d(\mu \times \nu)(x,y) = \\ &= \iint \chi_E(yx)g(x^{-1})d\nu(y)d\mu(x) = \\ &= \int g(x^{-1})\nu(Ex^{-1})d\mu(x).\end{aligned}$$

Since $g(x^{-1})\nu(Ex^{-1}) = f(x)$, the desired result follows from a comparison of the extreme terms of this chain of equalities. ∎

Theorem B. *If μ and ν are two measures such that $(X,\mathbf{S},\mu)$ and $(X,\mathbf{S},\nu)$ are measurable groups, and if E in $\mathbf{S}$ is such that $0 < \nu(E) < \infty$, then, for every F in $\mathbf{S}$, $\mu(E)\nu(F) = \nu(E)\mu(F)$.*

We remark that this result is indeed a uniqueness theorem; it asserts that, for every F in $\mathbf{S}$, $\mu(F) = c\nu(F)$, where c is the non negative finite constant $\frac{\mu(E)}{\nu(E)}$, i.e. that μ and ν coincide to within a multiplicative constant.

Proof. Let f be the characteristic function of F. Since Theorem A is true, in particular, if the two measures μ and ν are both equal to ν, we have

$$\int f(x)d\nu(x) = \nu(E)\int \frac{f(y^{-1})}{\nu(Ey)}\,d\nu(y).$$

Multiplying by $\mu(E)$ and applying Theorem A, we obtain

$$\mu(E)\int f(x)d\nu(x) = \nu(E)\int f(x)d\mu(x). \quad ∎$$

Theorem C. *If μ and ν are regular Haar measures on a locally compact topological group X, then there exists a positive finite constant c such that $\mu(E) = c\nu(E)$ for every Borel set E.*

Proof. If $\mathbf{S}_0$ is the class of all Baire sets in X, then $(X,\mathbf{S}_0,\mu)$ and $(X,\mathbf{S}_0,\nu)$ are measurable groups and hence, by Theorem B, $\mu(E) = c\nu(E)$ for every Baire set E, with a non negative finite constant c; the fact that c is positive may be inferred by choosing E to be any bounded open Baire set. Any two regular Borel measures (such as μ and $c\nu$) which coincide on Baire sets coincide also on all Borel sets; (cf. 52.H). ∎

(1) The Haar measure of the multiplicative group of all non zero real numbers is absolutely continuous with respect to Lebesgue measure; what is its Radon–Nikodym derivative?

(2) If X and Y are two locally compact groups with Haar measures μ and ν respectively, and if λ is a Haar measure in $X \times Y$, then, on the class of all Baire sets in $X \times Y$, λ is a constant multiple of $\mu \times \nu$.

(3) The metric transitivity established in 59.4 may be used to prove the uniqueness theorem for measurable groups with a finite measure. Suppose

first that μ and ν are left invariant measures such that $\nu \ll \mu$; then there exists a non negative integrable function f such that

$$\nu(E) = \int_E f(x) d\mu(x)$$

for every measurable set E. It follows that

$$\nu(yE) = \int_{yE} f(x) d\mu(x) = \int_E f(y^{-1}x) d\mu(x),$$

and hence, since ν is left invariant, that $f(x) = f(y^{-1}x)$ $[\mu]$. If $N_t = \{x: f(x) < t\}$, then

$$\mu(yN_t - N_t) = \mu(\{x: f(y^{-1}x) < t\} - \{x: f(x) < t\}) = 0,$$

and hence, for each real number t, either $\mu(N_t) = 0$ or $\mu(N_t') = 0$. Since this implies that f is a constant a.e. $[\mu]$, it follows that $\nu = c\mu$. To treat the general, not necessarily absolutely continuous, case, replace μ by $\mu + \nu$. Just as in 59.4, these considerations may be extended to apply to not necessarily finite cases also.

(4) If $(X,\mathbf{S},\mu)$ is a measurable group and if E and F are measurable sets, then there exist two sequences $\{x_n\}$ and $\{y_n\}$ of elements of X and a sequence $\{A_n\}$ of measurable sets such that (a) the sequences $\{x_nA_n\}$ and $\{y_nA_n\}$ are disjoint sequences of subsets of E and F respectively, and (b) at least one of the two measurable sets,

$$E_0 = E - \bigcup_{n=1}^{\infty} x_nA_n \quad \text{and} \quad F_0 = F - \bigcup_{n=1}^{\infty} y_nA_n,$$

has measure zero. (Hint: if either E or F has measure zero, the assertion is trivial. If both E and F have positive measure, apply 59.E to find x_1, y_1, A_1 so that $\mu(A_1) > 0$, $x_1A_1 \subset E$, $y_1A_1 \subset F$. If either $E - x_1A_1$ or $F - y_1A_1$ has measure zero, the assertion is true; if not, then 59.E may be applied again, and the argument may be repeated by countable but possibly transfinite induction.)

Since this result is valid for all left invariant measures, it may be used to give still another proof of the uniqueness theorem. It may be shown that if μ and ν are both left invariant measures, then the correspondence which, for every measurable set E, assigns $\mu(E)$ to $\nu(E)$, is an unambiguously defined, one to one correspondence between the set of all values of μ and the set of all values of ν. A more detailed, but not particularly difficult, examination of this correspondence yields the uniqueness theorem.

(5) Let μ be a regular Haar measure on a locally compact group X. Since, for each x in X, the set function μ_x, defined for every Borel set E by $\mu_x(E) = \mu(Ex)$, is also a regular Haar measure, it follows from the uniqueness theorem that $\mu(Ex) = \Delta(x)\mu(E)$, where $0 < \Delta(x) < \infty$.

(5a) $\Delta(xy) = \Delta(x)\Delta(y)$; $\Delta(e) = 1$.

(5b) If x is in the center of X, then $\Delta(x) = 1$.

(5c) If x is a commutator, and hence, more generally, if x is in the commutator subgroup of X, then $\Delta(x) = 1$.

(5d) The function Δ is continuous. (Hint: let C be a compact set of positive measure and let ϵ be any positive number. By regularity, there exists a bounded open set U such that $C \subset U$ and $\mu(U) \leq (1+\epsilon)\mu(C)$. If V is a neighborhood of e such that $V = V^{-1}$ and $CV \subset U$, and if $x \in V$, then

$$\Delta(x)\mu(C) = \mu(Cx) \leq \mu(U) \leq (1+\epsilon)\mu(C)$$

and

$$\frac{\mu(C)}{\Delta(x)} = \mu(Cx^{-1}) \leq \mu(U) \leq (1+\epsilon)\mu(C),$$

so that $\frac{1}{1+\epsilon} \leq \Delta(x) \leq 1+\epsilon$.)

(5e) The results (5a) and (5d) yield another proof of the identity of left and right invariant measures on a compact group X, since they imply that $\Delta(X)$ is a compact subgroup of the multiplicative group of positive real numbers.

(5f) For every Borel set E, $\mu(E^{-1}) = \int_E \frac{1}{\Delta(x)}\, d\mu(x)$. (Hint: by the uniqueness theorem for *right* invariant measures, $\mu(E^{-1}) = c\int_E \frac{1}{\Delta(x)}\, d\mu(x)$ for some positive finite constant c. This implies that $\int f(x^{-1})d\mu(x) = c\cdot\int \frac{f(x)}{\Delta(x)}\, d\mu(x)$ for every integrable function f. Replacing $f(x)$ by $f(x^{-1})$, writing $g(x^{-1}) = f(x^{-1})/\Delta(x)$, and applying the last written equation to g in place of f, yields the result that

$$\frac{1}{c}\int g(x^{-1})d\mu(x) = c\cdot\int g(x^{-1})d\mu(x).)$$

(5g) If $\Gamma(x)$ is the right handed analog of $\Delta(x)$, i.e. if, for a right invariant measure ν, Γ is defined by $\nu(xE) = \Gamma(x)\nu(E)$, then $\Gamma(x) = \frac{1}{\Delta(x)}$.

(6) A **relatively invariant** measure on a locally compact group X is a Baire measure ν, not identically zero, such that for each fixed x in X the measure ν_x, defined by $\nu_x(E) = \nu(xE)$, is a constant non zero multiple of ν. A necessary and sufficient condition that ν be relatively invariant is that $\nu(E) = \int_E \phi(y)d\mu(y)$, where μ is Haar measure and ϕ is a continuous representation of X in the multiplicative group of positive real numbers. (Hint: if ϕ is non negative, continuous, and such that $\phi(xy) = \phi(x)\phi(y)$, and if $\nu(E) = \int_E \phi(y)d\mu(y)$, then

$$\nu(xE) = \int_{xE} \phi(y)d\mu(y) = \int_E \phi(xy)d\mu(y) =$$

$$= \int_E \phi(x)\phi(y)d\mu(y) = \phi(x)\nu(E).$$

If, conversely, $\nu(xE) = \phi(x)\nu(E)$, then it follows (cf. (5)) that $\phi(xy) = \phi(x)\phi(y)$ and ϕ is continuous. Consequently $\tilde{\mu}(E) = \int_E \phi(y^{-1})d\nu(y)$ may be formed, and by the uniqueness theorem $\tilde{\mu} = \mu$.)

(7) If μ is a σ-finite, left invariant measure on the class $\mathbf{S}_0$ of Baire sets in a locally compact group X, then μ is a constant multiple of the Baire contraction of Haar measure, and hence, in particular, μ is finite on compact sets. (Hint: if μ is not identically zero, then $(X,\mathbf{S}_0,\mu)$ is a measurable group.)

Chapter XII

MEASURE AND TOPOLOGY IN GROUPS

§ 61. TOPOLOGY IN TERMS OF MEASURE

In the preceding chapter we showed that in every locally compact group it is possible to introduce a left invariant Baire measure (or a left invariant, regular Borel measure) in an essentially unique manner. In this chapter we shall show that there are very close connections between the measure theoretic and the topological structures of such a group. In particular, in this section we shall establish some of the many results whose total effect is the assertion that not only is the measure determined by the topology, but that, conversely, all topological concepts may be described in measure theoretic terms. Throughout this section we shall assume that

> X is a locally compact topological group, μ is a regular Haar measure on X, and $\rho(E,F) = \mu(E \Delta F)$ for any two Borel sets E and F.

Theorem A. *If E is a Borel set of finite measure and if $f(x) = \rho(xE,E)$, for every x in X, then f is continuous.*

Proof. If $\epsilon > 0$, then, because of the regularity of μ, there exists a compact set C such that $\rho(E,C) < \frac{\epsilon}{4}$ and there exists an open Borel set U containing C such that $\rho(U,C) < \frac{\epsilon}{4}$. Let V be a neighborhood of e such that $V = V^{-1}$ and $VC \subset U$. If $y^{-1}x \in V$, then $x^{-1}y \varepsilon V$ and therefore

$$\begin{aligned}\rho(xC,yC) &= \mu(xC - yC) + \mu(yC - xC) = \\ &= \mu(y^{-1}xC - C) + \mu(x^{-1}yC - C) \leqq \\ &\leqq 2\mu(VC - C) \leqq 2\mu(U - C) < \frac{\epsilon}{2}.\end{aligned}$$

It follows that

$$\begin{aligned}&| \rho(xE,E) - \rho(yE,E) | \leqq \\ &\leqq \rho(xE,yE) \leqq \rho(xE,xC) + \rho(xC,yC) + \rho(yC,yE) < \epsilon. \quad \blacksquare\end{aligned}$$

Theorem A implies that for every Borel set E of finite measure, and for every positive number ϵ, the set $\{x: \rho(xE,E) < \epsilon\}$ is open. Our next result shows that there are enough open sets of this kind.

Theorem B. *If U is any neighborhood of e, then there exist a Baire set E of positive, finite measure and a positive number ϵ such that $\{x: \rho(xE,E) < \epsilon\} \subset U$.*

Proof. Let V be a neighborhood of e such that $VV^{-1} \subset U$ and let E be a Baire set of positive, finite measure such that $E \subset V$. If ϵ is such that $0 < \epsilon < 2\mu(E)$, then

$$\{x: \rho(xE,E) < \epsilon\} \subset \{x: xE \cap E \neq 0\} = EE^{-1} \subset VV^{-1} \subset U. \quad \blacksquare$$

Theorems A and B together imply that the class of all sets of the form $\{x: \rho(xE,E) < \epsilon\}$ is a base at e, and hence that it is indeed possible to describe all topological concepts in measure theoretic terms. To illustrate in detail how such descriptions are made, we proceed to give a measure theoretic characterization of boundedness.

Theorem C. *A necessary and sufficient condition that a set A be bounded is that there exist a Baire set E of positive, finite measure and a number ϵ, $0 \leqq \epsilon < 2\mu(E)$, such that*

$$A \subset \{x: \rho(xE,E) \leqq \epsilon\}.$$

Proof. In order to prove the sufficiency of the condition, we shall show that if E is a Baire set of positive, finite measure, and if $0 \leqq \epsilon < 2\mu(E)$, then the set $\{x: \rho(xE,E) \leqq \epsilon\}$ is bounded. Let δ be a positive number such that $4\delta < 2\mu(E) - \epsilon$, and let C be a compact subset of E such that $\mu(E) - \delta < \mu(C)$. It follows that

$$\rho(xC,C) \leqq \rho(xC,xE) + \rho(xE,E) + \rho(E,C) < 2\delta + \rho(xE,E)$$

and hence that

$$\{x: \rho(xE,E) \leqq \epsilon\} \subset \{x: \rho(xC,C) \leqq \epsilon + 2\delta\}.$$

Since $\epsilon + 2\delta < 2\mu(C)$, it follows that

$$\{x: \rho(xE,E) \leqq \epsilon\} \subset \{x: \mu(xC \cap C) \neq 0\} \subset CC^{-1}.$$

To prove the necessity of the condition, let C be a compact set such that $A \subset C$ and let D be a compact set of positive measure. Suppose that the Baire set E of positive, finite measure is selected so that $E \supset C^{-1}D \cup D$. Since $D \subset E$, and since, for x in C, $D \subset xC^{-1}D \subset xE$, it follows that, for x in C, $D \subset xE \cap E$. This implies that

$$A \subset C \subset \{x: D \subset xE \cap E\} \subset \{x: \rho(xE,E) \leqq \epsilon\},$$

where $\epsilon = 2(\mu(E) - \mu(D))$. ∎

(1) The analogs of Theorems A, B, and C with $\mu(xE \cap F)$ in place of $\rho(xE,E)$ are true, where E and F are Baire sets of positive finite measure.

(2) If, for a fixed Borel set E of finite measure, $f(x) = xE$, then f is a continuous function from X to the metric space of measurable sets of finite measure.

(3) If E is any Borel set of positive measure, then there exists a neighborhood U of e such that $U \subset EE^{-1}$.

(4) X is separable if and only if the metric space of measurable sets of finite measure is separable.

(5) If E is any bounded Borel set, and if, for every x in X and every bounded neighborhood U of e,

$$f_U(x) = \frac{\mu(E \cap Ux)}{\mu(Ux)},$$

then f_U converges in the mean (and therefore in measure) to χ_E as $U \to e$. In other words, for every positive number ϵ there exists a bounded neighborhood V of e such that if $U \subset V$, then $\int |f_U - \chi_E| d\mu < \epsilon$. This result may be called the **density theorem** for topological groups. (Hint: let V be a neighborhood of e such that if $y \in V$, then $\rho(yE,E) < \frac{\epsilon}{2}$. If $U \subset V$ and if F is any Borel set, then

$$\frac{\epsilon}{2} > \frac{1}{\mu(U)} \int_U \int_F |\chi_E(yx) - \chi_E(x)| d\mu(x) d\mu(y) \geqq$$

$$\geqq \left| \int_F d\mu(x) \int_U \frac{1}{\mu(U)} \chi_{Ex^{-1}}(y) d\mu(y) - \int_F \chi_E(x) d\mu(x) \int_U \frac{d\mu(y)}{\mu(U)} \right| =$$

$$= \left| \int_F (f_U(x) - \chi_E(x)) d\mu(x) \right|.$$

[Recall that $\frac{\mu(A)}{\mu(B)} = \frac{\mu(Ax)}{\mu(Bx)}$ for any Borel sets A and B and every x in X; cf. 60.5.] The desired conclusion follows upon applying this result first to

$$F = \{x: f_U(x) - \chi_E(x) > 0\}$$

and then to $F = \{x: f_U(x) - \chi_E(x) < 0\}$.)

(6) If ν is any finite signed measure on the class of all Baire sets in X, then (cf. 17.3) there exists a Baire set N_ν such that $\nu(E) = \nu(E \cap N_\nu)$ for every Baire set E. If λ and ν are any two such finite signed measures, their **convolution** $\lambda * \nu$ is defined, for every Baire set E, by

$$(\lambda * \nu)(E) = \int_{N_\lambda \times N_\nu} \chi_E(xy) d(\lambda \times \nu)(x, y).$$

If λ and ν are the indefinite integrals (with respect to Haar measure μ) of the integrable functions f and g respectively, then $\lambda * \nu$ is the indefinite integral of h, where

$$h(y) = \int f(x) g(x^{-1}y) d\mu(x).$$

(7) If λ and ν are finite signed measures (cf. (6)), then

$$(\lambda * \nu)(E) = \int_{N_\lambda} \nu(x^{-1}E) d\lambda(x).$$

If the group X is abelian, then $\lambda * \nu = \nu * \lambda$.

(8) If λ and ν are finite measures on the class of all Baire sets of a locally compact, σ-compact, abelian group X, then $\int \lambda(xE) d\nu(x) = \int \nu(xE^{-1}) d\lambda(x)$. (Hint: if $\bar{\nu}(E) = \nu(E^{-1})$, then

$$\int \lambda(xE) d\nu(x) = \int \lambda(x^{-1}E) d\bar{\nu}(x) \quad \text{and} \quad \int \nu(xE^{-1}) d\lambda(x) = \int \bar{\nu}(x^{-1}E) d\lambda(x);$$

the desired result follows from the relation $\lambda * \bar{\nu} = \bar{\nu} * \lambda$.)

(9) If f and g are two bounded, continuous, monotone functions on the real line, then (cf. 25.4)

$$\int_a^b f dg + \int_a^b g df = f(b)g(b) - f(a)g(a),$$

i.e. the usual equation for integration by parts is valid. (Hint: let λ and ν be the measures induced by f and g respectively, and apply (8) with $E = \{x: -\infty < x < 0\}$.)

§ 62. WEIL TOPOLOGY

We have seen that every locally compact group in which measurability is interpreted in the sense of Baire is a measurable group, and that, moreover, the topology of the group is uniquely determined by its measure theoretic structure. In this section we shall treat the converse problem: is it possible to introduce a natural topology in a measurable group so that it becomes a locally compact topological group? We shall see that the answer is essentially affirmative; we proceed to the precise description of the details.

Throughout this section we shall work with a fixed measurable group $(X,\mathbf{S},\mu)$; as usual, we shall write $\rho(E,F) = \mu(E \Delta F)$ for any two measurable sets E and F. We shall denote by $\mathbf{A}$ the class of all sets of the form EE^{-1}, where E is a measurable set of positive, finite measure, and by $\mathbf{N}$ the class of all sets of the form $\{x: \rho(xE,E) < \epsilon\}$, where E is a measurable set of positive, finite measure and ϵ is a real number such that $0 < \epsilon < 2\mu(E)$.

Theorem A. *If $N = \{x: \rho(xE,E) < \epsilon\} \varepsilon \mathbf{N}$, then every measurable set F of positive measure contains a measurable subset G of positive, finite measure such that $GG^{-1} \subset N$.*

Proof. It is sufficient to treat the case in which F has finite measure. If $T(x,y) = (yx,y)$, then $T(E \times F)$ is a measurable set of finite measure in $X \times X$ and hence there exists a set A in $X \times X$ such that A is a finite union of measurable rectangles and

$$\frac{\epsilon}{4}\mu(F) > \rho(T(E \times F),A) =$$

$$= \iint |\chi_{T(E\times F)}(x,y) - \chi_A(x,y)|\,d\mu(x)d\mu(y) \geqq$$

$$\geqq \int_F \int |\chi_E(y^{-1}x) - \chi_A(x,y)|\,d\mu(x)d\mu(y).$$

If we write $C = \left\{y : \int |\chi_E(y^{-1}x) - \chi_A(x,y)|\,d\mu(x) \geqq \frac{\epsilon}{2}\right\}$, then it follows that $\mu(F \cap C) \leqq \frac{1}{2}\mu(F)$ and hence that

$$\mu(F - C) \geqq \tfrac{1}{2}\mu(F) > 0.$$

If $y \,\varepsilon\, F - C$, then

$$\rho(yE, A^y) = \int |\chi_E(y^{-1}x) - \chi_A(x,y)|\,d\mu(x) < \frac{\epsilon}{2}.$$

Since A is a finite union of measurable rectangles, there are only a finite number of distinct sets of the form A^y; we denote them by $A_1, \cdots, A_n$. What we have proved may now be expressed by the relation

$$F - C \subset \bigcup_{i=1}^{n} \left\{y : \rho(yE, A_i) < \frac{\epsilon}{2}\right\}.$$

Since $\frac{\epsilon}{2} < \mu(E) = \mu(yE)$ it follows from 59.F that each of the sets $\left\{y : \rho(yE, A_i) < \frac{\epsilon}{2}\right\}$ is measurable, and therefore, since $\mu(F - C) > 0$, that at least one of them intersects $F - C$ in a set of positive measure. We select an index i such that if

$$G_0 = (F - C) \cap \left\{y : \rho(yE, A_i) < \frac{\epsilon}{2}\right\},$$

then $\mu(G_0) > 0$. It is clear that G_0 is a measurable set of positive, finite measure and that $G_0 \subset F$. If $y_1 \,\varepsilon\, G_0^{-1}$ and $y_2 \,\varepsilon\, G_0^{-1}$, then

$$\rho(y_1y_2^{-1}E, E) = \rho(y_2^{-1}E, y_1^{-1}E) \leqq$$
$$\leqq \rho(y_2^{-1}E, A_i) + \rho(y_1^{-1}E, A_i) < \epsilon,$$

so that $G_0^{-1}G_0 \subset N$. We have proved, in other words, the existence of a set G_0 satisfying all the requirements of the theorem except that instead of $GG^{-1} \subset N$ we have $G_0^{-1}G_0 \subset N$. If we apply this result to F^{-1} in place of F, and if we denote the set so obtained by G^{-1}, then the set G satisfies all the requirements without exception. ∎

Theorem A asserts in particular that every set in **N** contains a set in **A**. We shall also need the following result which goes in the opposite direction.

Theorem B. *If* $A = EE^{-1} \varepsilon \mathbf{A}$ *and* $0 < \epsilon < 2\mu(E)$, *and if*

$$N = \{x: \rho(xE,E) < \epsilon\},$$

then $N \varepsilon \mathbf{N}$ *and* $N \subset A$.

Proof. It is trivial that $N \varepsilon \mathbf{N}$; to see that $N \subset A$, observe that $N \subset \{x: xE \cap E \neq 0\} = EE^{-1}$. ∎

Theorem C. *If* $N = \{x: \rho(xE,E) < \epsilon\} \varepsilon \mathbf{N}$, *then* N *is a measurable set of positive measure. If* $\mu(E^{-1}) < \infty$, *then* $\mu(N) < \infty$.

Proof. Since $N = \left\{x: \mu(xE \cap E) > \mu(E) - \frac{\epsilon}{2}\right\}$, the measurability of N follows from 59.F. To prove that $\mu(N) > 0$, we apply Theorem A. If G is a measurable set of positive measure such that $GG^{-1} \subset N$, then, in particular, $Gy^{-1} \subset N$ for any y in G. The last assertion of the theorem follows from the relations

$$\left(\mu(E) - \frac{\epsilon}{2}\right)\mu(N) \leq \int_N \mu(xE \cap E)d\mu(x) \leq$$

$$\leq \int \mu(xE \cap E)d\mu(x) = \mu(E)\mu(E^{-1}). \quad ∎$$

Theorem D. *If* A *and* B *are any two sets of* $\mathbf{A}$, *then there exists a set* C *in* $\mathbf{A}$ *such that* $C \subset A \cap B$.

Proof. Let E and F be measurable sets of positive, finite measure such that $A = EE^{-1}$ and $B = FF^{-1}$. By 59.E there exists a measurable set G of positive, finite measure and there exist two elements x and y in X such that

$$Gx \subset E \quad \text{and} \quad Gy \subset F.$$

If $C = GG^{-1}$, then $C \varepsilon \mathbf{A}$ and

$$C = (Gx)(Gx)^{-1} \subset A \quad \text{and} \quad C = (Gy)(Gy)^{-1} \subset B. \quad ∎$$

Before introducing the promised topology in X, we need to define one more concept. We recall that our definition of a measurable group was motivated by the continuity properties of topological groups and ignored entirely the separation axiom which is an essential part of the definition of a topological group. One way of phrasing the relevant separation axiom is this: if an element x of the group is different from e, then there exists a neighborhood U of e such that $x \varepsilon' U$. Guided by these considerations and 61.A and 61.B, we shall say that a measurable group X is **separated** if whenever an element x of the group is different from e, then there exists a measurable set E of positive, finite measure such that $\rho(xE,E) > 0$.

Theorem E. *If X is separated, and if the class $\mathbf{N}$ is taken for a base at e, then, with respect to the induced topology, X is a topological group.*

We shall refer to this topology of the measurable group X as the **Weil topology.**

Proof. We shall verify that $\mathbf{N}$ satisfies the conditions (a), (b), (c), (d), and (e) of § 0.

Suppose that x_0 is an element of X, $x_0 \neq e$, and that E is a measurable set of positive, finite measure such that $\rho(x_0E,E) > 0$. If ϵ is a positive number such that $0 < \epsilon < \rho(x_0E,E)$, then $\epsilon < 2\mu(E)$. It follows that if $N = \{x: \rho(xE,E) < \epsilon\}$, then $N \varepsilon \mathbf{N}$, and, clearly, $x_0 \varepsilon' N$.

If N and M are in $\mathbf{N}$, then by Theorem A, there exist sets A and B in $\mathbf{A}$ such that $A \subset N$ and $B \subset M$. By Theorem D, there exists a set C in $\mathbf{A}$ such that $C \subset A \cap B$; by an application of Theorem B, we obtain a set K in $\mathbf{N}$ such that

$$K \subset C \subset A \cap B \subset N \cap M.$$

If $N = \{x: \rho(xE,E) < \epsilon\}$, we write $M = \left\{x: \rho(xE,E) < \frac{\epsilon}{2}\right\}$.

If x_0 and y_0 are any two elements of M, then

$$\begin{aligned}\rho(x_0y_0^{-1}E,E) &\leq \rho(y_0^{-1}E,E) + \rho(x_0^{-1}E,E) = \\ &= \rho(y_0E,E) + \rho(x_0E,E) < \epsilon,\end{aligned}$$

and therefore $MM^{-1} \subset N$.

If $N \varepsilon \mathbf{N}$ and $x \varepsilon X$, then by Theorem A, there exists a measurable set E of positive, finite measure such that $EE^{-1} \subset N$.

Applying Theorem B to the set $(xE)(xE)^{-1}$ in **A**, we may find a set M in **N** such that

$$M \subset (xE)(xE)^{-1} = xEE^{-1}x^{-1} \subset xNx^{-1}.$$

If, finally, $N = \{x: \rho(xE,E) < \epsilon\} \in \mathbf{N}$ and if $x_0 \in N$, then $\rho(x_0E,E) < \epsilon$. Since $\epsilon < 2\mu(E)$, it follows that $\epsilon - \rho(x_0E,E) < 2\mu(x_0E)$ and hence that if

$$M = \{x: \rho(xx_0E,x_0E) < \epsilon - \rho(x_0E,E)\},$$

then $M \in \mathbf{N}$. Since

$$Nx_0^{-1} = \{xx_0^{-1}: \rho(xE,E) < \epsilon\} = \{x: \rho(xx_0E,E) < \epsilon\},$$

we have, for every x in M,

$$\rho(xx_0E,E) \leqq \rho(xx_0E,x_0E) + \rho(x_0E,E) <$$
$$< (\epsilon - \rho(x_0E,E)) + \rho(x_0E,E) = \epsilon.$$

This implies that $x \in Nx_0^{-1}$ and hence that $Mx_0 \subset N$. ∎

Theorem F. *If X is a separated, measurable group, then X is locally bounded with respect to its Weil topology. If a measurable set E has a non empty interior, then $\mu(E) > 0$; if a measurable set E is bounded, then $\mu(E) < \infty$.*

Proof. Let N_0 be an arbitrary set of finite measure in **N** (see Theorem C), and let M_0 be a set in **N** such that $M_0M_0^{-1} \subset N_0$. We shall prove that M_0 is bounded. If M_0 is not bounded, then there exists a set N in **N** and a sequence $\{x_n\}$ of elements in M_0 such that

$$x_{n+1} \epsilon' \bigcup_{i=1}^{n} x_iN, \quad n = 1, 2, \cdots.$$

By Theorem A, there exists a measurable set E of positive, finite measure such that $E \subset M_0^{-1}$ and $EE^{-1} \subset N$. Since the condition on $\{x_n\}$ implies that the sequence $\{x_nE\}$ is disjoint, and since $x_nE \subset M_0M_0^{-1} \subset N_0$, it follows that $\mu(N_0) = \infty$. Since this contradicts Theorem C, we have proved the first assertion of the theorem.

The fact that a measurable set with a non empty interior has positive measure follows from Theorem C; the last assertion is a consequence of Theorem C and the fact that, by definition, a bounded set may be covered by a finite number of left translations of any set in **N**. ▮

Theorem F is in a certain sense the best possible result in this direction. If, however, we make use of the fact that every locally bounded group may be viewed as a dense subgroup of a locally compact group, then we may reformulate our results in a somewhat more useful way. We do this in Theorem H; first we prove an auxiliary result concerning arbitrary (i.e. not necessarily right or left invariant) Baire measures in locally compact groups.

Theorem G. *If μ is any Baire measure in a locally compact topological group X, and if Y is the set of all those elements y for which $\mu(yE) = \mu(E)$ for all Baire sets E, then Y is a closed subgroup of X.*

Proof. The fact that Y is a subgroup of X is trivial. To prove that Y is closed, let y_0 be any fixed element of $\bar{Y}$ and let C be any compact Baire set. If U is any open Baire set such that $y_0C \subset U$, then there exists a neighborhood V of e such that $Vy_0C \subset U$. Since Vy_0 is a neighborhood of y_0, it follows that there exists an element y in Y such that $y \in Vy_0$. Since $yC \subset Vy_0C \subset U$, it follows that

$$\mu(C) = \mu(yC) \leqq \mu(U),$$

and hence, by the regularity of μ, that $\mu(C) \leqq \mu(y_0C)$. Applying this conclusion to y_0^{-1} and y_0C in place of y_0 and C we obtain the reverse inequality, and hence the identity, $\mu(C) = \mu(y_0C)$ for all C. It follows that $\mu(E) = \mu(y_0E)$ for every Baire set E and hence that $y_0 \in Y$. ▮

By a **thick subgroup** of a measurable group we mean a subgroup which is a thick set; (cf. § 17).

Theorem H. *If $(X,\mathbf{S},\mu)$ is a separated, measurable group, then there exists a locally compact topological group $\hat{X}$ with a Haar measure $\hat{\mu}$ on the class $\hat{\mathbf{S}}$ of all Baire sets, such that X is a thick subgroup of $\hat{X}$, $\mathbf{S} \supset \hat{\mathbf{S}} \cap X$, and $\mu(E) = \hat{\mu}(\hat{E})$ whenever $\hat{E} \in \hat{\mathbf{S}}$ and $E = \hat{E} \cap X$.*

Proof. Let $\hat{X}$ be the completion of X in its Weil topology; i.e. $\hat{X}$ is a locally compact group containing X as a dense subgroup. Consider the class of all those subsets $\hat{E}$ of $\hat{X}$ for which $\hat{E} \cap X \varepsilon \mathbf{S}$. It is clear that this class is a σ-ring; in order to show that this σ-ring contains all Baire sets, we shall show that it contains a base for the topology of $\hat{X}$.

Suppose that $\hat{x}$ is any point of $\hat{X}$ and $\hat{U}$ is any neighborhood of $\hat{e}$ in $\hat{X}$; let $\hat{V}$ be a neighborhood of $\hat{e}$ such that $\hat{V}^{-1}\hat{V} \subset \hat{U}$. Since $\hat{V} \cap X$ is an open set in X, there exists a measurable open set W in X such that $W \subset \hat{V} \cap X$; since the topology of X is (by the definition of $\hat{X}$) the relative topology it inherits from $\hat{X}$, there exists an open set $\hat{W}$ in $\hat{X}$ such that $W = \hat{W} \cap X$. Since we may replace $\hat{W}$ by $\hat{W} \cap \hat{V}$, there is no loss of generality in assuming that $\hat{W} \subset \hat{V}$. Since X is dense in $\hat{X}$, there exists a point x in X such that $x \varepsilon \hat{x}\hat{W}^{-1}$; it follows that

$$\hat{x} \varepsilon x\hat{W} \subset \hat{x}\hat{W}^{-1}\hat{W} \subset \hat{x}\hat{V}^{-1}\hat{V} \subset \hat{x}\hat{U}.$$

If we define $\hat{\mu}$ by writing $\hat{\mu}(\hat{E}) = \mu(\hat{E} \cap X)$ for every $\hat{E}$ in $\hat{\mathbf{S}}$, then it is easy to verify that $\hat{\mu}$ is a Baire measure in $\hat{X}$. It follows from Theorem G, and the fact that $\hat{\mu}(x\hat{E}) = \hat{\mu}(\hat{E})$ whenever $x \varepsilon X$ and $\hat{E} \varepsilon \hat{\mathbf{S}}$, that $\hat{\mu}$ is left invariant. The uniqueness theorem implies therefore that $\hat{\mu}$ coincides on $\hat{\mathbf{S}}$ with a Haar measure in $\hat{X}$. It follows that if $\hat{E} \varepsilon \hat{\mathbf{S}}$ and $\hat{E} \cap X = 0$, then $\hat{\mu}(\hat{E}) = \mu(\hat{E} \cap X) = 0$, i.e. that X is thick in $\hat{X}$. ∎

(1) Let X be any locally compact topological group with a Haar measure μ on the class $\mathbf{S}$ of all Baire sets. If $\tilde{X} = X \times X$, if $\tilde{\mathbf{S}}$ is the class of all sets of the form $E \times X$, where $E \varepsilon \mathbf{S}$, and if $\tilde{\mu}(E \times X) = \mu(E)$, then $(\tilde{X}, \tilde{\mathbf{S}}, \tilde{\mu})$ is a measurable group which is not separated. To what extent is this example typical of non separated measurable groups in general?

(2) If X is a separated measurable group, then a set E is bounded with respect to the Weil topology of X if and only if there exists a measurable set A of positive, finite measure such that EA is contained in a measurable set of finite measure.

(3) Is Theorem G true for Borel measures?

(4) Is the subgroup Y described in Theorem G necessarily invariant?

(5) Under the hypotheses of Theorem G, write $f(x) = \mu(xE)$ for every x in X and every Baire set E. Is f a continuous function?

(6) The purpose of the following considerations is to give a non trivial example of a thick subgroup. Let X be the real line and consider the locally compact

topological group $X \times X$. A subset B of X is **linearly independent** if the conditions $\sum_{i=1}^{n} r_i x_i = 0$, $x_i \in B$, $i = 1, \cdots, n$, with *rational* numbers r_i, imply that $r_1 = \cdots = r_n = 0$.

(6a) If E is a Borel set of positive measure in $X \times X$ and if B is a linearly independent set in X, of cardinal number smaller than that of the continuum, then there exists a point (x,y) in E such that $B \cup \{x\}$ is a linearly independent set. (Hint: there exists a value of y such that E^y has positive measure and therefore has the cardinal number of the continuum.)

(6b) There exists a set C of points in $X \times X$ such that (i) $C \cap E \neq 0$ for every Borel set E of positive measure in $X \times X$, (ii) the set B of first coordinates of points of C is linearly independent, and (iii) C intersects every vertical line in at most one point. (Hint: well order the class of all Borel sets of positive measure in $X \times X$ and construct C, using (6a), by transfinite induction.)

(6c) A **Hamel basis** is a linearly independent set B in X with the property that for every x in X there exists a finite subset $\{x_1, \cdots, x_n\}$ of B and a corresponding finite set $\{r_1, \cdots, r_n\}$ of rational numbers such that $x = \sum_{i=1}^{n} r_i x_i$. The expression of x as a rational linear combination of elements of B is unique. Every linearly independent set is contained in a Hamel basis. (Hint: use transfinite induction or Zorn's lemma.)

(6d) By (6b) and (6c) there exists a set C of points in $X \times X$ having the properties (i), (ii), and (iii) described in (6b) and such that the set B of first coordinates of points of C is a Hamel basis. If $x = \sum_{i=1}^{n} r_i x_i$, where r_i is rational and $(x_i, y_i) \in C$, $i = 1, \cdots, n$, write $f(x) = \sum_{i=1}^{n} r_i y_i$. If $Z = \{(x,y): y = f(x)\}$ (i.e. if Z is the graph of f), then Z is a thick subgroup of $X \times X$.

§ 63. QUOTIENT GROUPS

Throughout this section we shall assume that

X is a locally compact topological group and μ is a Haar measure in X; Y is a compact, invariant subgroup of X, ν is a Haar measure in Y such that $\nu(Y) = 1$, and π is the projection from X onto the quotient group $\mathcal{X} = X/Y$.

While most of the important results of this section are valid for closed (but not necessarily compact) subgroups Y, we restrict our attention to the compact case because this will be sufficient for our purposes and because the proofs in this case are slightly simpler.

Theorem A. *If a compact set C is a union of cosets of Y and if U is an open set containing C, then there exists an open set $\mathcal{V}$ in $\mathcal{X}$ such that*

$$C \subset \pi^{-1}(\mathcal{V}) \subset U.$$

Proof. There is no loss of generality in assuming that U is bounded. If we write $X_0 = \overline{UY}$, then it follows that X_0 is compact. We assert that X_0 is, just as UY, a union of cosets of Y. To prove this, suppose that $x_1 \varepsilon X_0$ and $\pi(x_1) = \pi(x_2)$ (so that $x_1^{-1}x_2 \varepsilon Y$); we are to prove that $x_2 \varepsilon X_0$. If V is any neighborhood of x_2, then $Vx_2^{-1}x_1$ is a neighborhood of x_1 and therefore $UY \cap Vx_2^{-1}x_1 \neq 0$. Since $x_1^{-1}x_2 \varepsilon Y$, it follows that

$$UY \cap V = UYx_1^{-1}x_2 \cap Vx_2^{-1}x_1x_1^{-1}x_2 =$$
$$= (UY \cap Vx_2^{-1}x_1)x_1^{-1}x_2 \neq 0;$$

since V is arbitrary, this implies that $x_2 \varepsilon X_0$.

The fact that C is a union of cosets of Y implies that $\pi(X_0 - U) \cap \pi(C) = \pi((X_0 - U) \cap C) = 0$. Since the sets $\pi(X_0 - U)$ and $\pi(C)$ are compact, and since $\pi(U)$ is an open set containing $\pi(C)$, there exists an open set $\hat{V}$ in $\hat{X}$ such that

$$\pi(C) \subset \hat{V} \subset \pi(U) \subset \pi(X_0) \quad \text{and} \quad \hat{V} \cap \pi(X_0 - U) = 0.$$

If $x \varepsilon \pi^{-1}(\hat{V})$, so that $\pi(x) \varepsilon \hat{V}$, then $\pi(x) \varepsilon' \pi(X_0 - U)$ and therefore $x \varepsilon' X_0 - U$. Since, however, $x \varepsilon X_0$, it follows that $x \varepsilon U$ and therefore that $C \subset \pi^{-1}(\hat{V}) \subset U$. ▌

Theorem B. *If $\hat{C}$ is a compact subset of $\hat{X}$, then $\pi^{-1}(\hat{C})$ is a compact subset of X; if $\hat{E}$ is a Baire set [or a Borel set] in X, then $\pi^{-1}(\hat{E})$ is a Baire set [or a Borel set] in X.*

Proof. Suppose that $\mathbf{K}$ is an open covering of $\pi^{-1}(\hat{C})$. Since, for each $\hat{x}$ in $\hat{C}$, $\pi^{-1}(\{\hat{x}\})$ is a coset of Y, and therefore compact, it follows that $\mathbf{K}$ contains a finite subclass $\mathbf{K}(\hat{x})$ such that $\pi^{-1}(\{\hat{x}\}) \subset U(\hat{x}) = \bigcup \mathbf{K}(\hat{x})$. By Theorem A there is an open set $\hat{V}(\hat{x})$ in $\hat{X}$ such that

$$\pi^{-1}(\{\hat{x}\}) \subset V(\hat{x}) = \pi^{-1}(\hat{V}(\hat{x})) \subset U(\hat{x}).$$

Since $\hat{C}$ is compact, there exists a finite subset $\{\hat{x}_1, \cdots, \hat{x}_n\}$ of $\hat{C}$ such that $\hat{C} \subset \bigcup_{i=1}^n \hat{V}(\hat{x}_i)$; it follows that

$$\pi^{-1}(\hat{C}) \subset \bigcup_{i=1}^n V(\hat{x}_i) \subset \bigcup_{i=1}^n \bigcup \mathbf{K}(\hat{x}_i),$$

and hence that $\pi^{-1}(\hat{C})$ is compact.

The assertion concerning Baire sets and Borel sets follows from the preceding paragraph and the additional facts that the inverse image (under π) of a G_δ is a G_δ, and that the class of all those sets in $\hat{X}$ whose inverse image lies in a prescribed σ-ring is a σ-ring. ∎

Theorem B implies that if measurability in both X and $\hat{X}$ is interpreted in the sense of Baire, or else in the sense of Borel, then the transformation π is measurable. In other words, π^{-1} maps measurable sets satisfactorily; how does it map their measures?

Theorem C. *If $\hat{\mu} = \mu\pi^{-1}$, then $\hat{\mu}$ is a Haar measure in $\hat{X}$.*

Proof. The fact that $\hat{\mu}$ is finite on compact sets and positive on non empty open Borel sets follows from the fact that the inverse image (under π) of a compact set or a non empty open set is a compact set or a non empty open set, respectively. It remains only to prove that $\hat{\mu}$ is left invariant.

If $\hat{E}$ is a Borel set in $\hat{X}$ and $\hat{x}_0 \in \hat{X}$, let x_0 be an element of X such that $\pi(x_0) = \hat{x}_0$. If $x \in x_0\pi^{-1}(\hat{E})$, then (since π is a homomorphism) $\pi(x) \in \hat{x}_0\hat{E}$, so that

$$x_0\pi^{-1}(\hat{E}) \subset \pi^{-1}(\hat{x}_0\hat{E}).$$

If, conversely, $x \in \pi^{-1}(\hat{x}_0\hat{E})$, then $\pi(x) \in \hat{x}_0\hat{E}$ and therefore $\pi(x_0^{-1}x) = \hat{x}_0^{-1}\pi(x) \in \hat{E}$. This implies that $x_0^{-1}x \in \pi^{-1}(\hat{E})$ and hence that $x \in x_0\pi^{-1}(\hat{E})$. Since we have thus proved that

$$\pi^{-1}(\hat{x}_0\hat{E}) \subset x_0\pi^{-1}(\hat{E}),$$

it follows that

$$\hat{\mu}(\hat{x}_0\hat{E}) = \mu\pi^{-1}(\hat{x}_0\hat{E}) = \mu(x_0\pi^{-1}(\hat{E})) = \mu\pi^{-1}(\hat{E}) = \hat{\mu}(\hat{E}). \quad ∎$$

Theorem D. *If $f \in \mathfrak{L}_+(X)$ and if*

$$g(x) = \int_Y f(xy)d\nu(y),$$

then $g \in \mathfrak{L}_+(X)$ and there exists a (uniquely determined) function $\hat{g}$ in $\mathfrak{L}_+(\hat{X})$ such that $g = \hat{g}\pi$.

Proof. If $f_x(y) = f(xy)$, then the continuity of f implies the continuity, and hence integrability, of f_x on Y. Since f is uniformly continuous, to every positive number ϵ there corresponds a neighborhood U of e such that if $x_1x_2^{-1} \varepsilon U$, then $|f(x_1) - f(x_2)| < \epsilon$. If $x_1x_2^{-1} \varepsilon U$, then

$$(x_1y)(x_2y)^{-1} = x_1x_2^{-1} \varepsilon U$$

and therefore

$$|g(x_1) - g(x_2)| \leq \int_Y |f(x_1y) - f(x_2y)| d\nu(y) < \epsilon,$$

so that g is continuous. Since g is clearly non negative and since $\{x: g(x) \neq 0\} \subset \{x: f(x) \neq 0\} \cdot Y$, it follows that $g \varepsilon \mathfrak{L}_+(X)$.

If $\pi(x_1) = \pi(x_2)$, then $x_1^{-1}x_2 \varepsilon Y$ and it follows from the left invariance of ν that

$$g(x_1) = \int_Y f(x_1y)d\nu(y) = \int_Y f(x_1(x_1^{-1}x_2y))d\nu(y) = g(x_2).$$

Consequently writing $\hat{g}(\hat{x}) = g(x)$, whenever $\hat{x} = \pi(x)$, unambiguously defines a function $\hat{g}$ on $\hat{X}$; clearly $g = \hat{g}\pi$. Since (39.A), for every open subset M of the real line,

$$\{\hat{x}: \hat{g}(\hat{x}) \varepsilon M\} = \pi(\{x: g(x) \varepsilon M\}),$$

the continuity of $\hat{g}$ follows from the openness of π. Since π maps the bounded set $\{x: g(x) \neq 0\}$ on a bounded subset of $\hat{X}$, it follows that $\hat{g} \varepsilon \mathfrak{L}_+(\hat{X})$; the uniqueness of $\hat{g}$ is a consequence of the fact that π maps X *onto* $\hat{X}$. ∎

Theorem E. *If C is a compact Baire set in X and if $g(x) = \nu(x^{-1}C \cap Y)$, then there exists a (uniquely determined) Baire measurable and integrable function $\hat{g}$ on $\hat{X}$ such that $g = \hat{g}\pi$. If C is a union of cosets of Y, then $\int \hat{g}d\hat{\mu} = \mu(C)$.*

Proof. Let $\{f_n\}$ be a decreasing sequence of functions in $\mathfrak{L}_+(X)$ such that $\lim_n f_n(x) = \chi_C(x)$ for every x in X. If

$$g_n(x) = \int_Y f_n(xy)d\nu(y), \quad n = 1, 2, \cdots,$$

then $\{g_n\}$ is a decreasing sequence of functions in $\mathfrak{L}_+(X)$ (cf.

Theorem D), and hence (by, for instance, the bounded convergence theorem)

$$\lim_n g_n(x) = \int_Y \chi_C(xy)d\nu(y) = \int_Y \chi_{x^{-1}C}(y)d\nu(y) =$$

$$= \nu(x^{-1}C \cap Y) = g(x)$$

for every x in X.

By Theorem D, for each positive integer n there is a function $\hat{g}_n$ in $\mathcal{L}_+(\hat{X})$ such that $g_n = \hat{g}_n\pi$. Since the sequence $\{\hat{g}_n\}$ is decreasing, we may write $\hat{g}(\hat{x}) = \lim_n \hat{g}_n(\hat{x})$; clearly $g = \hat{g}\pi$. Since (39.C)

$$\int \hat{g}d\hat{\mu} = \int g d\mu = \int \nu(x^{-1}C \cap Y)d\mu(x),$$

and since $\{x: \nu(x^{-1}C \cap Y) \neq 0\} \subset \{x: x^{-1}C \cap Y \neq 0\} = CY$, the integrability of g follows from the finiteness of ν.

If, finally, C is a union of cosets of Y, then

$$x^{-1}C \cap Y = \begin{cases} Y & \text{if} \quad x \in C, \\ 0 & \text{if} \quad x \in' C, \end{cases}$$

and therefore

$$\int \hat{g}d\hat{\mu} = \int g d\mu = \int \nu(x^{-1}C \cap Y)d\mu(x) = \mu(C). \quad \blacksquare$$

Theorem F. *If, for each Baire set E in X,*

$$g_E(x) = \nu(x^{-1}E \cap Y),$$

then there exists a (uniquely determined) Baire measurable function $\hat{g}_E$ on $\hat{X}$ such that $g_E = \hat{g}_E\pi$.

Proof. We observe first that (by the definition of topology in Y) the set $x^{-1}E \cap Y$ is always a Baire set in Y and, consequently, that $g_E(x)$ is always defined.

If we denote by $\mathbf{E}$ the class of all those sets E for which the desired conclusion is valid, then, by Theorem E, it follows that every compact Baire set belongs to $\mathbf{E}$. Since the elementary properties of the (finite) measure ν imply that $\mathbf{E}$ is closed under the formation of proper differences, finite disjoint unions, and monotone unions and intersections, it follows that **E** contains all Baire sets. ▮

Theorem G. *If, for each Baire set E in X, $\hat{g}_E$ is the unique Baire measurable function on $\hat{X}$ for which*

$$\hat{g}_E(\pi(x)) = \nu(x^{-1}E \cap Y) = g_E(x)$$

for every x in X, then

$$\int \hat{g}_E d\hat{\mu} = \mu(E)$$

for every Baire set E.

Proof. Write $\lambda(E) = \int \hat{g}_E d\hat{\mu} = \int \nu(x^{-1}E \cap Y) d\mu(x)$ for every Baire set E in X. Since $\lambda(C)$ is finite for every compact Baire set C (Theorem E) and since λ is clearly non negative, we see that λ is a Baire measure in X. If $x_0 \in X$, then

$$\lambda(x_0E) = \int g_{x_0E}(x) d\mu(x) = \int \nu(x^{-1}x_0E \cap Y) d\mu(x) =$$

$$= \int \nu((x_0^{-1}x)^{-1}E \cap Y) d\mu(x) = \int g_E(x_0^{-1}x) d\mu(x) =$$

$$= \int g_E(x) d\mu(x) = \lambda(E),$$

so that λ is left invariant. It follows from the uniqueness theorem that $\lambda(E) = c\mu(E)$ for a suitable constant c. Since, by Theorem E, $\lambda(C) = \mu(C)$ whenever C is a compact Baire set which is a union of cosets of Y, and since there exist such sets with $\mu(C) > 0$, it follows that $c = 1$. ▮

§ 64. THE REGULARITY OF HAAR MEASURE

The purpose of this section is to prove that every Haar measure is regular. Throughout this section, up to the statement of the final, general result, we shall assume that

X is a locally compact and σ-compact topological group, and μ is a left invariant Baire measure in X which is not identically zero (and which, therefore, is positive on all non empty open Baire sets).

It is convenient to introduce an auxiliary concept; by an **invariant σ-ring** we shall mean a σ-ring $\mathbf{T}$ of Baire sets such that if $E \varepsilon \mathbf{T}$ and $x \varepsilon X$, then $xE \varepsilon \mathbf{T}$. Since the class of all Baire sets is an invariant σ-ring, and since the intersection of any collection of invariant σ-rings is itself an invariant σ-ring, we may define the invariant σ-ring **generated** by any class $\mathbf{E}$ of Baire sets as the intersection of all invariant σ-rings containing $\mathbf{E}$.

Theorem A. *If* $\mathbf{E}$ *is a class of Baire sets and* $\mathbf{T}$ *is the invariant σ-ring generated by* $\mathbf{E}$, *then* $\mathbf{T}$ *coincides with the σ-ring* $\mathbf{T}_0$ *generated by the class* $\{xE: x \varepsilon X, E \varepsilon \mathbf{E}\}$.

Proof. Since $xE \varepsilon \mathbf{T}$ for every x in X and every E in $\mathbf{E}$, it follows that $\mathbf{T}_0 \subset \mathbf{T}$; it is sufficient therefore to prove that $\mathbf{T}_0$ is invariant. Let x_0 be any fixed element of X. The class of all those Baire sets F for which $x_0F \varepsilon \mathbf{T}_0$ is a σ-ring; since, for every x in X and every E in $\mathbf{E}$, $x_0(xE) = (x_0x)E \varepsilon \mathbf{T}_0$, it follows that this σ-ring contains $\mathbf{T}_0$. We have proved, in other words, that if $F \varepsilon \mathbf{T}_0$, then $x_0F \varepsilon \mathbf{T}_0$. ∎

Theorem B. *If* $\mathbf{E}$ *is a countable class of Baire sets of finite measure and* $\mathbf{T}$ *is the invariant σ-ring generated by* $\mathbf{E}$, *then the metric space* $\mathfrak{J}$ *of all sets of finite measure in* $\mathbf{T}$ (*with the metric* ρ *defined by* $\rho(E,F) = \mu(E \Delta F)$) *is separable.*

Proof. Since every subspace of a separable metric space is separable, it is sufficient to prove that there exists a σ-ring $\mathbf{T}_0$ of Baire sets such that $\mathbf{T} \subset \mathbf{T}_0$ and such that $\mathbf{T}_0$ has a countable set of generators of finite measure; (cf. 40.B). Since X is a Baire set, it follows that $X \times E$ is a Baire set in $X \times X$ for each E in $\mathbf{E}$. If we write, as before, $S(x,y) = (x,xy)$, then $S(X \times E)$ is also a Baire set in $X \times X$ for each E in $\mathbf{E}$. Consequently there exists, for each E in $\mathbf{E}$, a countable class $\mathbf{R}_E$ of rectangles of finite measure such that $S(X \times E) \varepsilon \mathbf{S}(\mathbf{R}_E)$. If we denote by $\mathbf{T}_0$ the σ-ring generated by the class of all sides of all rectangles in all

$\mathbf{R}_E$, $E \in \mathbf{E}$, then clearly

$$S(X \times E) \in \mathbf{T}_0 \times \mathbf{T}_0$$

for every E in $\mathbf{E}$. Since every section of a set in $\mathbf{T}_0 \times \mathbf{T}_0$ belongs to $\mathbf{T}_0$, it follows that, for every x in X and every E in $\mathbf{E}$,

$$xE = x(X \times E)_x = (S(X \times E))_x \in \mathbf{T}_0,$$

and hence (by Theorem A) that $\mathbf{T} \subset \mathbf{T}_0$. ▮

Theorem C. *If* $\mathbf{T}$ *is an invariant* σ*-ring, if* f *is a function in* $\mathfrak{L}$ *which is measurable* ($\mathbf{T}$), *and if* y *in* X *is such that* $\rho(yE,E) = 0$ *for every* E *in* $\mathbf{T}$, *then* $f(y^{-1}x) = f(x)$ *for every* x *in* X.

Proof. If E is any set of finite measure in $\mathbf{T}$, then

$$0 = \rho(yE,E) = \int | \chi_{yE}(x) - \chi_E(x) | d\mu(x) =$$

$$= \int | \chi_E(y^{-1}x) - \chi_E(x) | d\mu(x).$$

It follows that

$$\int | g(y^{-1}x) - g(x) | d\mu(x) = 0$$

for every integrable simple function g which is measurable ($\mathbf{T}$), and hence, by approximation, that $\int | f(y^{-1}x) - f(x) | d\mu(x) = 0$. Since the integrand of the last written integral belongs to $\mathfrak{L}_+$, the desired conclusion follows from 55.B. ▮

Theorem D. *If* $\mathbf{T}$ *is an invariant* σ*-ring generated by its sets of finite measure and containing at least one bounded set of positive measure, if* $\mathbf{E}$ *is a class of sets dense in the metric space of sets of finite measure in* $\mathbf{T}$, *and if*

$$Y = \{y\colon \rho(yE,E) = 0, \quad E \in \mathbf{E}\},$$

then Y *is a compact, invariant subgroup of* X.

Proof. If $Y_0 = \{y\colon \rho(yE,E) = 0, E \in \mathbf{T}\}$, then clearly $Y_0 \subset Y$. On the other hand if E_0 is a set of finite measure in $\mathbf{T}$ and ϵ is a positive number, then there exists a set E in $\mathbf{E}$ such that $\rho(E_0,E) < \frac{\epsilon}{2}$. It follows that if $y \varepsilon Y$, then

$$0 \leqq \rho(yE_0,E_0) \leqq \rho(yE_0,yE) + \rho(yE,E) + \rho(E,E_0) < \epsilon.$$

Since ϵ is arbitrary, this implies that $y \varepsilon Y_0$, and hence that $Y = Y_0$.

If y_1 and y_2 are in Y and E is in $\mathbf{T}$, then

$$0 \leqq \rho(y_1^{-1}y_2E,E) \leqq \rho(y_1^{-1}y_2E,y_2E) + \rho(y_2E,E).$$

Since $y_2E \varepsilon \mathbf{T}$ and $\rho(y_1^{-1}y_2E,y_2E) = \rho(y_2E,y_1y_2E)$, it follows that $y_1^{-1}y_2 \varepsilon Y$, so that Y is indeed a subgroup of X. If $y \varepsilon Y$, $x \varepsilon X$, and $E \varepsilon \mathbf{T}$, then $xE \varepsilon \mathbf{T}$ and therefore $\rho(x^{-1}yxE,E) = \rho(yxE,xE) = 0$, so that Y is invariant.

If E_0 is a bounded set of positive measure in $\mathbf{T}$, then the fact that $\rho(yE_0,E_0) = 0$ for every y in Y implies that $yE_0 \cap E_0 \neq 0$. It follows that $y \varepsilon E_0E_0^{-1}$ and hence that Y is contained in the bounded set $E_0E_0^{-1}$. To prove, finally, that Y is closed (and therefore compact) we observe that

$$Y = \bigcap_{E \varepsilon \mathbf{E}} \{y: \rho(yE,E) = 0\};$$

the desired result follows from 61.A. ∎

Theorem E. *If E is any Baire set in X, then there exists a compact, invariant Baire subgroup Y of X such that E is a union of cosets of Y.*

Proof. Let $\{C_i\}$ be a sequence of compact Baire sets, of which at least one has positive measure, such that $E \varepsilon \mathbf{S}(\{C_i\})$. For each i, let $\{f_{ij}\}$ be a decreasing sequence of functions in $\mathfrak{L}_+(X)$ such that $\lim_j f_{ij}(x) = \chi_{C_i}(x)$ for every x in X. For each positive rational number r, the set $\{x: f_{ij}(x) \geqq r\}$ is a compact Baire set; let $\mathbf{T}$ be the invariant σ-ring generated by the class of all sets of this form. It follows from Theorem B that the metric space of all sets of finite measure in $\mathbf{T}$ is separable; let $\{E_n\}$ be a sequence which is dense in this metric space. If

$$Y = \bigcap_{n=1}^{\infty} \{y: \rho(yE_n,E_n) = 0\},$$

then it follows from Theorem D that Y is a compact invariant subgroup of X and it follows from 61.A that Y is a Baire set.

Since each f_{ij} is measurable (**T**), Theorem C implies that $f_{ij}(y^{-1}x) = f_{ij}(x)$ for every y in Y and every x in X. It follows that $\chi_{C_i}(y^{-1}x) = \chi_{C_i}(x)$, i.e. that $yC_i = C_i$, for every y in Y and every $i = 1, 2, \cdots$. Since, for each y in Y, the class of all those sets F for which $yF = F$ is a σ-ring, it follows that $yE = E$ for every y in Y. Hence $E = YE = \bigcup_{x \varepsilon E} Yx$; that is, E is a union of cosets of the invariant subgroup Y. ∎

Theorem F. *If $\{e\}$ is a Baire set, then X is separable.*

Proof. Let $\{U_n\}$ be a sequence of bounded open sets such that $\{e\} = \bigcap_{n=1}^{\infty} U_n$; we have seen before that there is no loss of generality in assuming that

$$\overline{U}_{n+1} \subset U_n, \quad n = 1, 2, \cdots.$$

There exists a sequence $\{C_i\}$ of compact sets such that $X = \bigcup_{i=1}^{\infty} C_i$; since each C_i is compact there exists, for each i and n, a finite subset $\{x_{ij}^{(n)}\}$ of C_i such that $C_i \subset \bigcup_j x_{ij}^{(n)} U_n$. We shall prove that the countable class $\{x_{ij}^{(n)} U_n\}$ is a base.

We prove first that if U is any neighborhood of e, then there exists a positive integer n such that $e \varepsilon U_n \subset U$. Indeed since

$$\{e\} = \bigcap_n U_n = \bigcap_n \overline{U}_n \quad \text{and} \quad e \varepsilon U,$$

it follows that

$$\bigcap_n (\overline{U}_n - U) = (\bigcap_n \overline{U}_n) - U = 0.$$

Since $\{\overline{U}_n - U\}$ is a decreasing sequence of compact sets with an empty intersection, it follows that $U_n - U (\subset \overline{U}_n - U)$ is empty for at least one value of n.

Suppose now that x is any element of X and V is any neighborhood of x. Since $x^{-1}V$ is a neighborhood of e, there exists a neighborhood U of e such that $U^{-1}U \subset x^{-1}V$, and, by the preceding paragraph, there is a positive integer n such that $e \varepsilon U_n \subset U$. Since $x \varepsilon \bigcup_{i=1}^{\infty} C_i$, there is a value of i such that $x \varepsilon C_i$ and therefore there is a value of j such that $x \varepsilon x_{ij}^{(n)} U_n$. Since the last written relation implies that $x_{ij}^{(n)} \varepsilon x U_n^{-1}$, we have

$$x \varepsilon x_{ij}^{(n)} U_n \subset x U_n^{-1} U_n \subset x U^{-1} U \subset x x^{-1} V = V. \quad ∎$$

Theorems E and F together yield the following startling and useful result.

Theorem G. *If E is any Baire set in X, then there exists a compact, invariant subgroup Y of X such that E is a union of cosets of Y, and such that the quotient group X/Y is separable.*

Proof. By Theorem E, there exists a compact, invariant Baire subgroup Y such that E is a union of cosets of Y. If $\{U_n\}$ is a sequence of open sets such that $Y = \bigcap_{n=1}^{\infty} U_n$, then, for each positive integer n, there exists an open set $\hat{U}_n$ in the quotient group $\hat{X} = X/Y$ such that

$$Y \subset \pi^{-1}(\hat{U}_n) \subset U_n,$$

where π is the projection from X onto $\hat{X}$; (cf. 63.A). It follows that $Y = \bigcap_{n=1}^{\infty} \pi^{-1}(\hat{U}_n)$ and hence that $\{\hat{e}\} = \bigcap_{n=1}^{\infty} \hat{U}_n$; the separability of $\hat{X}$ is now implied by Theorem F. ∎

Theorem H. *Every Haar measure in X is completion regular.*

Proof. It is sufficient to prove that if U is any bounded open set, then there exists a Baire set E contained in U such that $U - E$ may be covered by a Baire set of measure zero. Given U, we select the Baire set E $(\subset U)$ so that $\mu(E)$ is maximal; by Theorem G there exists a compact invariant subgroup Y of X such that E is a union of cosets of Y and such that the quotient group $\hat{X}(= X/Y)$ is separable.

Let π be the projection from X onto $\hat{X}$, and write $F = \pi^{-1}\pi(U - E)$; we shall prove that F is a Baire set of measure zero. The fact that E is a union of cosets of Y implies that $\pi(U - E) = \pi(U) - \pi(E)$; since $\pi(U)$ is an open set in a separable space, $\pi(U)$ is a Baire set in $\hat{X}$; (cf. 50.E). It follows from the relation $F = \pi^{-1}\pi(U) - E$ that F is indeed a Baire set.

Since the Baire open sets of X form a base, corresponding to each point x in $U - E$ there is a Baire open set $V(x)$ such that $x \in V(x) \subset U$. Since $\{\pi(V(x)) : x \in U - E\}$ is an open covering of $\pi(U - E)$, it follows from the separability of $\hat{X}$ that there exists a sequence $\{x_i\}$ of points in $U - E$ such that

$$\pi(U - E) \subset \bigcup_i \pi(V(x_i)).$$

Since $\pi(U - E) = \pi(U) - \pi(E)$, we have

$$\pi(U - E) \subset (\bigcup_i \pi(V(x_i))) - \pi(E) = \bigcup_i \pi(V(x_i) - E).$$

It follows from these considerations that it is sufficient, in order to complete the proof of the theorem, to prove that

$$\mu(\pi^{-1}(\pi(V - E))) = 0$$

for every Baire open set V contained in U; we turn therefore to the proof of this result. (Observe that the reasoning, used above to show that $\pi^{-1}\pi(U - E)$ is a Baire set, may also be applied to V in place of U.)

If V is a Baire open set contained in U, then it follows from the maximal property of E that $\mu(V - E) = 0$. If ν is a Haar measure in Y such that $\nu(Y) = 1$, and if we write $\hat{\mu} = \mu\pi^{-1}$ and $g(x) = \nu(x^{-1}(V - E) \cap Y)$, then (63.G) there exists a (non negative) Baire measurable function $\hat{g}$ on $\hat{X}$ such that $g = \hat{g}\pi$ and such that

$$0 = \mu(V - E) = \int \hat{g}\, d\hat{\mu} =$$

$$= \int g\, d\mu \geqq \int_{\pi^{-1}\pi(V-E)} \nu(x^{-1}(V - E) \cap Y)\, d\mu(x) \geqq 0.$$

We have

$$x^{-1}(V - E) \cap Y = (x^{-1}V \cap Y) - (x^{-1}E \cap Y).$$

If $x \in V$, then $e \in x^{-1}V \cap Y$ and if $x \varepsilon' E$, then $x^{-1}E \cap Y = 0$, so that if $x \in V - E$, then $x^{-1}(V - E) \cap Y$ is a non empty open subset of Y. It follows that if $x \in \pi^{-1}\pi(V - E)$, so that $\pi(x) = \pi(x_0)$ for some x_0 in $V - E$, then

$$g(x) = \hat{g}(\pi(x)) = \hat{g}(\pi(x_0)) = g(x_0) > 0,$$

and therefore, by 25.D, $\mu(\pi^{-1}\pi(V - E)) = 0$. ▮

Theorem I. *If X is an arbitrary (not necessarily σ-compact) locally compact topological group, and if μ is a left invariant Borel measure in X, then μ is completion regular.*

Proof. Given any Borel set E in X, there exists a σ–compact full subgroup Z of X such that $E \subset Z$. By Theorem H, μ on Z is completion regular and therefore there exist two sets A and B in Z which are Baire subsets of Z and for which

$$A \subset E \subset B \quad \text{and} \quad \mu(B - A) = 0.$$

Since Z is both open and closed in X, A and B are also Baire subsets of X. ∎

REFERENCES

(The numbers in brackets refer to the bibliography that follows.)

§ 0. (1)–(7): [7] and [15]. Topology: [1, Chapters I and II] and [42, Chapter I]. Metric spaces: [41]. Tychonoff's theorem: [14]. Topological groups: [58, Chapters I, II, and III] and [73, Chapter I]. Completion of topological groups: [72].

§ 4. Rings and algebras: [27]. Semirings: [52].

§ 5. Lattices: [6]. (3): [52, p. 70].

§ 6. (2): [61, p. 85].

§ 7. (5): [52, pp. 77–78].

§ 9. (10): [57, p. 561].

§ 11. Outer measures and measurability: [12, Chapter V]. Metric outer measures: [61, pp. 43–47]; cf. also [10].

§ 12. Hausdorff measure: [29, Chapter VII].

§ 17. Thick sets: [3, p. 108]. Theorem A and (1): [19, pp. 109–110].

§ 18. (10): [51, pp. 602–603] and [20, pp. 91–92]. Distribution functions: [16].

§ 21. Egoroff's theorem: [61, pp. 18–19].

§ 26. (7): [63].

§ 29. Jordan decomposition, (3): [61, pp. 10–11].

§ 31. Radon–Nikodym theorem: [56, p. 168], [61, pp. 32–36], and [74].

§ 37. (4): [12, pp. 340–349].

§ 38. Theorem B: [64]; cf. also [32].

§ 39. [25].

§ 40. Boolean rings: [6, Chapter VI] and [68]. (8): [49]. (12): [59] and [60]. (15a): [21] and [66]. (15b): [68] and [69]. (15c): [47].

§ 41. Theorem C: [22]. (2): [50, pp. 85–87]. (1), (2), (3), and (4): [11], [24], and [44]. Theorem C and (7): [48]; cf. also [8].

§ 42. Hölder's and Minkowski's inequalities: [26, pp. 139–143 and pp. 146–150]. Function spaces: [5]. (3): [54, p. 130].

§ 43. Point functions and set functions: [28]. Theorem E: [28, p. 338 and p. 603].

§ 44. [23], [45], and [67].

§ 46. [71]. Kolmogoroff's inequality: [37, p. 310]. Series theorems: [35], [37], [38].

§ 47. Law of large numbers: [30] and [39]. Normal numbers: [9, p. 260].

§ 48. Conditional probabilities and expectations: [40, Chapter V]. (4): [18] and [20, p. 96].

§ 49. Theorem A: [40, pp. 24–30]. Theorem B: [43, pp. 129–130]. (3): [20, p. 92] and [65].

§ 51. Borel sets and Baire sets: [33] and [36].

§ 52. [55]. (10): [17].

§ 54. Regular contents: [2].

§ 55. Lusin's theorem: [61, p. 72] and [62].

§ 56. Linear functionals: [5, p. 61] and [31, p. 1008].

§ 58. [73, pp. 33–34]. Circles covering plane sets: [34].

§ 59. [73, pp. 140–149].

§ 60. [13], [46], [53].

§ 61. Integration by parts: [61, p. 102] and [70].

§ 62. (6): [36, p. 93].

§ 63. [4]; cf. also [73, pp. 42–45].

§ 64. [33].

BIBLIOGRAPHY

1. P. ALEXANDROFF and H. HOPF, *Topologie*, Berlin, 1935.

2. W. AMBROSE, *Lectures on topological groups* (unpublished), Ann Arbor, 1946.

3. W. AMBROSE, *Measures on locally compact topological groups*, Trans. A.M.S. **61** (1947) 106–121.

4. W. AMBROSE, *Direct sum theorem for Haar measures*, Trans. A.M.S. **61** (1947) 122–127.

5. S. BANACH, *Théorie des opérations linéaires*, Warszawa, 1932.

6. G. BIRKHOFF, *Lattice theory*, New York, 1940.

7. G. BIRKHOFF and S. MACLANE, *A survey of modern algebra*, New York, 1941.

8. A. BISCHOF, *Beiträge zur Carathéodoryschen Algebraisierung des Integralbegriffs*, Schr. Math. Inst. u. Inst. Angew. Math. Univ. Berlin **5** (1941) 237–262.

9. E. BOREL, *Les probabilités dénombrables et leurs applications arithmétiques*, Rend. Circ. Palermo **27** (1909) 247–271.

10. N. BOURBAKI, *Sur un théorème de Carathéodory et la mesure dans les espaces topologiques*, C. R. Acad. Sci. Paris **201** (1935) 1309–1311.

11. K. R. BUCH, *Some investigations of the set of values of measures in abstract space*, Danske Vid. Selsk. Math.-Fys. Medd. **21** (1945) No. 9.

12. C. CARATHÉODORY, *Vorlesungen über reelle Funktionen*, Leipzig–Berlin, 1927.

13. H. CARTAN, *Sur la mesure de Haar*, C. R. Acad. Sci. Paris **211** (1940) 759–762.

14. C. CHEVALLEY and O. FRINK, *Bicompactness of Cartesian products*, Bull. A.M.S. **47** (1941) 612–614.

15. R. COURANT, *Differential and integral calculus*, London–Glasgow, 1934.

16. H. CRAMÉR, *Random variables and probability distributions*, Cambridge, 1937.

17. J. DIEUDONNÉ, *Un exemple d'espace normal non susceptible d'une structure uniforme d'espace complet*, C. R. Acad. Sci. Paris **209** (1939) 145–147.

18. J. DIEUDONNÉ, *Sur le théorème de Lebesgue–Nikodym* (III), Ann. Univ. Grenoble **23** (1948) 25–53.

19. J. L. DOOB, *Stochastic processes depending on a continuous parameter*, Trans. A.M.S. **42** (1937) 107–140.

20. J. L. DOOB, *Stochastic processes with an integral-valued parameter*, Trans. A.M.S. **44** (1938) 87–150.

21. O. FRINK, *Representations of Boolean algebras*, Bull. A.M.S. **47** (1941) 755–756.

22. P. R. HALMOS and J. v. NEUMANN, *Operator methods in classical mechanics*, II, Ann. Math. **43** (1942) 332–350.

23. P. R. HALMOS, *The foundations of probability*, Amer. Math. Monthly **51** (1944) 493–510.

24. P. R. HALMOS, *The range of a vector measure*, Bull. A.M.S. **54** (1948) 416–421.

25. P. R. HALMOS, *Measurable transformations*, Bull. A.M.S., **55** (1949) 1015–1034.

26. G. H. HARDY, J. E. LITTLEWOOD, and G. PÓLYA, *Inequalities*, Cambridge, 1934.

27. F. HAUSDORFF, *Mengenlehre* (zweite Auflage), Berlin–Leipzig, 1927.

28. E. W. HOBSON, *The theory of functions of a real variable and the theory of Fourier's series* (vol. I, third edition), Cambridge, 1927.

29. W. HUREWICZ and H. WALLMAN, *Dimension theory*, Princeton, 1941.

30. M. KAC, *Sur les fonctions indépendantes* (I), Studia Math. **6** (1936) 46–58.

31. S. KAKUTANI, *Concrete representation of abstract (M)–spaces*, Ann. Math. **42** (1941) 994–1024.

32. S. KAKUTANI, *Notes on infinite product measure spaces*, I, Proc. Imp. Acad. Tokyo **19** (1943) 148–151.

33. S. KAKUTANI and K. KODAIRA, *Über das Haarsche Mass in der lokal bikompakten Gruppe*, Proc. Imp. Acad. Tokyo **20** (1944) 444–450.

34. R. KERSHNER, *The number of circles covering a set*, Am. J. Math. **61** (1939) 665–671.

35. A. KHINTCHINE and A. KOLMOGOROFF, *Über Konvergenz von Reihen, deren Glieder durch den Zufall bestimmt werden*, Mat. Sbornik **32** (1925) 668–677.

36. K. KODAIRA, *Über die Beziehung zwischen den Massen und Topologien in einer Gruppe*, Proc. Phys.-Math. Soc. Japan **23** (1941) 67–119.

37. A. KOLMOGOROFF, *Über die Summen durch den Zufall bestimmter unabhängiger Grössen*, Math. Ann. **99** (1928) 309–319.

38. A. KOLMOGOROFF, *Bemerkungen zu meiner Arbeit "Über die Summen zufälliger Grössen,"* Math. Ann. **102** (1930) 484–488.

39. A. KOLMOGOROFF, *Sur la loi forte des grandes nombres*, C. R. Acad. Sci. Paris **191** (1930) 910–912.

40. A. KOLMOGOROFF, *Grundbegriffe der Wahrscheinlichkeitsrechnung*, Berlin, 1933.

41. C. KURATOWSKI, *Topologie*, Warszawa–Lwów, 1933.

42. S. LEFSCHETZ, *Algebraic topology*, New York, 1942.

43. P. LÉVY, *Théorie de l'addition des variables aléatoires*, Paris, 1937.

44. A. LIAPOUNOFF, *Sur les fonctions-vecteurs complétement additives*, Bull. Acad. Sci. URSS **4** (1940) 465–478.

45. A. LOMNICKI, *Nouveaux fondements du calcul des probabilités*, Fund. Math. **4** (1923) 34–71.

46. L. H. LOOMIS, *Abstract congruence and the uniqueness of Haar measure*, Ann. Math. **46** (1945) 348–355.

47. L. H. LOOMIS, *On the representation of σ–complete Boolean algebras*, Bull. A.M.S. **53** (1947) 757–760.

48. D. MAHARAM, *On homogeneous measure algebras*, Proc. N.A.S. **28** (1942) 108–111.

49. E. MARCZEWSKI, *Sur l'isomorphie des mesures séparables*, Colloq. Math. **1** (1947) 39–40.

50. K. MENGER, *Untersuchungen über allgemeine Metrik*, Math. Ann. 100 (1928) 75–163.

51. J. v. NEUMANN, *Zur Operatorenmethode in der klassischen Mechanik*, Ann. Math. **33** (1932) 587–642.

52. J. v. NEUMANN, *Functional operators*, Princeton, 1933–1935.

53. J. v. NEUMANN, *The uniqueness of Haar's measure*, Mat. Sbornik **1** (1936) 721–734.

54. J. v. NEUMANN, *On rings of operators*, III, Ann. Math. **41** (1940) 94–161.

55. J. v. NEUMANN, *Lectures on invariant measures* (unpublished), Princeton, 1940.

56. O. NIKODYM, *Sur une généralisation des intégrales de M. J. Radon*, Fund. Math. **15** (1930) 131–179.

57. J. C. OXTOBY and S. M. ULAM, *On the existence of a measure invariant under a transformation*, Ann. Math. **40** (1939) 560–566.

58. L. PONTRJAGIN, *Topological groups*, Princeton, 1939.

59. S. SAKS, *On some functionals*, Trans. A.M.S. **35** (1933) 549–556.

60. S. SAKS, *Addition to the note on some functionals*, Trans. A.M.S. **35** (1933) 965–970.

61. S. SAKS, *Theory of the integral*, Warszawa–Lwów, 1937.

62. H. M. SCHAERF, *On the continuity of measurable functions in neighborhood spaces*, Portugaliae Math. **6** (1947) 33–44.

63. H. SCHEFFÉ, *A useful convergence theorem for probability distributions*, Ann. Math. Stat. **18** (1947) 434–438.

64. E. SPARRE ANDERSEN and B. JESSEN, *Some limit theorems on integrals in an abstract set*, Danske Vid. Selsk. Math.-Fys. Medd. **22** (1946) No. 14.

65. E. SPARRE ANDERSEN and B. JESSEN, *On the introduction of measures in infinite product sets*, Danske Vid. Selsk. Math.-Fys. Medd. **25** (1948) No. 4.

66. E. R. STABLER, *Boolean representation theory*, Amer. Math. Monthly **51** (1944) 129–132.

67. H. STEINHAUS, *Les probabilités dénombrables et leur rapport à la théorie de la mesure*, Fund. Math. **4** (1923) 286–310.

68. M. H. STONE, *The theory of representations for Boolean algebras*, Trans. A.M.S. **40** (1936) 37–111.

69. M. H. STONE, *Applications of the theory of Boolean rings to general topology*, Trans. A.M.S. **41** (1937) 375–481.

70. G. TAUTZ, *Eine Verallgemeinerung der partiellen Integration; uneigentliche mehrdimensionale Stieltjesintegrale*, Jber. Deutsch. Math. Verein. **53** (1943) 136–146.

71. E. R. VAN KAMPEN, *Infinite product measures and infinite convolutions*, Am. J. Math. **62** (1940) 417–448.

72. A. WEIL, *Sur les espaces a structure uniforme et sur la topologie générale*, Paris, 1938.

73. A. WEIL, *L'intégration dans les groupes topologiques et ses applications*, Paris, 1940.

74. K. YOSIDA, *Vector lattices and additive set functions*, Proc. Imp. Acad. Tokyo **17** (1941) 228–232.

LIST OF FREQUENTLY USED SYMBOLS

(References are to the pages on which the symbols are defined)

INDEX

Graduate Texts in Mathematics

(continued from page ii)

131 LAM. A First Course in Noncommutative Rings.
132 BEARDON. Iteration of Rational Functions.
133 HARRIS. Algebraic Geometry: A First Course.
134 ROMAN. Coding and Information Theory.
135 ROMAN. Advanced Linear Algebra.
136 ADKINS/WEINTRAUB. Algebra: An Approach via Module Theory.
137 AXLER/BOURDON/RAMEY. Harmonic Function Theory. 2nd ed.
138 COHEN. A Course in Computational Algebraic Number Theory.
139 BREDON. Topology and Geometry.
140 AUBIN. Optima and Equilibria. An Introduction to Nonlinear Analysis.
141 BECKER/WEISPFENNING/KREDEL. Gröbner Bases. A Computational Approach to Commutative Algebra.
142 LANG. Real and Functional Analysis. 3rd ed.
143 DOOB. Measure Theory.
144 DENNIS/FARB. Noncommutative Algebra.
145 VICK. Homology Theory. An Introduction to Algebraic Topology. 2nd ed.
146 BRIDGES. Computability: A Mathematical Sketchbook.
147 ROSENBERG. Algebraic K-Theory and Its Applications.
148 ROTMAN. An Introduction to the Theory of Groups. 4th ed.
149 RATCLIFFE. Foundations of Hyperbolic Manifolds.
150 EISENBUD. Commutative Algebra with a View Toward Algebraic Geometry.
151 SILVERMAN. Advanced Topics in the Arithmetic of Elliptic Curves.
152 ZIEGLER. Lectures on Polytopes.
153 FULTON. Algebraic Topology: A First Course.
154 BROWN/PEARCY. An Introduction to Analysis.
155 KASSEL. Quantum Groups.
156 KECHRIS. Classical Descriptive Set Theory.
157 MALLIAVIN. Integration and Probability.
158 ROMAN. Field Theory.
159 CONWAY. Functions of One Complex Variable II.
160 LANG. Differential and Riemannian Manifolds.
161 BORWEIN/ERDÉLYI. Polynomials and Polynomial Inequalities.
162 ALPERIN/BELL. Groups and Representations.
163 DIXON/MORTIMER. Permutation Groups.
164 NATHANSON. Additive Number Theory: The Classical Bases.
165 NATHANSON. Additive Number Theory: Inverse Problems and the Geometry of Sumsets.
166 SHARPE. Differential Geometry: Cartan's Generalization of Klein's Erlangen Program.
167 MORANDI. Field and Galois Theory.
168 EWALD. Combinatorial Convexity and Algebraic Geometry.
169 BHATIA. Matrix Analysis.
170 BREDON. Sheaf Theory. 2nd ed.
171 PETERSEN. Riemannian Geometry.
172 REMMERT. Classical Topics in Complex Function Theory.
173 DIESTEL. Graph Theory. 2nd ed.
174 BRIDGES. Foundations of Real and Abstract Analysis.
175 LICKORISH. An Introduction to Knot Theory.
176 LEE. Riemannian Manifolds.
177 NEWMAN. Analytic Number Theory.
178 CLARKE/LEDYAEV/STERN/WOLENSKI. Nonsmooth Analysis and Control Theory.
179 DOUGLAS. Banach Algebra Techniques in Operator Theory. 2nd ed.
180 SRIVASTAVA. A Course on Borel Sets.
181 KRESS. Numerical Analysis.
182 WALTER. Ordinary Differential Equations.
183 MEGGINSON. An Introduction to Banach Space Theory.
184 BOLLOBAS. Modern Graph Theory.
185 COX/LITTLE/O'SHEA. Using Algebraic Geometry.
186 RAMAKRISHNAN/VALENZA. Fourier Analysis on Number Fields.
187 HARRIS/MORRISON. Moduli of Curves.
188 GOLDBLATT. Lectures on the Hyperreals: An Introduction to Nonstandard Analysis.
189 LAM. Lectures on Modules and Rings.
190 ESMONDE/MURTY. Problems in Algebraic Number Theory.
191 LANG. Fundamentals of Differential Geometry.
192 HIRSCH/LACOMBE. Elements of Functional Analysis.
193 COHEN. Advanced Topics in Computational Number Theory.
194 ENGEL/NAGEL. One-Parameter Semigroups for Linear Evolution Equations.
195 NATHANSON. Elementary Methods in Number Theory.
196 OSBORNE. Basic Homological Algebra.
197 EISENBUD/HARRIS. The Geometry of Schemes.
198 ROBERT. A Course in p-adic Analysis.
199 HEDENMALM/KORENBLUM/ZHU. Theory of Bergman Spaces.
200 BAO/CHERN/SHEN. An Introduction to Riemann–Finsler Geometry.

201 HINDRY/SILVERMAN. Diophantine Geometry: An Introduction.
202 LEE. Introduction to Topological Manifolds.
203 SAGAN. The Symmetric Group: Representations, Combinatorial Algorithms, and Symmetric Function. 2nd ed.
204 ESCOFIER. Galois Theory.
205 FÉLIX/HALPERIN/THOMAS. Rational Homotopy Theory.
206 MURTY. Problems in Analytic Number Theory.
Readings in Mathematics
207 GODSIL/ROYLE. Algebraic Graph Theory.